CAMBRIDGE MONOGRAPHS ON
MECHANICS AND APPLIED MATHEMATICS

General Editors
G.K. BATCHELOR, F.R.S.
Professor of Fluid Dynamics at the University of Cambridge
J.W. MILES
Professor of Applied Mechanics and Geophysics, University of California, La Jolla

THEORY OF LAMINAR FLAMES

T0296391

Theory of Laminar Flames

J.D. BUCKMASTER

Professor of Mathematics and
Professor of Theoretical and Applied Mechanics,
University of Illinois

G.S.S. LUDFORD

Professor of Applied Mathematics, Cornell University

CAMBRIDGE UNIVERSITY PRESS

Cambridge

London New York New Rochelle

Melbourne Sydney

CAMBRIDGE UNIVERSITY PRESS
Cambridge, New York, Melbourne, Madrid, Cape Town, Singapore, São Paulo, Delhi

Cambridge University Press
The Edinburgh Building, Cambridge CB2 8RU, UK

Published in the United States of America by Cambridge University Press, New York

www.cambridge.org
Information on this title: www.cambridge.org/9780521239295

First published 1982
This digitally printed version 2008

A catalogue record for this publication is available from the British Library

Library of Congress Catalogue Card Number: 81-21573

ISBN 978-0-521-23929-5 hardback
ISBN 978-0-521-09192-3 paperback

CONTENTS

Preface ix

1 Governing equations of combustion

1 Approach 1
2 Continuum theory of a mixture of reacting species 2
3 The Arrhenius factor 5
4 Differential mass diffusion; equality of c_{pi} and of m_i 7
5 The combustion and constant-density approximations 11
6 Constant properties: nondimensionalization 13
7 Equidiffusion: Shvab–Zeldovich variables 15
8 Activation-energy asymptotics 17

2 The premixed plane flame

1 Models of mixtures 20
2 Cold-boundary difficulty 21
3 Anchored flames 23
4 Asymptotic analysis 25
5 Flat burners 28
6 Near-surface, surface, and remote flames 31
7 Reactions of arbitrary order 33
8 The two-reactant model 34

3 Perturbations: SVFs and NEFs

1 Prologue 38
2 Modifications of the plane premixed flame for $\mathscr{L} \neq 1$ 39
3 Theory of perturbations 41
4 Steady heat loss: the excess-enthalpy flame 43
5 SVFs 47
6 Nonuniform ducts 50
7 An elementary flame holder 51
8 NEFs 53

4 Steady burning of a linear condensate

1 Responses 58
2 Experimental results; extinction 60
3 Solid pyrolysis under adiabatic conditions 61
4 Radiation from the surface 64
5 Background radiation 67
6 True nature of effective extinction 71
7 Liquid evaporation under adiabatic conditions 73
8 Radiative exchange at the surface; distributed heat exchange 76

5 Unsteady burning of a linear condensate

1 The problem 79
2 Solid pyrolysis 80
3 Liquid evaporation 86
4 Response to an impinging acoustic wave 87
5 Amplification 90
6 Stability of anchored flames 92

6 Spherical diffusion flames

1 Diffusion flames 95
2 Steady combustion for $\mathscr{K}=\mathscr{L}=1$; D-asymptotics 98
3 The nearly adiabatic flame for $\mathscr{K}$, $\mathscr{L}\neq 1$ and its stability 103
4 General ignition and extinction analyses for $\mathscr{K}=\mathscr{L}=1$ 108
5 Remarks on the middle branch of the S-response 112
6 Other responses 113
7 The burning fuel drop 115
8 Further results on stability 117

7 Cylindrical and spherical premixed flames

1 Cylindrical flames 120
2 Planar character 120
3 Near-surface, surface, and remote flames 123
4 Spherical flames; D-asymptotics 124
5 Ignition and extinction 127
6 Other aspects of responses 130

8 Multidimensional theory of premixed flames

1 The flame as a hydrodynamic discontinuity 136
2 Slow variation and near-equidiffusion 139
3 The basic equation for SVFs 142
4 Flame stretch 146
5 The basic equations for NEFs 149
6 Reduction to Stefan problems 151

9 Burners

1 Features of a tube flame 154
2 Hydrodynamic considerations 156

3 SVF tips 159
4 NEF tips 163
5 Quenching by a cold surface 168
6 Further hydrodynamic considerations 170

10 Effects of shear and strain

1 Nonuniformities 175
2 Response of NEFs to simple shear 176
3 Response of NEFs to simple strain 179
4 More general nonuniformity 185
5 Effect of slowly varying shear 187
6 Effect of slowly varying strain 189

11 Stability

1 Scope 193
2 Slowly varying perturbations of the plane wave 196
3 Buoyancy and curvature 199
4 Stability of plane NEFs 202
5 Cellular flames 206
6 Hydrodynamic effects 209
7 Curved cellular flames 214
8 Delta-function models and the right stability boundary 218

12 Ignition and explosion

1 Synopsis 221
2 Spontaneous combustion (auto-ignition) 222
3 Adiabatic explosion 227
4 Explosion with heat loss 232
5 Ignition by heat flux 238
6 Auto-ignition and explosion of separated reactants 241

Text references 248
Further references 256
Citation Index 261
Subject Index 263

PREFACE

Existing combustion books are primarily phenomenological in the sense that explanation, where provided, is usually set in an intuitive framework; when mathematical modeling is employed it is often obscured by *ad hoc* irrational approximation, the emphasis being on the explanation of existing experimental results. It is hardly necessary to add that the philosophy underlying such texts is scientifically legitimate and that they will undoubtedly stay in the mainstream of combustion science for many years to come. Nevertheless, we are of the opinion that there is need for texts that treat combustion as a mathematical science and the present work is an attempt to meet that need in part.

In this monograph we describe, within a mathematical framework, certain basic areas of combustion science, including many topics rightly covered by introductory graduate courses in the subject. Our treatment eschews sterile rigor inappropriate for a subject in which the emphasis has been physical, but we are deeply concerned with maintaining clear links between the mathematical modeling and the analytical results; irrational approximation is carefully avoided. All but the most fastidious of readers will be satisfied that the mathematical conclusions are correct, except for slips of the pen.

Although the material covered inevitably reflects our special interests and personal perspectives, the entire discussion is connected by a singular perturbation procedure known as activation-energy asymptotics. The description of reacting systems characterized by Arrhenius kinetics can be simplified when the activation energy is large, corresponding to an extreme sensitivity to temperature. The notion is an old one; certainly the Russian combustion school was aware of it by the '40s. But its full power is only achieved within the formal framework of modern singular perturbation theory. Developments of this nature are quite recent. Although the earliest example is possibly a 1962 study of P. A. Blythe on the flow of a reacting gas through a Laval nozzle, it was not until the '70s, following a seminal paper by W. B. Bush and F. E. Fendell (on the laminar flame speed) and the call by F. A. Williams in the 1971 *Annual Review of Fluid Mechanics*, that extensive applications to combustion theory were undertaken.

Development has been increasingly rapid. A few years ago this monograph would not have been possible; but now there are more than enough results in a wide range of topics to justify the connected account attempted here. Our hope is that, by collecting many of the results together in a single volume with unified and self-contained presentation, other applied mathematicians will be encouraged to enter the field; for, although activation-energy asymptotics is no panacea, many exciting discoveries undoubtedly lie ahead.

Our discussion is concerned only with these modern developments; we have made no attempt to trace the history of the subject before the advent of activation-energy asymptotics. No disparagement is intended. On the contrary, this new tool has been used on problems that were largely identified, formulated, and discussed by the giants of the past; those who developed the material used in this monograph stood firmly on broad shoulders. Our failure to trace the earlier work stems from the conviction that it is better left to someone more scholarly than we are, someone who has been immersed in the subject longer than we have. (Similar remarks apply to our neglect of parallel problems in other areas, such as chemical reactor theory.)

Our treatment of the physical aspects of the subject requires some comment. The neophyte is well advised to read a traditional text in conjunction with ours, since phenomenology plays only a minor role in our presentation. The description of a physical phenomenon by clear-cut mathematical analysis is applied mathematics at its best and this ultimate justification of our efforts is not neglected in the discussion. But we are never in danger of letting the existing physical world be both judge and jury for the mathematical results. Where the results do not presently appear to have convincing physical counterpart, we say so; but no topic has been omitted for lack of such a counterpart. The value of rational deductions from physically plausible models is not always immediately apparent and applied mathematics is ultimately most fruitful when permitted a great deal of independence.

Reaction-diffusion equations arise copiously elsewhere, but activation-energy asymptotics has been largely restricted to combustion. In biochemical contexts, temperature effects (when present) are of a different character. For chemical reactors, problems involving large activation energy are far less common than for combustion. While significant results for chemical reactors have been obtained by activation-energy asymptotics, and more will be, it seems unlikely that the coverage will ever be as extensive as for combustion.

This monograph is not a review of laminar flames for large activation energy but an essay on the subject. The text only mentions works that helped us in writing a connected account of the topics chosen. To recognize other contributions and to give an idea of the wealth of results that has been obtained, we have compiled a list of further references. Substantial use of activation-energy asymptotics was a necessary qualification; in particular, treatments of thermal ignition in the style of Frank-Kamenetskii were excluded. Such a list is bound to be incomplete, especially since no serious attempt was made to search the literature. Any omissions are due to carelessness or ignorance, but not to malice. Only published work available to us by the end of 1981 is included; in particular, theses do not appear.

The notation has been kept as simple as possible. Rather than conscripting exotic symbols, we have made latin and greek letters do multiple duty; the meaning of a symbol should be clear from context. We have also been economical in the designation of and reference to chapters, sections, figures, and displays (so as not to litter the page with numbers): a decimal is used only for reference to another chapter. When part of a multiple display is intended, the letters *a*, *b*,... are added to indicate the first, second,... part. Finally, the notation used uniformly throughout the text for coefficients in asymptotic expansions is explained on page 19.

The task of writing this book has been lightened and made more pleasant by the enthusiastic encouragement we have received from so many friends and colleagues. We particularly want to thank and identify those who provided direct technical help, all of which was invaluable; they are

T. S. Chang	H. V. McConnaughey
P. Clavin	D. M. Michelson
J. George	D. Mikolaitis
R. D. Janssen	I. Müller
J. T. Jenkins	A. Nachman
A. K. Kapila	A. K. Sen
D. R. Kassoy	G. I. Sivashinsky
C. K. Law	D. S. Stewart
A. Liñán	F. A. Williams

For bringing the material into the form of a final typescript we are indebted to Julia Dethier; her editorial touch is evident everywhere.

The constant support of our research by the US Army Research Office provided the climate in which such an endeavour could be undertaken. We salute Dr. Jagdish Chandra of ARO-Mathematics, whose confidence seldom seemed to waver over the years.

The monograph was started six years ago during a sabbatical at the University of Queensland, arranged by A. F. Pillow. Solid progress was made during a joint leave at the Mathematics Research Center–Wisconsin in the Fall of 1977, arranged by B. Noble. The work was finished early this year, but we have tried to up-date the material as the printing process slowly ran its course.

J. D. Buckmaster
Vancouver B.C., Canada

G. S. S. Ludford
Cambridge, England

December, 1981

1

GOVERNING EQUATIONS OF COMBUSTION

1 Approach

The development of the equations governing combustion involves derivation of the equations of motion of a chemically reacting gaseous mixture and judicious simplification to render them tractable while retaining their essential characteristics. A rigorous derivation requires a long apprenticeship in either kinetic theory or continuum mechanics. (Indeed, the general continuum theory of reacting mixtures is only now being perfected.) We choose instead a plausible, but potentially rigorous, derivation based on the continuum theory of a mixture of fluids, guided by experience with a single fluid. *Ad hoc* arguments, in particular the inconsistent assumption that the mixture itself is a fluid for the purpose of introducing certain constitutive relations, will not be used.

Treating the flow of a reacting mixture as an essentially isobaric process, the so-called combustion approximation, is a safe simplification under a wide range of circumstances if detonations are excluded. But the remaining simplifications, designed as they are solely to make the equations tractable, should be accepted tentatively. They are always revocable should faulty predictions result; for that reason they are explained carefully. Nevertheless, whosoever is primarily interested in solving nontrivial combustion problems, as we are, can have the same confidence in the final equations as is normally placed in the equations of a non-Newtonian fluid, for example.

These final equations retain most of the complexity of a compressible, heat-conducting, viscous fluid; but diffusion of the species, and source terms representing the chemical reaction, have been added. This complexity has usually been fought with irrational approximation and computers. Combustion processes, however, tend by their very nature to have large activation energies. It is, therefore, more appropriate to analyze the problem rationally by localizing the reaction in layers (or flames), which can be done through activation-energy asymptotics, as is shown in section 8. Such an approach was scarcely mentioned before publication of the review article of Williams (1971), which contains the first suggestion of so forming a mathematical theory of combustion. The classic paper of Bush & Fendell (1970) did come earlier and Frank-Kamenetskii (1969, p. 375) did introduce approximations based on large activation energy more than forty years ago, but Williams marked the start of a clear path through the complex and fascinating problems of combustion, as is shown by the wealth of papers that have appeared since.

Our immediate intention is to reach a convincing set of equations and a method of attack, so as to come to grips with the subject of laminar flames.

2 Continuum theory of a mixture of reacting species

The mixture has density ρ and is considered to be made up of N fluids whose separate densities are $\rho Y_i (i=1, 2, \ldots, N)$. Here the Y_i are mass fractions (or concentrations), with

$$\sum_{i=1}^{N} Y_i = 1. \tag{1}$$

If $\boldsymbol{v}_i$ is the velocity of the ith fluid, i.e. species, the balance of mass requires

$$\partial(\rho Y_i)/\partial t + \nabla\cdot(\rho Y_i \boldsymbol{v}_i) = \dot{\rho}_i, \tag{2}$$

where $\dot{\rho}_i$ is the rate of production of species i (mass per unit volume) by the chemical reactions so that

$$\sum_{i=1}^{N} \dot{\rho}_i = 0. \tag{3}$$

Summation of the equations over all species yields the familiar continuity equation

$$\partial\rho/\partial t + \nabla\cdot(\rho \boldsymbol{v}) = 0, \tag{4}$$

where

$$\boldsymbol{v} = \sum_{i=1}^{N} Y_i \boldsymbol{v}_i \tag{5}$$

is the mass-average velocity of the mixture. Each balance equation (2) can now be recast by subtracting Y_i times the overall equation (4) and introducing the diffusion velocities

$$\boldsymbol{V}_i = \boldsymbol{v}_i - \boldsymbol{v}; \tag{6}$$

this gives

$$\rho(\partial Y_i/\partial t + \boldsymbol{v}\cdot\nabla Y_i) = \dot{\rho}_i - \nabla\cdot(\rho Y_i \boldsymbol{V}_i). \tag{7}$$

The momentum balances of the separate fluids will not be given since they play no role in the sequel. But Truesdell (1965) shows that if they are added the result is the single-fluid balance

$$\rho(\partial \boldsymbol{v}/\partial t+\boldsymbol{v}\cdot\nabla\boldsymbol{v})=\nabla\cdot\Sigma+\rho\boldsymbol{g}, \tag{8}$$

provided no momentum is created by the chemical reactions. Here Σ is the sum of the stresses in the individual fluids plus stresses due to diffusion of the species and $\boldsymbol{g}$ is the gravity force, assumed to be the only force external to the mixture. Interactions between the species (see also the discussion of Fick's law (19)), which give additional external forces on each individual fluid, sum to zero.

We shall also forego writing the energy balances for the separate fluids, but Truesdell shows their sum gives the single-fluid balance

$$\rho(\partial U/\partial t+\boldsymbol{v}\cdot\nabla U)=\Sigma:\nabla\boldsymbol{v}+\nabla\cdot\boldsymbol{q}. \tag{9}$$

provided no energy is created by the chemical reactions and the work of the interaction forces is negligible. (For the forces mentioned at the end of this section the work is nonlinear in the (small) $\boldsymbol{V}_i$.) Here U is the sum of the separate internal energies plus kinetic energies of diffusion, while $\boldsymbol{q}$ is the sum of the separate energy fluxes plus fluxes due to diffusion. Note that the work of the gravity force has not been neglected: it is $\Sigma_{i=1}^{N}(\rho Y_i)\boldsymbol{g}\cdot\boldsymbol{V}_i=0$ by virtue of the result

$$\sum_{i=1}^{N} Y_i\boldsymbol{V}_i=0, \tag{10}$$

which follows from the definition (6).

The new variables $\dot{\rho}_i$, $\boldsymbol{V}_i$, Σ, U, and $\boldsymbol{q}$ introduced by these balances have to be related to the primitive variables by constitutive equations. Continuum mechanics does not supply such relations, but rather judges them for consistency with certain general principles after they have been proposed. For reacting mixtures such judgements are still being made, but the linear equations we propose are acceptable (Bowen 1976). With the exception of those for $\dot{\rho}_i$ and $\boldsymbol{V}_i$ they result from experience with a single fluid. The coefficients (50) that will appear are, for the moment, to be considered functions of the primitive variables.

The primitive variables will be Y_i, $\boldsymbol{v}$, ρ, and p; but in formulating the constitutive equations it is convenient to introduce the temperature T, which is assumed to be the same for all fluids. If each species is a perfect gas its partial pressure is

$$p_i=R(\rho Y_i)T/m_i, \tag{11}$$

and the consequence of Dalton's law is

$$p=R\rho T\sum_{i=1}^{N}(Y_i/m_i), \tag{12}$$

which determines the common temperature. Here R is the gas constant and m_i is the mass of a molecule of species i. The separate internal energies of the fluids are then functions of T alone, which may be written

$$U_i=h_i-p_i/(\rho Y_i), \tag{13}$$

where the enthalpies are

$$h_i=h_i^0+\int_{T^0}^{T}c_{pi}(T)\,\mathrm{d}T. \tag{14}$$

Here h_i^0 is the heat of formation of the species i at some standard temperature T^0 and the c_{pi} are the specific heats at constant pressure. If we neglect the kinetic energies of diffusion because they are nonlinear in the V_i, the internal energy of the mixture is now

$$U=\sum_{i=1}^{N}Y_ih_i-p/\rho. \tag{15}$$

Next the energy flux $\boldsymbol{q}$ is specified by requiring that the energy flux in each individual fluid is due entirely to heat conduction, so that

$$\boldsymbol{q}=\lambda\nabla T-\rho\sum_{i=1}^{N}Y_ih_iV_i, \tag{16}$$

where λ, the coefficient of thermal conductivity of the mixture, is the sum of the species coefficients. Each of the diffusion fluxes is that of a single fluid moving relative to the mixture with velocity V_i when kinetic energy and deviation of the stress tensor (17) from $-p_i\mathbf{I}$ are neglected, the latter anticipating the combustion approximation (section 5) that makes $\boldsymbol{v}_i$ small. Such a $\boldsymbol{q}$ neglects, in particular, radiative transfer and the Dufour effect (heat flux due to concentration gradients).

The separate stresses are assumed to be Newtonian, i.e.

$$\boldsymbol{\Sigma}_i=-(p_i+\tfrac{2}{3}\kappa_i\nabla\cdot\boldsymbol{v}_i)\mathbf{I}+\kappa_i[\nabla\boldsymbol{v}_i+(\nabla\boldsymbol{v}_i)^T], \tag{17}$$

where $\mathbf{I}$ is the unit tensor and bulk viscosity has been neglected (for simplicity only). We shall take the remaining coefficients $\kappa_i=Y_i\kappa$; this corresponds to assuming that the intrinsic viscosities of the fluids are all equal. (Otherwise gradients of V_i must be neglected in comparison with those of $\boldsymbol{v}$ to arrive at the result (18).) Since stresses due to diffusion and terms $V_i\nabla Y_i$ are nonlinear, the separate stresses sum to give

$$\boldsymbol{\Sigma}=-(p+\tfrac{2}{3}\kappa\nabla\cdot\boldsymbol{v})\mathbf{I}+\kappa[\nabla\boldsymbol{v}+(\nabla\boldsymbol{v})^T] \tag{18}$$

by virtue of the result (10).

The thermomechanical constitution is completed by equations for the diffusion velocities V_i; we adopt Onsager's (1945) generalization of Fick's (1855) law

$$\rho Y_i V_i = -\sum_{j=1}^{N} \mu_{ij} \nabla Y_j, \tag{19}$$

making each diffusive mass flux a linear combination of the concentration gradients. The result (10) now reads

$$\sum_{i=1}^{N} \sum_{j=1}^{N} \mu_{ij} \nabla Y_j = \sum_{j=1}^{N-1} \left[\sum_{i=1}^{N} (\mu_{ij} - \mu_{iN}) \right] \nabla Y_j = 0$$

when the relation (1) is used to eliminate ∇Y_N. Since the remaining gradients are independent of each other, it follows that the coefficients μ_{ij} must satisfy

$$\sum_{i=1}^{N} \mu_{ij} = \sum_{i=1}^{N} \mu_{iN} \quad \text{for } j = 1, 2, \ldots, N-1. \tag{20}$$

An equivalent of the law (19), in which the gradients are expressed as linear combinations of the diffusion velocities, can be derived from the separate momentum balances by making assumptions about the interaction forces between the species (Williams 1965, p. 416). An implicit new assumption in such arguments is that the accelerations $\partial V_i/\partial t$ are negligible. They only occur in unsteady problems and then produce a finite velocity of propagation of diffusion effects (Müller 1977): such an effect appears to be of no great importance in combustion. The generalized Fick's law (which neglects the Soret effect of a diffusive flux due to temperature gradients) is the only nonchemical constitutive equation not derived from experience with a single fluid.

Finally we come to the $\dot{\rho}_i$, which require a discussion of chemical reactions.

3 The Arrhenius factor

For simplicity we shall first consider one-step combustion, where a single unopposed chemical reaction is involved. If N_i is the number of molecules of species i per unit volume, then by definition

$$\dot{\rho}_i = m_i \dot{N}_i. \tag{21}$$

The chemical reaction may be described in terms of the m_i as the mass balance

$$\sum_{i=1}^{N} \nu_i m_i = \sum_{i=1}^{N} \lambda_i m_i, \tag{22}$$

where ν_i and λ_i are called the stoichiometric coefficients, each a non-negative integer; ν_i is zero if the species i is not a reactant, while λ_i is zero if it is not a product. Both are zero for an inert. Since these coefficients represent the relative numbers of molecules consumed or produced by the reaction we may write

$$\dot{N}_i = (\lambda_i - \nu_i)\omega, \tag{23}$$

where $\omega(>0)$ measures the rate at which the reaction is proceeding. Clearly the requirement (3) is now a consequence of the definition (21) and the mass balance (22).

For example, consider the exothermic reaction

$$2NO + Cl_2 \rightarrow 2NOCl \tag{24}$$

by which nitrosyl chloride is formed from nitric oxide and chlorine. If the three species are numbered

$$1: NO; \quad 2: Cl_2; \quad 3: NOCl;$$

then the stoichiometric integers are

$$\nu_1 = 2,\ \nu_2 = 1,\ \nu_3 = 0; \quad \lambda_1 = 0,\ \lambda_2 = 0,\ \lambda_3 = 2;$$

and

$$\dot{N}_1 = -2\omega, \quad \dot{N}_2 = -\omega, \quad \dot{N}_3 = 2\omega$$

are the rates of consumption (negative) and production (positive) of the molecules. Note that the reaction (24) is written

$$NO + NO + Cl_2 \rightarrow NOCl + NOCl \tag{25}$$

when single molecules are used.

It is, therefore, ω that must be related to the primitive variables; a common assumption is that it is proportional to the number density (concentration) $\rho Y_j/m_j$ of each reactant, where the species j is counted ν_j times as a reactant (see the single-molecule form (25) of the reaction). Thus,

$$\omega = k \prod_{j=1}^{N} (\rho Y_j/m_j)^{\nu_j} = k\rho^{\nu} \prod_{j=1}^{N} (Y_j/m_j)^{\nu_j}, \quad \text{where } \nu = \sum_{j=1}^{N} \nu_j \tag{26}$$

and the factor k is assumed to depend only on the temperature. Since experimental determinations of k are often done under isothermal conditions, it is called the rate constant; but we shall avoid the term since the temperature dependence of k is an essential feature in our analysis. The Arrhenius (1889) law

$$k = BT^{\alpha} e^{-E/RT} \tag{27}$$

will be taken for its variation with temperature, where B, α (equals $\frac{1}{2}$ theoretically), and E are constants. The latter is known as the activation energy since E/R is roughly the temperature below which k is relatively small.

Thus we now have

$$\dot{\rho}_i = \mu_i \omega, \tag{28}$$

where

$$\mu_i = m_i(\lambda_i - \nu_i), \quad \omega = BT^{\alpha} e^{-E/RT} \rho^{\nu} \prod_{j=1}^{N} (Y_j/m_j)^{\nu_j}. \tag{29}$$

More complex reactions are first split up into elementary reaction steps that may contain reversals of each other (opposing reactions). For each elementary reaction a formula (28) holds and for the complete reaction $\dot{\rho}_i$ is the sum of all such terms. (A pair of opposing reactions may be in equilibrium, i.e. the corresponding terms cancel and thereby furnish an algebraic relation between mass fractions.) The central question of reaction kinetics is to determine these elementary steps and their parameters B, α, and E. This monograph will be restricted to a single elementary reaction. Equilibrium is then reached only when one of the reactants is exhausted, i.e.

$$Y_i = 0 \quad \text{for some } i. \tag{30}$$

In practice, a combustion reaction may be a complex network of elementary steps, so that attempts are made to model the overall reaction with a smaller number of terms of the form (28), preferably one. The exponents ν_i are then released from being the stoichiometric coefficients of the overall reaction and ν is not necessarily their sum. On the other hand, the μ_i retain their meaning since they still represent the proportions in which the species take part in the reaction. The free constants B, α, E, ν, and ν_i are determined experimentally; of course, the ν_i need no longer be integers.

4 Differential mass diffusion; equality of c_{pi} and of m_i

Once the various parameters (constants and functions) that appear in the constitutive equations (12), (15), (16), (18), (19), and (28) are given, there are $N+6$ equations (1), (4), (7), (8), and (9) for the $N+5$ unknowns Y_i, $\mathbf{v}$, ρ, and p. Equation (1) is consistent with the remainder since they conserve $\Sigma_{i=1}^{N} Y_i$ whenever the relations (20) hold between the μ_{ij}; initial or boundary conditions will then give it the value 1. The system is very complicated and must be simplified if it is to be analyzed in detail.

The law (19) leads to terms

$$\sum_{j=1}^{N} \nabla \cdot (\mu_{ij} \nabla Y_j) \tag{31}$$

in the mass balances (7). To simplify these terms while retaining different diffusion properties for the species, it is tempting to set

$$\mu_{ij}=0 \quad \text{for } i\neq j \tag{32}$$

and allow the μ_{ii} to change with i (Emmons 1971). The conditions (20), however, would be violated: they require

$$\mu_{ii}=\mu_{NN}=\mu(\text{say}) \quad \text{for all } i. \tag{33}$$

Conclusions (32) and (33) are reached quite generally by Williams's theory when the so-called binary diffusion coefficients are equal, i.e. whenever the interactions between pairs of species are identical for momentum.

The same goal, however, can effectively be reached by taking the values (32) for all i except the last, and

$$\mu_{Nj}=\mu_{NN}-\mu_{jj} \quad \text{for } j\neq N \tag{34}$$

with the μ_{jj} different. The conditions (20) are now satisfied and the terms (31) simplify to

$$\nabla\cdot(\mu_{ii}\nabla Y_i) \quad \text{for } i\neq N. \tag{35}$$

Only the Nth equation remains complicated but, since Y_N can be otherwise calculated from $1-\Sigma_{i=1}^{N-1} Y_i$, that does not matter. Such a scheme can be justified from Williams's formulas when the first $N-1$ species are scarce compared to the last, a condition often roughly approximated in combustion where, for example, a stoichiometric air/propane mixture contains 76% by volume of the inerts nitrogen and argon.

In short, we shall generally employ the values (32) and (33) but, when differential diffusion cannot be ignored, the μ_{ii} for $i=1, 2, \ldots, N-1$ will be given different values. The results will then apply to a mixture of reactants and products highly diluted in an inert. Note that a common value of the μ_{ii} can still depend on the Y_i and, hence, take different values in different locations. This can give the appearance of differential diffusion when, for example, there are two reactants ($i=1, 2$) that do not coexist. Where $Y_1=0$ its equation is irrelevant and μ_{22} there may differ from μ_{11} in the region where $Y_2=0$ with irrelevant equation. Such is the case in the burning of a fuel droplet treated by Kassoy & Williams (1968). They use different diffusion coefficients for the fuel and oxidant equations but do not apply the equations in the same place: the combustion field is divided into two parts, in each of which there is either no fuel or no oxidant. There is no question of an abundant inert.

We come now to the energy balance (9). First note that the individual c_{pi} disappear from U in favor of

$$c_p(T, Y_i)=\sum_{i=1}^{N} Y_i c_{pi}. \tag{36}$$

Thus, we have immediately

$$U=\sum_{i=1}^{N} Y_i h_i^0 + I - p/\rho, \tag{37}$$

where

$$I(T, Y_i)=\sum_{i=1}^{N} Y_i \int_{T^0}^{T} c_{pi}\,\mathrm{d}T=\int_{T^0}^{T} c_p\,\mathrm{d}T, \tag{38}$$

the last integral being taken with the Y_i fixed, i.e. for constant composition. Adoption of $\mu_{ij}=\mu\delta_{ij}$ ensures that the same happens to $\boldsymbol{q}$: we have

$$\begin{aligned}\rho\sum_{i=1}^{N} Y_i h_i V_i &= \rho\sum_{i=1}^{N} Y_i h_i^0 V_i + \rho\sum_{i=1}^{N}\left(\int_{T^0}^{T} c_{pi}\,\mathrm{d}T\right) Y_i V_i\\ &=\rho\sum_{i=1}^{N} Y_i h_i^0 V_i - \mu\sum_{i=1}^{N}\left(\int_{T^0}^{T} c_{pi}\,\mathrm{d}T\right)\nabla Y_i\\ &=\rho\sum_{i=1}^{N} Y_i h_i^0 V_i - \mu\left(\nabla I - \sum_{i=1}^{N} Y_i c_{pi}\nabla T\right)\\ &=\rho\sum_{i=1}^{N} Y_i h_i^0 V_i + \mu(c_p\nabla T - \nabla I).\end{aligned} \tag{39}$$

Next note the form that the energy balance now takes. The total contribution to the equation from the heats of formation h_i^0, when placed on the right side, is

$$-\sum_{i=1}^{N} h_i^0[\rho(\partial Y_i/\partial t - \boldsymbol{v}\cdot\nabla Y_i)+\nabla\cdot(\rho Y_i V_i)] = -\sum_{i=1}^{N} h_i^0 \dot{\rho}_i \tag{40}$$

according to the mass balances (7). Also, the total contribution from heat conduction and terms containing c_p, when placed on the left side, is

$$\rho(\partial I/\partial t + \boldsymbol{v}\cdot\nabla I) + \nabla\cdot[\mu(c_p\nabla T - \nabla I) - \lambda\nabla T]. \tag{41}$$

The remaining terms, when placed on the right side, contribute

$$\partial p/\partial t + \boldsymbol{v}\cdot\nabla p - \tfrac{2}{3}\kappa(\nabla\cdot\boldsymbol{v})^2 + \kappa[\nabla\boldsymbol{v} + (\nabla\boldsymbol{v})^T]:\nabla\boldsymbol{v} \tag{42}$$

according to the continuity equation (4). In the combustion approximation discussed later the velocity is small, implying also that the pressure is effectively a spatial constant, so that these last terms are ignored, except for $\partial p/\partial t$, which is a function of t alone.

It is tempting to take the Lewis number

$$\mathscr{L}=\lambda/\mu c_p \tag{43}$$

equal to 1; then the terms (41) contain I only. If the c_{pi}, however, are not all equal, c_p is a linear function (36) of the Y_i that μ/λ can only equal for very special mixtures, if at all. Only when the c_{pi} are all equal does c_p lose its dependence on the composition of the mixture, and then it is unnecessary to make any assumption about the Lewis number, as we shall see next.

We are, therefore, faced with taking all the c_{pi} equal, when c_p becomes a function of T alone:

$$c_p(T)=c_{pi} \quad \text{for all } i. \tag{44}$$

Since ∇I is then $c_p\nabla T$ the energy equation reads

$$\rho(\partial I/\partial t+\boldsymbol{v}\cdot\nabla I)-\nabla\cdot[(\lambda/c_p)\nabla I]=-\left[\sum_{i=1}^{N} h_i^0\mu_i\right]\omega+\partial p/\partial t, \tag{45}$$

where

$$I(T)=\int_{T^0}^{T} c_p\,\mathrm{d}T \tag{46}$$

is a function of T only and we have used the rate result (28). When considering specific enthalpy the species are distinguished only by their heats of formation; thus, equation (45) is that of a single fluid in which differences in these heats are liberated as reactants change into products. The species equations (7) take a similar form

$$\rho(\partial Y_i/\partial t+\boldsymbol{v}\cdot\nabla Y_i)-\nabla\cdot(\mu\nabla Y_i)=\mu_i\omega. \tag{47}$$

Even differential diffusion leads to the energy equation (45) when the specific heats are equal; the sum in which the c_{pi} first appear in the calculation (39) is again zero, now by virtue of the identity (10). When all but the Nth species are scarce, the specific heats need not be equal to obtain an equation of the form (45); then $Y_1, Y_2, \ldots, Y_{N-1}, V_N$ are small and the same sum is negligible compared to the heat-conduction term in the resulting equation. A corresponding approximation of I then leads to equation (45) with c_p replaced by c_{pN}. On the other hand, the species equations (47) are changed by differential diffusion (the terms $\mu\nabla Y_i$ being replaced by $\Sigma_{j=1}^{N}\mu_{ij}\nabla Y_j$).

Another complication is the presence of the mass fractions in the equation of state (12). They effectively disappear if the first $N-1$ species are scarce compared to the last; but the same simplification can only be achieved, in general, when all the molecular masses are the same, i.e. $m_i=m$ for all i. Then the law (12) becomes

$$p=R\rho T/m \tag{48}$$

and the mixture behaves like a perfect gas. Such is the case when the reaction changes a single molecule into a single molecule, but in practice the condition is never met exactly.

We may certainly expect the assumptions of equal specific heats and molecular masses to be valid for a mixture in which c_{pi} and m_i do not vary much from species to species. In practice, c_p and m would be given intermediate values, e.g. by using the definition (36) and

$$m=\left[\sum_{i=1}^{N} Y_i/m_i\right]^{-1} \tag{49}$$

for a characteristic set of Y_i. Possibly more serious is the consequence (33) of the assumption (32). Since Turing's seminal work (1953) it has been known that differential diffusion can have important effects in reacting systems, particularly in pattern formation. In combustion it is believed to play a role in the stability of near-stoichiometric mixtures (Markstein 1964, p. 75). Certainly such mixtures are sensitive to differential mass diffusion (section 2.8), while a difference between heat and mass diffusion has wide-ranging effects, in particular on stability (Chapter 11). Because so little is known of the subject, only passing reference will be made to differential mass diffusion; then the simple form (35) will be used. By contrast, a major part of our theory is concerned with different heat and mass diffusivities.

The coefficients

$$\kappa,\ \lambda,\ \mu, \quad \text{and } c_p \tag{50}$$

are functions of the basic variables Y_i, ρ, and p. Further simplifications will result from taking them constant, but we shall delay presenting the equations until those are made dimensionless later.

5 The combustion and constant-density approximations

While the simplifications introduced so far have the nature of idealizations that are simulated more or less by actual mixtures, the basic approximation of the theory characterizes combustion as a low-speed phenomenon in a chemically reacting, compressible mixture for which gravity is usually unimportant. More precisely, the kinetic energy is small compared to the thermal energy so that a representative Mach number is small.

An immediate consequence (in the absence of gravity) is that spatial variations in pressure are also small, so we may set

$$p=p_c(t) \tag{51}$$

everywhere except in the momentum equation (8). (The notation indicates that the pressure can be controlled, for example from the surroundings.) This conclusion is reached from the momentum equation, which shows that fractional changes in pressure at any time are proportional to the square of the Mach number (see section 6). The pressure can still vary in time, but when it does, the variations will be considered to be about some constant value $\bar{p}_c$. We shall be concerned with at most small oscillations about a mean value (Chapter 5), which may then be taken for $\bar{p}_c$.

Buoyancy would appear to be an important factor in certain combustion problems, and then $\boldsymbol{g}$ should be retained in the momentum equation; the conclusion (51) is not affected provided the Froude number is moderate, a realistic restriction. Often, buoyancy is neglected everywhere, as in the usual discussions of fuel-drop burning, and although the flow field (which is not of primary interest) is then poorly described, remarkably accurate predictions of combustion characteristics are obtained (see Williams 1965, pp. 48 and 232). The reason may lie in the absence of gravity from the energy equation (9) and in an overall insensitivity of its solutions to inaccurate determinations of the convection $\rho\boldsymbol{v}$. Buoyancy will only be introduced for one aspect of stability (sections 11.3, 11.6), which is insufficient justification for carrying the more general equations throughout.

It is tempting to think that the momentum equation can now be discarded because its only role is the calculation of these small 'flow-induced' variations in pressure. But that would leave $N+2$ equations for the $N+4$ unknowns Y_i, $\boldsymbol{v}$, and ρ. If these variations must be eliminated, the momentum equation should be replaced by its own curl (Helmholtz equation), which provides two independent scalar equations. We prefer to leave it intact and interpret p as the small, flow-induced difference in pressure from its value $p_c(t)$.

If the reaction rate ω is zero (as it is in whole regions for the asymptotic approach we shall describe in section 8), the five equations (4), (8), and (45) for the five unknowns $\boldsymbol{v}$, p, and ρ become uncoupled from the species equations. They can be recognized as the equations of motion of a viscous, heat-conducting, compressible fluid, for which few exact solutions are known. The combustion approximation has admittedly been applied but such modification leaves them only slightly more tractable. A nonzero ω can only make analytical solution more difficult; this accounts for the heavy reliance on numerical methods in the combustion literature.

Most progress can be made for steady conditions, since only the mass flux $\rho\boldsymbol{v}$ is needed to make the energy equation (45) and species equations (47) a determinate set for T and Y_i. For certain geometries the mass flux is obvious *a priori*, while for others it can be approximated locally. The most important example of the former is a uniform flux in the x-direction, for which

$$\rho\boldsymbol{v}=M\boldsymbol{i} \quad \text{with } M \text{ constant.} \tag{52}$$

It can be used when T varies with x alone, but otherwise needs justification. If T also varies radially from the x-axis, then so do $\rho = p_c m/RT$ and the speed $v = M/\rho$. Such a unidirectional velocity field is acceptable to the momentum equation only in special circumstances, if at all. Burke & Schumann (1928), in tackling one of the earliest combustion problems, ignored this inconsistency.

Justification comes when the variations in temperature are due entirely to the combustion, and heat release is small compared to the thermal energy of the mixture. Since the density variations are correspondingly small, the velocity field is that of a constant-density fluid to leading order. The undisturbed values of ρ and $\rho \mathbf{v}$ may be used in equations (45) and (47) to determine the leading terms in the mass fractions and the perturbation of the otherwise constant temperature. In effect, Burke and Schumann introduced the constant-density approximation, a rational scheme valid even for unsteady problems. The general treatment of near-equidiffusional flames (section 3.8), on which a large part of the theory is based, depends on this approximation (section 8.5).

6 Constant properties: nondimensionalization

The coefficients (50) will now be taken constant.

The starting point for making the variables dimensionless is the unit of temperature, which comes from the factor

$$-\sum_{i=1}^{N} h_i^0 \mu_i = \sum_{i=1}^{N} h_i^0(\nu_i - \lambda_i) m_i = \nu \bar{m} Q \text{ (say)} \quad \text{with } \bar{m} = \sum_{i=1}^{N} \nu_i m_i/\nu, \tag{53}$$

on the right side of the energy equation (45). Here $\bar{m}$ is the stoichiometric average mass and Q is the excess of the total heat of formation of the reactants over that of the products per unit mass of the reactants (or products), i.e. the heat per unit mass released by the reaction at the standard temperature T^0. (When all the c_{pi} are equal, it is also the heat released at any temperature.) Combustion is only concerned with exothermic reactions; then $Q(>0)$ is called the (specific) heat of combustion. The unit of temperature is taken to be Q/c_p, so that the perfect gas law becomes Charles's law

$$T = p_c/\rho \tag{54}$$

if pressure is measured in units of $\bar{p}_c$ and density in units of $\rho_c = c_p \bar{p}_c m/QR$.

The pressure $\bar{p}_c$ will be varied from 0 to ∞, so that the choice of density unit ensures that ρ stays finite. The same should hold for the mass flux, but there is no such combination of existing parameters (excluding the inappropriate rate coefficient B). We accordingly take a representative mass flux M_r and refer the velocity to M_r/ρ_c. A diffusion length λ/c_pM_r and a diffusion time $\rho_c\lambda/c_pM_r^2$ can then be defined as units, recognizing that these are the scales for spreading the heat generated by the combustion. Finally, pressure variations are referred to M_r^2/ρ_c, which is proportional to the square of the Mach number (in accordance with section 5).

In these units the governing equations become

$$\partial\rho/\partial t+\nabla\cdot(\rho\boldsymbol{v})=0, \tag{55}$$

$$\rho(\partial Y_i/\partial t+\boldsymbol{v}\cdot\nabla Y_i)-\mathscr{L}^{-1}\nabla^2 Y_i=\alpha_i\Omega, \tag{56}$$

$$\rho(\partial\boldsymbol{v}/\partial t+\boldsymbol{v}\cdot\nabla\boldsymbol{v})=-\nabla p+\mathscr{P}[\nabla^2\boldsymbol{v}+\tfrac{1}{3}\nabla(\nabla\cdot\boldsymbol{v})], \tag{57}$$

$$\rho(\partial T/\partial t+\boldsymbol{v}\cdot\nabla T)-\nabla^2 T=\Omega+\beta\partial p_c/\partial t, \tag{58}$$

where the dimensionless rate of heat release per unit volume is

$$\Omega=\mathscr{D}\,e^{-\theta/T}\prod_{j=1}^{N}Y_j^{\nu_j}\quad\text{with }\mathscr{D}\equiv DM_r^{-2}. \tag{59}$$

Here $\mathscr{P}=c_p\kappa/\lambda$ is the Prandtl number; $\mathscr{L}$ has the definition (43); the quantities characterizing the chemical reaction are

$$\alpha_i=\mu_i/\nu\bar{m}=(\lambda_i-\nu_i)m_i/\nu\bar{m},\quad \theta=Ec_p/QR,$$

$$D(T)=\left(B\nu\bar{m}m^\nu\lambda Q^{\alpha-\nu}c_p^{\nu-\alpha-1}R^{-\nu}\bar{p}_c^\nu\prod_{j=1}^{N}m_j^{-\nu_j}\right)T^\alpha\rho^\nu; \tag{60}$$

and $\beta=(\gamma-1)/\gamma$, where $\gamma=1/(1-mc_p/R)$ is the ratio of specific heats. The perfect gas law (48) was obtained by approximating the molecular masses with m, but here it is important to distinguish between the m_i. Otherwise the α_i do not correctly represent the transformations that are taking place between the species.

When the geometry provides a characteristic length a, the non-dimensionalization is usually done differently. Then $\lambda/\rho_c c_p a$, $\rho_c c_p a^2/\lambda$ and $\lambda^2/\rho_c c_p^2 a^2$ are taken as units of velocity, time and pressure variation, respectively, to give the same equations (55)–(58), except that now M_r is missing from the definition (59) of Ω and

$$D(T)=\left(B\nu\bar{m}m^\nu\lambda^{-1}Q^{\alpha-\nu}c_p^{\nu-\alpha+1}R^{-\nu}a^2\bar{p}_c^\nu\prod_{j=1}^{N}m_j^{-\nu_j}\right)T^\alpha\rho^\nu. \tag{61}$$

In effect, M_r has been replaced by the new unit of mass flux λ/c_pa.

For any species that is solely a reactant or product, α_i is respectively negative or positive; also

$$\sum_{i=1}^{N}\alpha_i=0, \tag{62}$$

since the μ_i sum to zero according to the mass balance (22). The variations in $D(T)$ are unimportant and its value at a characteristic temperature T_* will be taken (see section 8). The constant

$$D = D(T_*) \tag{63}$$

is called the Damköhler number, even though the definition (60) gives it the dimensions of M_r^2; it represents the pressure $\bar{p}_c$. The term would be better applied to $\mathscr{D}$ which, as the ratio of a diffusion time to a reaction time under either definition of D, measures the rapidity of the reaction. Finally, θ is the dimensionless activation energy.

When the reactants and products are distinct, so that we may take

$$\lambda_i = 0,\ \nu_i \neq 0 \text{ for } i = 1, 2, \ldots, n \text{ (say)}, \quad \nu_i = 0 \text{ for } i = n+1, n+2, \ldots, N; \tag{64}$$

the equations for the products and inerts can be solved after the rest. Thus, the first n of the species equations (56) contain only the reactant mass fractions, as does the energy equation (58); together with the equations of continuity (55) and momentum (57) they form $n+5$ equations for the $n+5$ unknowns $Y_1, Y_2, \ldots, Y_n, \boldsymbol{v}, \rho$ and p. The coefficients on the right sides of these reactant equations have the property

$$\sum_{i=1}^{n} \alpha_i = -\sum_{i=1}^{n} m_i \nu_i / m\nu = -1. \tag{65}$$

Finally we mention the modifications when differential diffusion of the simple form (35) is considered: the species equation $i = N$ is dropped in favor of $Y_N = 1 - \Sigma_{i=1}^{N-1} Y_i$ and $\mathscr{L}$ is replaced by

$$\mathscr{L}_i = \lambda / \mu_{ii} c_p \tag{66}$$

in the remaining equations. It is worth remembering that the resulting modifications of equations (54)–(58) even hold for unequal c_{pi}, provided that $\mathscr{L}_i$, $\mathscr{D}$, and units are calculated with $c_p = c_{pN}$, and there is an inert diluent.

7 Equidiffusion: Shvab–Zeldovich variables

When there is equidiffusion of mass and heat, i.e.

$$\mathscr{L} = 1, \tag{67}$$

the convection–diffusion operators on the left sides of equations (56) and (58) are identical, so that

$$(\rho\partial/\partial t + \rho\boldsymbol{v}\cdot\nabla - \nabla^2)(Y_i - \alpha_i T) = 0 \quad \text{for all } i. \tag{68}$$

The reaction terms have been eliminated to show that, except for an inert,

$$H_i = T - Y_i/\alpha_i \tag{69}$$

is convected and diffused as in a reactionless mixture. Such variables may be attributed to Shvab and Zeldovich (see Williams 1965, p. 10). Note that

$$\sum_{i=1}^{N} \alpha_i H_i = -1 \tag{70}$$

according to the requirement (62), so that only $N-1$ of the Shvab–Zeldovich variables are independent.

When no reactant is a product, only $i=1, 2, \ldots, n$ need be considered (section 6) and then the H_i can be interpreted as conditional enthalpies. Thus, $v\bar{m}Q/v_i m_i = -Q/\alpha_i$ is the heat release per unit mass of the ith species and $(-Q/\alpha_i)\rho Y_i$ is, therefore, the heat that would be released if the ith species could be exhausted by the reaction (i.e. if there were enough of the other species present). It follows that $c_p T - QY_i/\alpha_i$ is the (dimensional) enthalpy per unit volume available under those conditions; and the H_i may be called (dimensionless) conditional enthalpies. For a single reactant, therefore,

$$H_1 = T + Y_1 \tag{71}$$

is the available enthalpy of the mixture.

The H_i can often be determined first, leaving only the temperature equation (58), with the Y_i replaced by $H_i + \alpha_i T$ in Ω, to be solved. (The $N-1$ linear combinations $\alpha_j Y_i - \alpha_i Y_j$ effect a similar reduction even when $\mathscr{L} \neq 1$, but leave two equations to be solved.) Considerable simplification can, therefore, be obtained by adopting the value (67).

However, such simplification is unnecessary in a theory based on activation-energy asymptotics. Except in certain flame sheets, Ω is required to be zero so that only convection–diffusion equations are to be solved in general. Inside the sheets the convection terms are negligible and the operators are once more identical (except for the factor $\mathscr{L}^{-1}$): local Shvab–Zeldovich variables can be formed for an arbitrary value of the Lewis number. The excuse for the value (67) is solely to eliminate an awkward symbol and whenever there is a significant difference (as in unsteady problems) a general value will be adopted.

These convection–diffusion equations have a particularly simple form when conditions are steady. If the (dimensionless) energy flux

$$\boldsymbol{E} = \rho T \boldsymbol{v} - \nabla T \tag{72}$$

is introduced, then

$$\nabla \cdot \boldsymbol{E} = 0 \tag{73}$$

in view of the continuity equation (55). Similarly, the (dimensionless) species flux

$$\boldsymbol{J}_i = \rho Y_i \boldsymbol{v}_i = \rho Y_i(\boldsymbol{v} + \boldsymbol{V}_i) = \rho \boldsymbol{v} Y_i - \mathscr{L}^{-1} \nabla Y_i, \tag{74}$$

i.e. the mass-flux fraction, satisfies

$$\nabla \cdot \boldsymbol{J}_i = 0. \tag{75}$$

Equation (73) even holds when conditions are unsteady, provided p_c is constant, by virtue of equation (54). (Of course, these are nothing more than conservation equations.) In one dimension integration is immediate. Even when a reaction is present, $\boldsymbol{E}$ and $\boldsymbol{J}_i$ are useful as intermediate variables.

8 Activation-energy asymptotics

We are for the most part concerned with combustion for which the exothermic reaction increases the temperature severalfold over the supply temperature. A true measure of the size of the activation energy is then θ, if the nondimensional maximum temperature T_m is close to 1. If this is not the case, the appropriate measure is θ/T_m. For the reaction (24) the activation temperature E/R is about 2500 K in the range 273–523 K (Kontrat'ev 1972, p. 230) so that at the upper end of this range θ/T_m is about 5, not a very large number. For reactions of interest in combustion, activation temperatures of at least 20 000 K and flame temperatures of 2000 K are common, giving (conservatively) $\theta/T_m = 10$. The limit $\theta \to \infty$ promises, therefore, to be relevant and, indeed, has been found to give remarkably accurate results for $\theta = 12$ (Matalon, Ludford, & Buckmaster 1979). In this connection it is worth noting that the Arrhenius factor $e^{-12} \simeq 6 \times 10^{-6}$ and this is reduced by 90% if the maximum temperature is reduced by 16%. An essential characteristic of activation-energy asymptotics is to strongly limit the regions in which reaction is significant, and this is achieved with quite modest values of θ.

Taking the straightforward limit in the reaction term (59) leads to the trivial result $\Omega = 0$; reaction is completely eliminated from the discussion, which then has nothing to do with combustion. Nontrivial results can be obtained by taking a distinguished limit in which $\mathscr{D} \to \infty$ in an appropriate fashion with θ. Generally we shall write

$$\mathscr{D} = \mathscr{D}' e^{\theta/T_*} \tag{76}$$

so that

$$\Omega = \mathscr{D}' Y_1^{\nu_1} Y_2^{\nu_2} \ldots Y_N^{\nu_N} e^{(\theta/T_* - \theta/T)}, \tag{77}$$

where T_* is a temperature characterizing the magnitude of $\mathscr{D}$. The exponential in $\mathscr{D}$ is the essence of the distinguished limit, $\mathscr{D}'$ only being allowed to depend algebraically on θ.

If T is smaller than T_*, the reaction term (59) is exponentially small in the limit, so that it may be neglected in regions that are not exponentially large for times that are not exponentially long. This is called the frozen limit, which may not hold in large regions or for long times because of cumulative effects.

If T is larger than T_*, the reaction term is exponentially large in the limit, unless (30) holds to all orders in θ^{-1}. The corresponding equation (56) then shows that, unless spatial or time derivatives are very large, Ω is again negligible (but now because one of the reactants has been consumed). This is called the equilibrium limit, in which the remaining reactants are still available but no reaction occurs.

If, at the instant considered, T is equal to T_* only at a point or on a surface, then the combustion field consists for the most part of regions where the reaction is negligible, either because it is frozen or because equilibrium prevails. Significant reaction will only occur in a small zone, called a hot-spot or flame sheet. For example, a combustion field containing no maximum temperature will be divided into an equilibrium region with $T > T_*$ and a frozen region with $T < T_*$ by a surface on which $T = T_*$. Only in the neighborhood of this surface where $T - T_* = O(\theta^{-1})$ is the chemical reaction significant. Herein lies the main achievement of activation-energy asymptotics: the reaction term is eliminated from most of the combustion field, where its incorporation would only add to the mathematical complexities inherent in the description of a compressible flow, and is retained only where special circumstances simplify the nonreactive terms. Note that only values of $D(T)$ close to $D(T_*)$ will appear in the analysis, which reinforces the assertion in section 6 that its variations are unimportant. Accordingly, T_* will be taken as the characteristic temperature introduced in defining the Damköhler number (63).

The theory we shall present is based on the systematic exploitation of the limit $\theta \to \infty$ as outlined, with modifications and extensions as appropriate. Success in covering a wide range of combustion problems, however, should not blind the reader to those aspects that are eliminated by restriction to reactions with extreme sensitivity to temperature. In particular, the termination step in chain reactions has a weak dependence on temperature (see Kapila (1978) who treats a spatially homogeneous problem) and, therefore, cannot be confined to a zone. Nevertheless, activation-energy asymptotics can be used to simplify the other steps (Liñán 1971). There are also phenomena, such as stable flame propagation in thermites, that certainly depend on θ being finite: the sensitivity of the self-heating in the limit does not accurately represent the process for finite θ. Even then activation-energy asymptotics can provide a model (section 11.8) which, though technically speaking irrational (Van Dyke 1975, p. 2), does exhibit the phenomenon.

To avoid repetition a uniform notation for asymptotic expansions will be used throughout, namely

$$v = v_0 + \theta^{-1} v_1 + \theta^{-2} v_2 + \dots \tag{78}$$

for the generic dependent variable v. The coefficients v_0, v_1, v_2, etc. will be called the leading term, (first) perturbation, second perturbation, etc. Only the variables on which these coefficients depend need be specified, changing as they do from region to region. Thus, an expansion will only be written when it differs from (78); but the independent variables will always be identified. In all cases the notation $+\dots$ is used (when a more accurate estimate is not needed) to denote a remainder that is of smaller order in θ than the preceding term. The same notation will be used for parameters; however, since their expansion coefficients are constants, no specification of variables is needed.

Similar conventions will be used when a small parameter other than θ^{-1} is involved or, in the case of $+\dots$, when asymptotic expansions in a variable are being considered. Numerical subscripts such as 0, 1, 2 will henceforth be used exclusively for terms in asymptotic expansions.

2

THE PREMIXED PLANE FLAME

1 Models of mixtures

In general, two-reactant flames can be classified as diffusion or premixed. In a premixed flame the reactants are constituents of a homogeneous mixture that burns when raised to a sufficiently high temperature. In a diffusion flame the reactants are of separate origin; burning occurs only at a diffusion-blurred interface.

Both kinds of flames can be produced by a Bunsen burner. If the air hole is only partly open, so that a fuel-rich mixture of gas and air passes up the burner tube, a thin conical sheet of flame (typically blue-green in color) stands at the mouth; this is a premixed flame. Any excess gas escaping downstream mixes by diffusion with the surrounding atmosphere and burns as a diffusion flame (typically blue-violet). If the air hole is closed, only the diffusion flame is seen; if all the oxygen is removed from the atmosphere (but not from the air entering the hole), only the premixed flame is observed. (Yellow coloring due to carbon-particle luminosity may also be seen.)

A premixed flame can also be obtained by igniting a combustible mixture in a long uniform tube; under the right conditions, a flame propagates down the tube as a (more or less) steady process, called a deflagration wave. The Bunsen burner brings such a wave to rest by applying a counterflow and stabilizes it with appropriate velocity and thermal gradients.

The reactions that occur in almost all flames (including those on Bunsen burners) are extremely complicated and often poorly understood. The overall combustion process can, however, be represented by a single one-step irreversible reaction:

$$\text{fuel} + \text{oxidant} \rightarrow \text{products}.$$

Such a model preserves the essential features that fuel and oxidant are consumed and heat is liberated, but still involves two mass fractions.

If fuel and oxidant are present in the fresh mixture in exactly the right proportion, both are completely consumed in the flame. Then it is reasonable to expect them to act like a single reactant that fuses into a product, the reaction being treated as one-step and irreversible:

$$\text{fuel/oxidant} \rightarrow \text{product}.$$

Only one mass fraction is now involved. A similar model is plausible when either fuel or oxidant is present in excess – the fuel in a rich flame, the oxidant in a lean one. Only the active portion of the excess component is linked with the deficient one; the remainder is considered to be part of the products and inerts.

This model, which is certainly valid for a single decomposing reactant, is widely used for premixed flames. It yields results that, with suitable interpretation, apply to a two-reactant model in which differential diffusion is negligible, at least for large activation energy. The relation between the two models is clarified in section 8; meanwhile, the simpler, one-reactant model will be considered. Of course, the model applies directly to problems in which there is a single decomposing reactant, such as an alkyl nitrate or nitrite. Other decomposing organic compounds used in combustion are azomethane, ethyl azide, nitroglycol and nitromethane; the inorganic compounds include hydrazine, hydrogen peroxide, nitrous oxide, and ozone.

We begin with the basic problem of combustion theory: to describe the unlimited, plane, steady solution of the governing equations when conditions upstream are specified. The problem is analogous to the shock wave of ordinary compressible flow but, dissimilarly, it is ill-posed as stated (section 2) and does not necessarily represent local behavior in more general circumstances (Chapter 8).

2 Cold-boundary difficulty

If a combustible mixture in a long, straight, uniform tube is ignited at an open end, a flame moves steadily toward the other end, separating unburnt from burnt gas. An observer moving with the flame sees a constant flux M of unburnt gas toward himself from the left (say) and an equal flux of burnt gas away to the right, the value of M depending on pressure. It is, therefore, reasonable to suppose that the equations governing steady plane combustion have a solution for which the mass fractions and temperature tend to given values far to the left (i.e. upstream) and to constant values (zero for one mass fraction) far to the right. (Only through the exhaustion of one reactant does the combustion cease.)

Take unit Lewis number and consider a single reactant whose molecules decompose in one step; then $n=1$, $\lambda_1=0$, $\alpha_1=-1$ (see section 1.6) if the reactant is not a product. (More complicated reactions will be treated in section 7.) Then the corresponding Shvab–Zeldovich variable is the available enthalpy (1.71), and it satisfies

$$\mathfrak{L}(H, 1)=0 \quad \text{for} -\infty<x<\infty, \tag{1}$$

where the subscript 1 has been dropped and

$$\mathfrak{L}(Y, \mathscr{L})\equiv(\mathscr{L}^{-1}\,\mathrm{d}^2/\mathrm{d}x^2-\mathrm{d}/\mathrm{d}x)Y. \tag{2}$$

The only bounded solutions are constants, so that the available enthalpy is conserved, i.e.

$$H\equiv T+Y=T_{\mathrm{f}}+Y_{\mathrm{f}}\equiv H_{\mathrm{f}} \tag{3}$$

and Y is, therefore, known in terms of T. The subscript f denotes the fresh (unburnt) state upstream; likewise b will be used for the burnt state downstream. Note that there may be either products, or inerts, or both in the fresh mixture. If M_{r} is taken to be M, the remaining equation is

$$\mathfrak{L}(T, 1)=-\mathscr{D}Y\mathrm{e}^{-\theta/T} \quad \text{with } \mathscr{D}\equiv DM^{-2} \tag{4}$$

and its solution must satisfy

$$T=T_{\mathrm{f}} \text{ at } x=-\infty, \quad T=H_{\mathrm{f}} \text{ at } x=+\infty \tag{5}$$

since Y is to vanish far to the right. The temperature far to the right will be denoted by

$$T_{\mathrm{b}}=H_{\mathrm{f}}, \tag{6}$$

an equation expressing the adiabatic nature of the combustion process; T_{b} is here equal to what is known as the adiabatic flame temperature. Once T is determined, the perfect gas law and the continuity and momentum equations are satisfied by $\rho=1/T$, $\boldsymbol{v}=T\boldsymbol{i}$, and $p=T_{\mathrm{b}}-T+\frac{4}{3}\mathscr{P}T'$ (p_{c} being identically 1 and the mass flux M having been used in the nondimensionalization). Here $\boldsymbol{i}$ is the unit vector in the x direction and the prime denotes differentiation. The goal is to determine M, from which the velocity upstream, known as the adiabatic flame speed, can be calculated.

The operator $\mathfrak{L}$ is of second order, whereas there are effectively three boundary conditions to be satisfied (that on the left should be counted twice, as we shall see later). Only if $\mathscr{D}$ has an appropriate value can we expect a solution. The term eigenvalue is used, but it should be noted that we are not dealing with an eigenvalue in the usual sense. Determining the eigenvalue gives the value of M corresponding to D, which represents the pressure $\bar{p}_{\mathrm{c}}$ of the gas in the tube (see the definition (1.60)). Note that D

has been taken constant, so that a characteristic temperature must still be chosen; in the asymptotic theory it is required to be T_* (section 1.8), which will turn out to be T_b.

It is not difficult to see that the problem as posed has no solution (von Kármán & Millán 1953; Hirschfelder, Curtiss, & Campbell 1953). As $x \to -\infty$ the left side of equation (4) would tend to zero, since both dT/dx and d^2T/dx^2 do so, but the right side would tend to the nonzero value $-\mathscr{D} Y_f e^{-\theta/T_f}$. Since T_f is less than T_b, according to the result (6), this is called the cold-boundary difficulty; much ingenuity and effort has been spent on it (see Williams 1965, Chapter 5).

The origin of the difficulty is clear on physical grounds. Our formulation requires the mixture to be reacting all the way in from $x = -\infty$, so that by the time finite x is reached the combustion would be complete. To overcome the difficulty, which is a result of idealizations, it is only necessary to avoid reaction at a finite rate over an infinite time; this has been done in two ways.

The first consists in modifying the reaction term to make it vanish as $T \to T_f$. A factor such as $(T - T_f)$ will achieve this, but creates another difficulty. Johnson (1963) has shown that, instead of no eigenvalue, there is a continuum of eigenvalues $0 < \mathscr{D} \leqslant \mathscr{D}_{max}$, irrespective of how strongly the factor vanishes, provided it is not identically zero in some neighborhood of T_f (as in the switch-on concept treated next). Such tinkering with the reaction term may be mathematically satisfying, but there is little else to recommend it. The reaction term as it stands has a sound theoretical and experimental footing; the modifications do not. They can be disqualified for depending on the problem at hand (through T_f), instead of properties of the reactant only.

A unique eigenvalue is obtained if the unit-step function $\mathscr{H}(T - T_s)$, with $T_f < T_s < T_b$, is used as a factor. When $\mathscr{H}$ is used in conjunction with the preceding factors, the eigenvalue tends to $\mathscr{D}_{max}$ from above as $T_s \to T_f$. When $\mathscr{H}$ is used alone, T_f is not involved in the reaction term, which therefore qualifies for consideration. Certainly it may be argued that a realistic reaction will have a switch-on temperature (von Kármán & Millán 1953). But the same effect can be achieved without the factor if only a certain idealization is removed, and that leads to the second way of resolving the difficulty.

3 Anchored flames

If we temporarily abandon the original question and set the left boundary at a finite value of x, which may be taken as the origin, then specifying how the reactant is supplied at $x = 0$ gives a properly posed problem. It provides three boundary conditions from which T, T', and the constant replacing H_f in the T, Y-relation (3) can be determined. The last mentioned is, of course, still the final temperature T_b. (Our remark about condition (5*a*) being counted twice comes from the provision of T and T' at the left boundary here.) We have removed the idealization of unboundedness on the left (but retained that of steadiness) and have thereby introduced the anchored flame, which occurs under a variety of circumstances.

For example, a liquid such as ethyl nitrate evaporates to form a gaseous reactant that burns by decomposition. If, therefore, $x<0$ is filled with the liquid maintained at a uniform temperature T_s, then the reactant is supplied by evaporation through its surface with

$$T=T_s, \quad dT/dx=L, \quad Y-dY/dx=1 \quad \text{at } x=0, \tag{7}$$

where L is the latent heat of evaporation at the temperature T_s measured in units of the heat of combustion. The second of these conditions ensures that heat conducted back from the flame is used to vaporize the liquid while, according to the definition (1.74), the third says that the whole mass flux is due to the reactant, i.e. none of the products or inerts is moving into or out of the liquid. Since $dT/dx=-dY/dx$, the last condition provides the value $Y=1-L$ at $x=0$, from which the constant value

$$H=1-L+T_s \tag{8}$$

is calculated. (The result can be obtained directly from conservation of enthalpy.)

The idea of a switch-on temperature fits into the same scheme. In the absence of reaction, $E=T-T'$ is conserved, according to equation (1.73), so that

$$T=T_f+(T_s-T_f)e^x \quad \text{for } x<0 \tag{9}$$

provided the switch-on temperature T_s is attained at $x=0$. The T, Y-relation (3) and this last result show that

$$T=T_s, \quad dT/dx=T_s-T_f, \quad H=H_f \quad \text{at } x=0, \tag{10}$$

from which one deduces (as expected) that the relation (3) continues to hold for $x>0$. For later reference note that the mass flux fraction of the reactant at $x=0$ is

$$J_s\equiv Y-dY/dx=Y_f>0, \tag{11}$$

its value everywhere in $x<0$ (see (1.75)).

Clearly the solution, and hence the eigenvalue, depends on how the reactant is supplied. This lack of universal significance disappears as θ becomes large, when the eigenvalue depends only on J_s and $T_b(=E_s+J_s)$, although the position of the flame sheet does not. The asymptotic value has been the focus of all analysis, and properly so; its determination can be attributed to Bush & Fendell (1970), even though they were concerned with the original unbounded problem of section 2.

4 Asymptotic analysis

The limit $\theta \to \infty$ which, in the present context, was first considered by Williams (1973) will now be treated in detail. The temperature increases from T_s at $x=0$, where the reactant is supplied, to T_b as $x \to \infty$. On physical grounds, we would expect this increase to be monotonic, corresponding to a steady decrease in the reactant concentration Y. If the T_* of section 1.8 is then taken to have some intermediate value we might anticipate that a flame sheet, located at $x=x_*$, separates a frozen region characterized by $T_s < T < T_*$ from an equilibrium region characterized by $T_* < T < T_b$. Taking T continuous across the sheet is consistent with the idea that the reaction term represents a simple source of heat and also a sink of the reactant. In both regions $\mathfrak{L}(T, 1)=0$.

The only solutions that are bounded as $x \to +\infty$ are constants, however, so that

$$T=T_b, \quad Y=0 \quad \text{for } x>x_* \tag{12}$$

and $T_*=T_b$. Since T_* cannot have an intermediate value, the question then arises of whether equilibrium must prevail behind the flame, consistent with (12). The answer lies in Ω which, if most of the reactant is to be consumed in a very thin region, must become unbounded in the limit $\theta \to \infty$ so as to have a finite integral. Hence the $\mathscr{D}'$ of section 1.8 must contain an algebraic factor in θ that becomes unbounded; it is this factor that forces equilibrium behind the flame sheet, there being nothing to balance an otherwise algebraically large reaction term. Moreover, based on the present assumption that Ω is proportional to Y, there is equilibrium to all orders in θ. (This is not necessarily so otherwise.) The appropriate algebraic factor will emerge in a natural way in seeking a limit equation to describe the structure of the flame sheet.

Up to the flame sheet the reaction must be frozen because there is reactant present, so that to all orders in θ we have

$$T=T_s+T_s'(e^x-1) \quad \text{for } 0<x<x_*, \tag{13}$$

where T_s and T_s' are values at $x=0$ determined by how the reactant is supplied. The flame sheet must lie at

$$x_*=\ln[1+(T_b-T_s)/T_s'] \tag{14}$$

if the temperature is to be continuous across it. Since $T_b=Y_s+T_s>T_s$, consistency requires $T_s' > 0$, i.e. whatever supplies the reactant must be a (conductive) heat sink. Both the vaporizing liquid and switch-on model of section 3 have this property. The T, Y-profiles are plotted in Figure 1, from which it is clear that the flame sheet is a source of heat and a sink of reactant, the reaction term behaving as a δ-function.

The temperature (13) decays from its value at the flame in an $O(1)$ multiple of the unit of distance λ/c_pM (see section 1.6), thereby delimiting a so-called preheat zone. The width of the zone is commonly called the flame thickness; we shall follow that usage even though the term more properly applies to the much thinner flame-sheet layer. The preheat zone is very thin, typically about 0.5 mm for a velocity of 50 cm/s, because the thermal conductivity of gases is small. Nevertheless, it is large compared to the thickness of shock waves, because the mass flux is so much smaller.

Completion of the asymptotic solution requires a discussion of the layer that must effect the changes in gradients from one side of the flame sheet to the other. Here the reaction term

$$\mathscr{D} Y \mathrm{e}^{-\theta/T} = \mathscr{D}' Y \mathrm{e}^{(\theta/T_* - \theta/T)} \tag{15}$$

is important, so that perturbations of T from $T_* = T_b$ must be $O(\theta^{-1})$. Accordingly, we take an expansion of the form (1.78) with

$$T_0 = T_b \tag{16}$$

and variable

$$\xi = \theta(x - x_*), \tag{17}$$

1 Profiles of T and Y, drawn for $\mathscr{L} = 1$ and $T_s = 0.25$, $T_s' = T_b = 1$. The flame sheet is at $x_* = 0.56$.

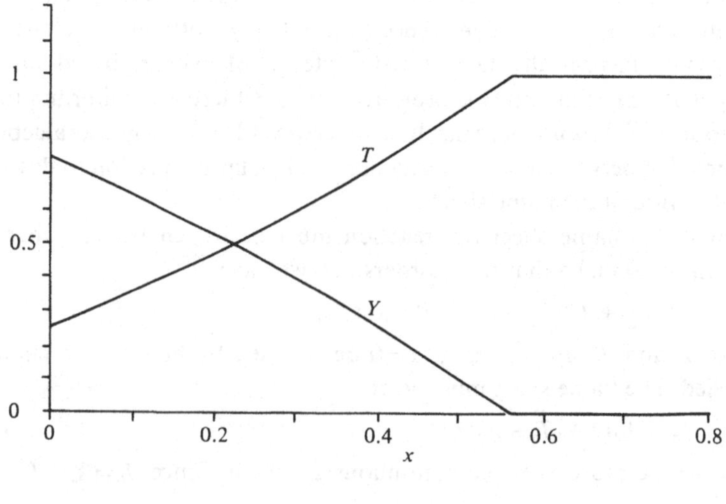

the coordinate scaling being dictated by the need to match gradients outside the layer. Then

$$d^2\phi/d\xi^2 = \tilde{\mathscr{D}}\phi\, e^{-\phi} \quad \text{with } \tilde{\mathscr{D}} = \mathscr{D}'/\theta^2, \quad \phi \equiv (1/T)_1 = -T_1/T_b^2 \tag{18}$$

is obtained as the structure equation, showing that the balance in the layer is between diffusion and chemical reaction with the convection playing no role. (Note that $\tilde{\mathscr{D}}$ is expected to be $O(1)$, which confirms that $\mathscr{D}' = O(\theta^2)$ is only algebraically large in θ (see section 1.8). Moreover, Ω is $O(\theta)$ in a zone of thickness $O(\theta^{-1})$ so that it is a δ-function.) To match with the solutions on either side of the flame sheet, the boundary conditions

$$\phi = -J_s\xi/T_b^2 + o(1) \quad \text{as } \xi \to -\infty, \quad \phi = o(1) \quad \text{as } \xi \to +\infty \tag{19}$$

must be applied. Here J_s is the mass-flux fraction (11). Note that the last condition cannot be replaced by the weaker requirement that ϕ tends to a nonzero constant, since the differential equation (18) would be violated. We conclude that the structure problem itself prohibits partial-burning solutions.

The first integral

$$(d\phi/d\xi)^2 = 2\tilde{\mathscr{D}}[1-(\phi+1)\, e^{-\phi}] \tag{20}$$

is obtained by using the boundary condition (19*b*). The remaining boundary condition will then be satisfied only if

$$\tilde{\mathscr{D}} = J_s^2/2T_b^4, \tag{21}$$

to which corresponds

$$\mathscr{D} \equiv DM^{-2} = J_s^2\theta^2\, e^{\theta/T_b}/2T_b^4. \tag{22}$$

A further integration using the same condition now yields the structure

$$J_s\xi/T_b^2 = \int_\phi^\infty \{[1-(\phi+1)\, e^{-\phi}]^{-\frac{1}{2}} - 1\}\, d\phi - \phi \tag{23}$$

(note again how the left boundary condition counts twice). The result (22) justifies our assertion that the eigenvalue $\mathscr{D}$ depends only on J_s and T_b and not all three values

$$T_s, \; Y_s, \quad \text{and } T_s' = -Y_s'. \tag{24}$$

For the evaporating liquid we find $x_* = \ln(L^{-1})$ and the formula (22) with $J_s = 1$. Clearly the solution is only valid if $L < 1$; there is, indeed, no solution for any value of θ when $L > 1$ since the boundary conditions (7) require $Y_s = 1 - L$. The vaporization must not require more heat than is available from the combustion, which is the only source when there is no conduction in from $x = +\infty$. This limitation does not hold for more dimensions, where there is a possibility of conduction from the surroundings. The position of the flame depends only on M (through the unit of length) and the temperature T_s of the liquid (through L). According to the definitions (1.60) and (1.63), determination of DM^{-2} enables M to be calculated for each $\bar{p}_c$. In that connection note that $\bar{p}_c$ is not independent of existing parameters: the Clausius–Clapeyron law for a vaporizing liquid, which has so far been ignored, determines the partial pressure $\bar{p}_c Y_s$ of the vapor from the temperature T_s of the liquid (see section 4.7).

For the switch-on reaction $x_* = \ln[(T_b - T_f)/(T_s - T_f)]$, which is always positive, and J_s has the value (11) in the eigenvalue (22). An apparent consequence is that a decrease in the amount of reactant in the fresh mixture has the effect of increasing the burning rate at the same pressure, which contradicts experience (Lewis & von Elbe 1961, p. 312). The flame temperature T_b, however, also depends on Y_f and, when that is taken into account, M decreases with Y_f except for values $O(\theta^{-1})$, where our analysis breaks down.

In the limit, the switch-on formulas provide a solution for the original unbounded problem (section 2), which is then well-posed. The function (13) is merely the continuation of the function (9), according to the determination (10*b*); together they form the solution in $-\infty < x < x_*$ where the reaction is frozen. Thus, T_s loses its role as the switch-on temperature, which is raised to T_* by the asymptotics, and merely becomes the temperature at the station from which x is measured. One achievement of activation-energy asymptotics is, therefore, to circumvent the improperly posed problem (3), (4), and (5). A glimmer of the idea shows through the 1970 paper of Bush and Fendell, but they are mainly concerned with modifying the reaction term to make a properly posed problem. (They also calculate the next approximation, which agrees remarkably well with numerical results for moderate values of θ.)

Resolution of the cold-boundary difficulty in the original unbounded problem can also be effected by removing the idealization of steadiness. In fact, the mixture ahead of any flame is reacting and the problem cannot be steady. When θ is large, however, the reaction is exponentially slow and so may be ignored for times that are not correspondingly large. Our asymptotic results for flames moving into unbounded mixtures apply for such times.

5 Flat burners

So far we have considered problems in which D is given and M is to be found. Our analysis, however, also applies when both are given but only two supply conditions are provided; the third is then determined *a posteriori*. For instance, consider a burner in which the reactant is injected into a tube at a given rate and temperature through a flat porous plug against a controlled pressure in $x>0$. The porous plug inhibits back-flow of the product, so that M, D, T_s, and J_s are prescribed (J_s being equal to 1 when pure reactant is supplied). The formula (22) for $\mathscr{D}$ then determines the third supply parameter T_b. From it the heat conducted back to the porous plug can be calculated as $T_s' = T_s + J_s - T_b$, and that heat must be removed by cooling coils if the temperature of the reactant is to be maintained at T_s. Such indirectly determined supply conditions were suggested by Hirschfelder & Curtiss (1949) as a means of overcoming the cold-boundary difficulty.

Since T_s' must be positive there is a maximum value of T_b, namely

$$T_a = T_s + J_s, \tag{25}$$

called the adiabatic flame temperature because none of the heat released is then lost to the burner. For given D, T_s, and J_s there is, therefore, a limiting value M_a of M, for which the flame blows off (i.e. $x_* \to \infty$). In terms of T_a and M_a we may write

$$x_* = \ln[J_s/(T_a - T_b)], \quad M/M_a = (T_b/T_a)^2 \, e^{(\theta/2T_a - \theta/2T_b)} \tag{26}$$

and, when T_b is within $O(\theta^{-1})$ of T_a, the pre-exponential factor may be omitted, to obtain a formula established empirically by Kaskan (1957).

Ferguson & Keck (1979) have considered the behavior of the flame as M decreases from M_a; we shall extend their discussion down to the value of M for which $T_b = T_s$ and the flame reaches the surface of the burner. Although T_b, and hence x_*, decrease monotonically with M, the (dimensional) stand-off distance of the flame does not. The reason lies in the unit of length λ/c_pM, which, by increasing, produces an opposing effect. If the fixed unit of length λ/c_pM_a is used, then the formula (26*a*) is replaced by

$$x_* = (T_a/T_b)^2 \, e^{(\theta/2T_b - \theta/2T_a)} \ln[J_s/(T_a - T_b)] \tag{27}$$

and, for almost all the range of M, the stand-off distance actually increases: the resulting minimum and maximum lie at $T_b = T_a - 2T_a{}^2/\theta \ln\theta + \cdots$ and $T_s + 2T_s{}^2/\theta + \cdots$, respectively.

Figure 2 shows how x_* varies with T_b (i.e. M) for the parameter values used by Ferguson and Keck. (Their formula lacks the factor T_a^2/T_b^2 due to its empirical derivation and, hence, describes the right side of the figure only.) As T_b decreases from T_a, the stand-off distance decreases very rapidly to a minimum (obtained experimentally by Ferguson and Keck) and then rises gradually to a maximum, after which it falls rapidly to zero for $T_b = T_s$. (When $T_b - T_s = O(\theta^{-1})$ the analysis must be modified as in section 7, since the burner intrudes into the reaction zone.)

2 Stand-off distance x_* (with fixed length unit) versus flame temperature T_b, drawn for $T_s = 0.25$, $J_s = 1$, $\theta = 12.5$: (*a*) complete range $T_s < T_b < T_a$; (*b*) detail near $T_b = T_s$; (*c*) detail near $T_b = T_a$.

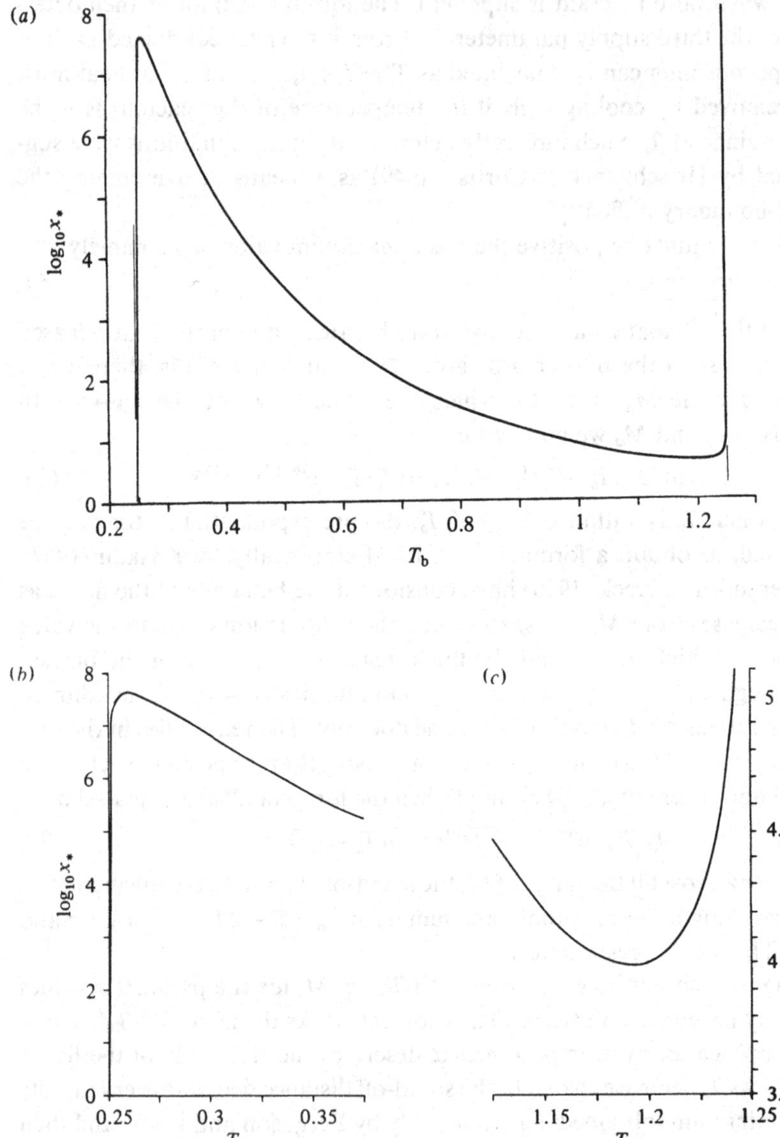

6 Near-surface, surface, and remote flames

Different values of M for the same supply parameters (24) give similar combustion fields, since only the length unit is affected. As $M \to \infty$ the latter tends to zero and the flame approaches the supply surface. An intense convective–diffusive zone of (dimensional) thickness $O(M^{-1})$ forms near the surface, bounded by a reaction zone whose (dimensional) thickness is $O(\theta^{-1}M^{-1})$. Such near-surface flames occur when D is correspondingly large, since $\mathscr{D}$ is fixed.

By contrast a true surface flame can be produced for any M if T_b can be made to approach T_s. For the switch-on reaction this can be effected by changing the upstream concentration and for the flat burner by adjusting the pressure; but for an evaporating liquid $T_b - T_s = 1 - L$ is fixed. Likewise, a remote flame can be obtained in the first two cases by making T_b approach the adiabatic flame temperature (25) but not for the last. Since $T_s' \to 0$, isothermal conditions are then obtained, according to equation (13). At both extremes the preceding analysis becomes invalid: either the boundary intrudes into the reaction zone and there is no frozen region between them, or the isothermal limit is not uniformly valid in the unbounded frozen region. In either case, surface flame or remote flame, the asymptotics must be reworked and we shall do this in a general manner.

The formula (14) predicts that x_* is $O(\theta^{-1})$, i.e. that intrusion has taken place, as soon as $T_b - T_s$ is $O(\theta^{-1})$. More precisely, if

$$T_* = T_s + \theta^{-1} t_s \quad \text{then } x_* = \theta^{-1} t_s / T_s'. \tag{28}$$

If these are inserted into the structure variable (17) and the corresponding expansion, the same structure equation (18*a*) and condition (19*b*) are obtained, but the condition (19*a*) is replaced by the pair

$$\phi = t_s/T_s^2, \quad \mathrm{d}\phi/\mathrm{d}\xi = -T_s'/T_s^2 \quad \text{at } \xi = -t_s/T_s'. \tag{29}$$

The integral (20) still applies but the determination (21) of $\tilde{\mathscr{D}}$ and the structure (23) do not. In their place we find

$$\tilde{\mathscr{D}} = J_s^2/2T_s^4[1 + (\phi_s - 1)\,\mathrm{e}^{\phi_s}] \quad \text{with } \phi_s = -t_s/T_s^2, \tag{30}$$

and

$$J_s(\xi + t_s/J_s) = T_s^2[1 + (\phi_s - 1)\,\mathrm{e}^{\phi_s}]^{\frac{1}{2}} \int_{\phi}^{-\phi_s} [1 - (\phi + 1)\,\mathrm{e}^{-\phi}]^{-\frac{1}{2}}\,\mathrm{d}\phi, \tag{31}$$

since $T_s' = -Y_s'$ is J_s when Y_s vanishes. As $t_s \to \infty$ these results merge with those for a flame standing clear of the boundary: $\tilde{\mathscr{D}}$ tends to the value (21) and the structure (31) to the structure (23). As $t_s \to 0$, $\tilde{\mathscr{D}} \to \infty$ and the structure behaves like

$$\phi = -\phi_s\,\mathrm{e}^{-(1 + J_s\xi/t_s)}, \tag{32}$$

which vanishes in the limit.

As $T_s' \to 0$ the flame recedes to infinity while T tends to T_s at any finite point. The solution (13), being based on a frozen reaction, continues to hold so long as T_s' is not too small; but when

$$T_s' = \varepsilon t_s' \quad \text{with } \varepsilon = e^{(\theta/T_b - \theta/T_s)} = o(\theta^{-1}), \tag{33}$$

it fails because the reaction can no longer be ignored. When changes with x are small the second derivative is even smaller than the first in the equation (4*a*), which becomes

$$dT/dx = \mathscr{D}'(T_b - T)\, e^{(\theta/T_b - \theta/T)} \tag{34}$$

representing a convective–reactive balance sometimes called an explosive regime, for a reason that will be evident. Integration shows that $T \to T_b$ asymptotically in x, but all we are interested in is the derivative $T_s' = \mathscr{D}'(T_b - T_s)\varepsilon$, which gives

$$\mathscr{D}' = t_s'/(T_b - T_s). \tag{35}$$

Thus $\mathscr{D}'$ varies from 0 to ∞ (the latter corresponding to $\tilde{\mathscr{D}} = \mathscr{D}'/\theta^2 = J_s^2/2T_b^4$) with t_s'.

The explosive regime does not admit a simple expansion scheme because the appropriate scale transformation varies from point to point. Under the transformation $t = \varepsilon x$ equation (34) becomes that governing the adiabatic explosion of a reactant (see section 12.3), for which three phases may be distinguished as the time t increases: induction, explosion, and relaxation. These phases will be encountered as we move downstream and, with the exception of the last, the formulas in section 12.3 describe them. The relaxation phase, characterized by temperatures within $O(\theta^{-1})$ of T_b, takes place over times that are $O(\varepsilon)$; for the corresponding $O(1)$ distances here the second derivative must be reinstated in equation (34). The resulting modification of that phase is straightforward, albeit numerical, but the details will be omitted.

The common feature of surface and remote flames is their ability to change $\tilde{\mathscr{D}}$ from its value (21). For the same M, the surface flame allows the pressure (represented by the Damköhler number) to range up to infinity and the remote flame allows it to range down to zero. The reverse is true of M for fixed pressure.

These features enable us to trace the position of the flame for the flat burner introduced in section 5 as the pressure is decreased from a sufficiently large value keeping the injection rate constant. Let D_s and D_∞ be the values given by the formula (22) when $T_b = T_s$ and $T_s + J_s$, respectively. So long as $D > D_s$ the flame remains at the plug, but moves away an ever-increasing distance as D decreases from D_s to D_∞. As D passes through

D_∞ and decreases further, the flame spreads out to form a very remote broad reaction zone. In practice the last stage will not be observed because heat loss through the tube wall, here neglected, will extinguish the flame before D_∞ is reached (see section 3.4).

7 Reactions of arbitrary order

The treatment so far has been based on a single reactant, each of whose molecules decomposes in one step, and a constant Damköhler number. Since the latter only enters into the determination of the flame structure, a D that varies with T is covered by taking its value for $T = T_*$. None of our results is changed. The decomposition of a compound, however, usually involves several elementary steps and then the order of the reaction by which it is approximated is found to be different from 1. It is, therefore, of interest to see how our results are affected by a general exponent ν_1 (Berman & Riazantsev 1972).

The order of the reaction only enters when the flame structure is considered so that in section 4 modifications first arise at equations (18), which are replaced by

$$d^2\phi/d\xi^2 = \tilde{\mathscr{D}}\phi^\nu e^{-\phi} \quad \text{with } \tilde{\mathscr{D}} = \mathscr{D}'/\theta^{\nu+1}. \tag{36}$$

Here the subscript on ν_1 has been dropped without fear of confusion with the old ν, which is concealed in $\mathscr{D}$. The first integral is now

$$(d\phi/d\xi)^2 = 2\tilde{\mathscr{D}}\gamma(\nu+1, \phi), \tag{37}$$

where γ is the incomplete gamma function, and the eigenvalue (21) is replaced by

$$\tilde{\mathscr{D}} = J_s^2/2\Gamma(\nu+1)T_b^4. \tag{38}$$

All else being equal (including the old ν) the ratio of the burning rate to its value for $\nu=1$ is $[\Gamma(\nu+1)]^{\frac{1}{2}}\theta^{\frac{1}{2}(1-\nu)}$, which changes from $\theta^{\frac{1}{2}}$ at $\nu=0$ to $0.94\theta^{0.27}$ at $\nu=0.46$ (the minimum of the numerical factor) and then to 1 at $\nu=1$, to $2.45\theta^{-1}$ at $\nu=3$.

Of more interest still is the structure

$$J_s\xi/T_b^2 = \Gamma^{\frac{1}{2}}(\nu+1)\int_\phi^\infty [\gamma^{-\frac{1}{2}}(\nu+1, \phi) - \Gamma^{-\frac{1}{2}}(\nu+1)]\, d\phi - \phi. \tag{39}$$

For $\nu > 1$ the integral diverges like $\phi^{\frac{1}{2}(1-\nu)}$ ($\log\phi$ for $\nu=1$) as $\phi \to 0$, so that the reactant is completely burnt only as $\xi \to +\infty$; but for $\nu < 1$ it converges, giving

$$J_s\xi/T_b^2 = \Gamma^{\frac{1}{2}}(\nu+1)\int_0^\infty [\gamma^{-\frac{1}{2}}(\nu+1, \phi) - \Gamma^{-\frac{1}{2}}(\nu+1)]\, d\phi. \tag{40}$$

At this value of ξ the slope $d\phi/d\xi$ is zero, as is also the curvature, so that there is a smooth transition to $\phi \equiv 0$ beyond (although higher derivatives are singular). In this context, the termination of the structure for $\nu < 1$ was first pointed out by Williams (1975). Similar modifications occur in the surface flame and remote flame (Buckmaster, Kapila, & Ludford 1976) but we shall not go into them.

Precise comparison with experiments of the eigenvalue (38) or the structure (39) is pointless considering the drastic simplification of the chemical kinetics. We see that both are sensitive to ν, the empirically determined reaction order. Nevertheless, the structures are qualitatively correct for actual flames and the predicted flame speeds have the correct order of magnitude for reasonable choices of the activation energy. In short, the simple model represented by equation (36) contains the essential physics in many circumstances.

On the other hand, there are qualitatively different phenomena that such a simple model cannot explain, so it is natural to consider how it can be modified. For the remainder of this chapter we shall consider a two-reactant model, to explain the behavior of mixtures near stoichiometry, i.e. mixtures containing the reactants in almost the correct proportions to leave only products (other than inerts) in the burnt state.

8 The two-reactant model

Extension to more than one reactant is of importance, since a pair of reacting species (usually called fuel and oxidant, even though oxygen may not be involved) occurs more commonly than a single decomposing species. Of special interest is the precise relation between the two models (see section 1); we shall find that the two-reactant model behaves like a one-reactant model except under near-stoichiometric conditions. Extension to an arbitrary number of species is due to Johnson & Nachbar (1963); for simplicity we shall treat two reactants, labeled X(oxidant) and Y(fuel) rather than Y_1 and Y_2, which combine to form a different product (see section 1.6). Certain aspects of the question, in somewhat greater generality, have been treated by Clarke (1975); by Carrier, Fendell, & Bush (1978); and by Mitani (1980). Here we shall follow Sen & Ludford (1979), who introduced differential diffusion (section 1.4) and subsequently (Ludford & Sen 1981, Sen & Ludford 1981*a*, *b*) made other improvements of the model.

Consider the unbounded flame. For simplicity, suppose $\nu_1 = \nu_2 = 1$ and $m_1 = m_2$; then $\alpha_1 = \alpha_2 = -\frac{1}{2}$ and the Shvab–Zeldovich variables (1.69) for the reactants are $G \equiv T + 2X$ and $H \equiv T + 2Y$, these being conditional enthalpies since neither reactant is a product. When the mixture is fuel-rich, G is the available enthalpy, a constant given by the result (41). Both G and H satisfy equation (1), of which the only solutions bounded at $x=+\infty$ are constants. Hence,

$$G=T_f+2X_f\equiv G_f,\quad H=T_f+2Y_f\equiv H_f \tag{41}$$

where, as usual, the subscript f denotes the fresh values at $x=-\infty$; so that we may write

$$X=\tfrac{1}{2}(G_f-T),\quad Y=\tfrac{1}{2}(H_f-T). \tag{42}$$

The reactant mass fractions are now known as functions of T; only the equation

$$\mathfrak{L}(T,1)=-\mathscr{D}XY\mathrm{e}^{-\theta/T}\quad \text{with } \mathscr{D}\equiv DM^{-2} \tag{43}$$

for the temperature remains.

The temperature behaves as if only one reactant were present. Suppose for the sake of argument that the fresh mixture is rich in fuel, i.e.

$$X_f<Y_f. \tag{44}$$

Then, as T increases from T_f, both mass fractions decrease but when X reaches zero, at the flame temperature

$$T_b=G_f, \tag{45}$$

the fuel fraction

$$Y_*=Y_f-X_f \tag{46}$$

is still positive. Rewriting equation (43) in the form

$$\mathfrak{L}(T,1)=-[\mathscr{D}Y/2](2X)\,\mathrm{e}^{-\theta/T}\quad \text{with } T+2X=G_f \tag{47}$$

now shows that we are dealing with a single reactant, the oxidant, of double concentration and a temperature-dependent Damköhler number. As explained at the beginning of section 7, the latter is accommodated by replacing the quantity in square brackets with its value at the flame temperature (45).

In short, the oxidant exhausts itself combining with the fuel, whose remaining portion (46) then passes beyond the flame to $x=+\infty$ without change, the process being described by our previous analysis. In particular, the result (21) leads to

$$\mathscr{D}=4J_s^2\theta^2\,\mathrm{e}^{\theta/T_b}/T_b^4Y_*\quad (J_s=X_f). \tag{48}$$

As X_f is increased so that the proportions of fuel and oxidant are brought closer to stoichiometry, T_b increases and so does the burning rate according to this formula, except very close to stoichiometry where the approach of Y_* to zero plays a role. However, the formula is then inapplicable; for when

$$Y_f = X_f + \theta^{-1} y_* \quad \text{with } y_* = O(1), \tag{49}$$

the asymptotic analysis clearly breaks down and the original temperature equation (43) must be reinstated together with the T, X, Y-relations (42). We now find that there is no factor in the burning rate tending to zero with y_*, the small amount of fuel that escapes burning.

Ahead of the flame sheet we still have

$$T = T_f + 2X_f e^x \quad \text{for } x < 0 \tag{50}$$

but the structure now satisfies

$$d^2\phi/d\xi^2 = \tilde{\mathscr{D}} X_1 Y_1 e^{-\phi} \quad \text{with } \tilde{\mathscr{D}} = \mathscr{D}'/\theta^3 T_b^2, \tag{51}$$

where

$$X_1 = \tfrac{1}{2} T_b^2 \phi, \quad Y_1 = y_* + \tfrac{1}{2} T_b^2 \phi. \tag{52}$$

Because the temperature profile (50) is the same, the boundary conditions

$$\phi = -2J_s \xi / T_b^2 + o(1) \quad \text{as } \xi \to -\infty, \quad \phi = o(1) \quad \text{as } \xi \to +\infty \tag{53}$$

still apply with

$$J_s = X_f = Y_f \text{ (to leading order).} \tag{54}$$

An integration similar to that in section 4 now gives $\tilde{\mathscr{D}} = 4J_s^2/T_b^6(y_* + T_b^2)$, so that

$$\mathscr{D} = 4J_s^2 \theta^3 e^{\theta/T_b}/T_b^4(y_* + T_b^2) \tag{55}$$

in agreement with the result (48) when $y_* = \theta Y_* \gg 1$.

We conclude that the burning rate is well defined as stoichiometry is approached and indeed

$$M = T_b^3 e^{-\theta/2T_b} D^{\frac{1}{2}}/2\theta^{\frac{3}{2}} J_s, \tag{56}$$

when $y_* = 0$. For such exact stoichiometry we are again dealing with a single reactant of double concentration but now with a reaction order of 2 and an eigenvalue $\frac{1}{4}\mathscr{D}$.

When the fresh mixture is fuel-lean, X and Y must be interchanged everywhere and $\theta^{-1} y_*$ interpreted as the amount of oxidant that escapes burning.

Experiments demonstrate (Fristrom & Westenberg 1965, p. 22) that, for fuels burning in oxidant–inert mixtures such as air, M has a maximum for a value of Y_f slightly larger than X_f; we shall show that such a maximum follows from the formula (55). In experiments with air or an air-like mixture the ratio of the inert to oxidant concentrations is held constant (e.g. air has a fixed nitrogen-to-oxygen mass ratio of 3.29) so that

$$N_f/X_f = b, \tag{57}$$

N_f being the mass fraction of the inert and b a positive constant. Thus,

$$(b+1)X_f + Y_f = 1, \tag{58}$$

since the mass fractions of the fresh mixture are subject to $N_f + X_f + Y_f = 1$, if it is assumed that no product species is present in the fresh mixture. In the formula (55), J_s and T_b may be replaced by their $O(1)$ values $(b+2)^{-1}$ and $T_{b0} = T_f + 2/(b+2)$, respectively, except in the exponential where the more accurate formula

$$T_b = T_{b0} - 2y_*/(b+2)\theta \tag{59}$$

must be used. Thus, M^2 is proportional to $(y_* + T_b^2)\exp[-2y_*/(b+2)T_b^2]$, which has a maximum at

$$y_* = \tfrac{1}{2}bT_b^2. \tag{60}$$

On the other hand, when the mixture is fuel-lean, M^2 is found to be proportional to $(y_* + T_b^2)\exp[-2(b+1)y_*/(b+2)T_b^2]$, which is monotonically decreasing with positive y_*: no further maximum is obtained.

We conclude that when N_f/X_f is held constant, the maximum burning rate is achieved only on the fuel-rich side, as given by (60). If, instead of keeping N_f/X_f fixed, the inert concentration N_f is held constant, so that $X_f + Y_f$ remains fixed, the maximum burning rate is attained at stoichiometry.

Other considerations can also play a role in practice. In the combustion of hydrocarbons in air the maximum flame temperature is not achieved at stoichiometry, as our theory demands, but on the fuel-rich side. This is a thermodynamic question which highlights a deficiency in the simple kinetic model that we have adopted. Combustion of lean hydrocarbons is characterized by more or less complete oxidation (i.e. the formation of carbon dioxide and water vapor) but, close to stoichiometry and on the rich side, incomplete oxidation also occurs with the formation of carbon monoxide. Our model, which ensures that the heat released is proportional to the amount of deficient component in the supply, cannot imitate this behavior, and accordingly cannot accurately predict the response to changes in mixture composition in the circumstances. It would be unreasonable to infer from this that it cannot be used to predict the behavior of rich hydrocarbon flames when the composition is fixed, however.

3

PERTURBATIONS: SLOWLY VARYING AND NEAR-EQUIDIFFUSIONAL FLAMES

1 Prologue

The propagation speed of a plane deflagration wave is extraordinarily sensitive to changes in the flame temperature. The result (2.22) shows that, for fixed D and J_s, an $O(1)$ change in T_b produces an exponentially large change in the burning rate. More modestly, an $O(\theta^{-1})$ change in T_b produces an $O(1)$ change in flame speed; it is perturbations of such a magnitude that concern us in this chapter.

Such a change may be engineered for the unbounded flame by an $O(\theta^{-1})$ change in T_f, an elementary example that is not of great importance either mathematically or physically. A much more interesting example is cooling by heat loss through the walls of a uniform duct along which the flame is traveling. It is well known that flames cannot propagate through very narrow passages (a key safety principle where explosive atmospheres are involved), and this can be adequately explained by such a heat-loss mechanism, as we shall see.

Perturbations of the same magnitude can also be produced by changes in the size of the duct that occur over distances $O(\theta)$. Now there will be slow variations in the combustion field developing on a time scale $O(\theta)$. Such slow variations can even be self-induced by residual perturbations of the initial conditions (on that time scale) in the absence of boundary perturbations. In all such cases an obvious conjecture is that the flame velocity is not close to the unperturbed value.

It is not so obvious that slowly varying flames (SVFs) behave quite differently accordingly as the Lewis number is smaller or greater than one. Moreover, $\mathscr{L} = 1$ is an exception; in particular, there are no self-induced variations. To develop the theory fully, it is necessary to re-do section 2.4 for a general Lewis number, and that is where we start. The bulk of the chapter introduces SVFs; the final section is devoted to near-equidiffusional flames (NEFs), which fill the gap around $\mathscr{L}=1$. Both have been unearthed using a perturbation theory that is proving increasingly useful in understanding the true role of such idealized solutions as the premixed plane flame. For SVFs the $O(\theta^{-1})$ flame temperature perturbations are due to the slowness of variations and for NEFs to the closeness of $\mathscr{L}$ to 1.

2 Modifications of the plane premixed flame for $\mathscr{L} \neq 1$

The analysis closely parallels that in section 2.4. The flame sheet is again located at $x=x_*$; beyond it the temperature is still constant and there is no reactant. The latter facts are based on the assumption that changes in $\mathscr{L}$ will only generate $O(1)$ changes in flame speed, so that $\mathscr{D}$ still has a factor θ^2 and equilibrium must prevail beyond the flame, i.e.

$$\mathfrak{L}(T, 1)= \mathfrak{L}(Y, \mathscr{L})=0, \quad (1)$$

the only bounded solutions being constants. Because there is no Shvab–Zeldovich variable, both T and Y must be carried, and we write

$$T= T_s + T_s'(e^x-1), \quad Y= Y_s + \mathscr{L}^{-1} Y_s'(e^{\mathscr{L}x}-1) \quad \text{for } 0<x<x_*, \quad (2)$$

where T_s, T_s', Y_s, and Y_s' are values at $x=0$ determined by how the reactant is supplied there. (Figure 2.1 still displays the general shape of these profiles.) Continuity at the flame sheet now leads to

$$x_*=\ln[1+(T_b-T_s)/T_s']=\mathscr{L}^{-1}\ln(1-\mathscr{L}Y_s/Y_s'). \quad (3)$$

The discussion of reactant supplies in section 2.3 showed that only three conditions are provided at $x=0$, so that we may expect Y_s' and T_b, for example, to be related to T_s, T_s', and Y_s. The result (3) provides one such relation and the second, namely

$$T_b= T_s + Y_s -(T_s' + \mathscr{L}^{-1} Y_s')=J_s + E_s, \quad (4)$$

comes from an enthalpy balance between the stations $x=0$ and ∞ (formally derived by integrating the linear combination of the equations for T and Y that eliminates the reaction term). The second relation enables the first to be rewritten as

$$T_s'= -\mathscr{L}^{-1} Y_s'(1-\mathscr{L}Y_s/Y_s')^{1-1/\mathscr{L}}. \quad (5)$$

It is clear that the reductions $T_b= T_s + Y_s$ and $T_s'= -Y_s'$ for $\mathscr{L}=1$ are recovered.

To ensure $0<x_*<\infty$ we must have $Y_s'<0$ and, hence, $T_s'>0$: whatever supplies the reactant must still be a (conductive) heat sink. Since the T-profile is monotonic, it follows that the temperature of the flame is always higher than that of the supply, something that is not obvious from the relations (4) and (5) when $\mathscr{L} \neq 1$.

Since $T_s + J_s$ is the final temperature when the reactant is allowed to decompose under adiabatic conditions (section 2.5), it is called the adiabatic flame temperature. The system, however, loses heat T_s' by conduction so that the actual flame temperature (4) is always lower.

Here expansions in the variable (2.17) lead to the structure equations

$$T_b^2 \, d^2\phi/d\xi^2 = \mathscr{L}^{-1} \, d^2 Y_1/d\xi^2 = \tilde{\mathscr{D}} Y_1 \, e^{-\phi}, \tag{6}$$

which, because of matching requirements, must be solved under the boundary conditions

$$\phi = -J_s\xi/T_b^2 + o(1), \quad Y_1 = -\mathscr{L} J_s \xi + o(1) \quad \text{as } \xi \to -\infty;$$
$$\phi, \, Y_1 = o(1) \quad \text{as } \xi \to +\infty, \tag{7}$$

where now

$$J_s = Y_s - \mathscr{L}^{-1} Y_s'. \tag{8}$$

We immediately conclude that there is a local Shvab–Zeldovich variable and that it stays constant through the layer, i.e.

$$T_b^2 \phi = \mathscr{L}^{-1} Y_1. \tag{9}$$

The first integral (2.20) is, therefore, replaced by

$$(d\phi/d\xi)^2 = 2\mathscr{L}\tilde{\mathscr{D}}[1 - (\phi + 1) e^{-\phi}] \tag{10}$$

and the result (2.22) by

$$\mathscr{D} \equiv DM^{-2} = J_s^2 \theta^2 \, e^{\theta/T_b} / 2\mathscr{L} T_b^4, \tag{11}$$

but the structure (2.23) still holds. The formula (11), as well as its correction of relative order θ^{-1}, was obtained by Bush & Fendell (1970).

Modifications in sections 2.5–2.7 that have to be made when $\mathscr{L} \neq 1$ are quite similar. Of more importance for our purpose is the form that the results take for the unbounded flame, our sole concern in the remainder of this section. By transferring the origin to the flame sheet and then letting $T_s' \to 0$, we find

$$T = \begin{cases} T_f + Y_f e^x \\ T_b = H_f \end{cases}, \quad Y = \begin{cases} Y_f(1 - e^{\mathscr{L}x}) \\ 0 \end{cases} \quad \text{for } x \lessgtr 0 \tag{12}$$

and the eigenvalue (11) with $J_s = Y_f$ again. Note that the adiabatic flame temperature is realized here because no heat is lost from, or gained by, the system.

Suppose now that the flame temperature is decreased by $\theta^{-1} T_b^2 \phi_*$, i.e. in terms of the original flame temperature T_b we have

$$T_* = T_b - \theta^{-1} T_b^2 \phi_*. \tag{13}$$

Then the formula (11) is equivalent to

$$M = M_r\, e^{-\phi_*/2} \quad \text{with } M_r = \sqrt{2\mathscr{L}D}\, T_b^2\, e^{-\theta/2T_b} / J_s \theta \tag{14}$$

a reference burning rate. This result has universal validity, i.e. it holds whatever the nature of the perturbation, steady or unsteady. The reason is that, in determining the eigenvalue, the perturbation only intrudes through the matching of ϕ at $\xi = +\infty$ (which leads to the exponential factor); matching ahead of the flame sheet is with $O(1)$ gradients that are unaffected by the perturbation. (Dimensional gradients, on the other hand, are significantly affected, due to changes in the mass flux M on which the length unit is based.)

3 Theory of perturbations

The equations that we shall discuss are:

$$\partial\rho/\partial t + \partial(\rho v)/\partial x = f/\theta, \tag{15}$$

$$\rho\partial Y/\partial t + \rho v \partial Y/\partial x - \mathscr{L}^{-1}\partial^2 Y/\partial x^2 = -\mathscr{D} Y e^{-\theta/T} + g/\theta, \tag{16}$$

$$\rho\partial T/\partial t + \rho v \partial T/\partial x - \partial^2 T/\partial x^2 = \mathscr{D} Y e^{-\theta/T} + h/\theta. \tag{17}$$

Here f, g, and h should be regarded, for the moment, as arbitrary functionals. For propagation into a quiescent gas,

$$Y, T, v \to Y_f, T_f, 0 \quad \text{as } x \to -\infty. \tag{18}$$

The units by which these equations have been made dimensionless are based on a representative mass flux, which will be taken as that through the flame sheet in the absence of perturbations, namely the M_r of equation (14); so that

$$\mathscr{D} = DM_r^{-2}. \tag{19}$$

To avoid an unnecessary factor in our results, we have changed the unit of density from ρ_c to ρ_c/T_f (with corresponding changes in other units). The unperturbed flame speed is then 1 and T_f appears in the equation of state, i.e.

$$\rho = T_f/T. \tag{20}$$

(In most circumstances, ρ_c is not characteristic of densities in the combustion field, because it is much too small. The change is a normalization.)

For finite θ, the flame speed is defined in a natural way only when the combustion field is steady: if the structure changes with time in all frames of reference, the location of the flame (and, hence, its speed) is not a precise concept. It is, therefore, customary to introduce an arbitrary definition, such as the speed with which the inflexion point of temperature moves

(Lewis & von Elbe 1961, p. 215). In the limit $\theta \to \infty$, however, a natural definition is the speed V of the sheet where all reaction is confined, whether the field is steady or not. Consistent with our previous remark, when discussing unsteady problems, V may be considered an $O(1)$ quantity, not to be expanded in θ as are the other unknowns. With $x_*(t)$ denoting the position of the sheet, we write

$$n = x - x_*(t) \tag{21}$$

for the distance ahead to obtain

$$\partial\rho/\partial t + \partial[\rho(V+v)]/\partial n = f/\theta, \tag{22}$$

$$\rho\partial Y/\partial t + \rho(V+v)\partial Y/\partial n - \mathscr{L}^{-1}\partial^2 Y/\partial n^2 = -\mathscr{D} Y e^{-\theta/T} + g/\theta, \tag{23}$$

$$\rho\partial T/\partial t + \rho(V+v)\partial T/\partial n - \partial^2 T/\partial n^2 = \mathscr{D} Y e^{-\theta/T} + h/\theta, \tag{24}$$

where $V = -\dot{x}_*(t)$ is the speed of the flame. The unperturbed flame corresponds to $V=1$; we are concerned with flames for which V stays $O(1)$. We may expect that to be the case when f, g, and h are all $O(1)$ and, when conditions are unsteady, $\partial/\partial t = O(\theta^{-1})$, i.e. for slowly varying flames. Integration of the continuity equation shows that V is equal to the dimensionless mass flux $\rho(V+v)$ everywhere, to leading order, so that the result (14a) now reads

$$V = e^{-\frac{1}{2}\phi_*}, \tag{25}$$

which shows that our task is to evaluate the change in flame temperature (represented by ϕ_*) due to the perturbations.

In this context the first calculation of ϕ_* was for a distributed heat loss, which was considered independently by Sivashinsky & Gutfinger (1975), by Buckmaster (1976), and by Joulin & Clavin (1976). Other perturbations (sections 5–7) were later treated by Buckmaster (1977).

We may calculate ϕ_* directly from the overall change in enthalpy of the mixture up to and including the flame sheet. If equations (23) and (24) are added to eliminate the reaction terms and then integrated from $n = -\infty$ to $0+$, we obtain

$$\int_{-\infty}^{0+} \rho(V+v)\frac{\partial H}{\partial n}\,dn = \left.\frac{\partial T}{\partial n}\right|_{0+} + \int_{-\infty}^{0}\left[-\rho\frac{\partial H}{\partial t} + \frac{g}{\theta} + \frac{h}{\theta}\right]dn$$

correct to $O(\theta^{-1})$, since Y vanishes behind the flame sheet (cf. section 2.4). Integrating by parts on the left side and using the continuity equation (22) now yields

$$(T_b - T_*)V = -\left.\frac{\partial T}{\partial n}\right|_{0+} + \int_{-\infty}^{0}\left[\frac{\partial}{\partial t}[\rho(Y-T_*)] + (T_* - H)\frac{f}{\theta} - \frac{g}{\theta} - \frac{h}{\theta}\right]dn$$

by virtue of the equation of state (20). Here $T_* = T(0+, t)$ may properly be called the flame temperature, while T_b is the adiabatic flame temperature

$T_f + Y_f$. Thus,

$$T_b^2\phi_* V = -\theta \frac{\partial T}{\partial n}\bigg|_{0+} + \int_{-\infty}^{0} \left[\frac{\partial}{\partial \tau}[\rho(Y - T_b)] + (T_b - H)f - g - h\right] dn, \quad (26)$$

where

$$\tau = t/\theta \quad (27)$$

represents a long time (first introduced by Sivashinsky (1974*a*)) and all terms are to be evaluated to leading order only. (The actual times involved can be quite short: a typical unit of time t is 0.2 ms, see section 2.4.)

The problem is reduced to calculating the right side of equation (26) in terms of V for the particular perturbation of interest. It then becomes an equation for the flame speed (the goal of our analysis) since the universal formula (25) shows the left side to be $-T_b^2 V \ln V^2$.

4 Steady heat loss: the excess-enthalpy flame

In practice, the mixture must be confined laterally; there is then the possibility of losing heat to the sides. For example, heat conduction through the wall of the tube introduced in section 2.2 will result in a heat loss proportional to $T - T_f$, if we assume that the temperature outside the tube at any point is that of the mixture at $x = -\infty$. Similarly, thermal radiation (as distinct from chemiluminescence) is present in combustion processes, although usually negligible, resulting in a loss proportional to $T^4 - T_f^4$. The importance of considering such effects comes from the observed inability of the mixture to ignite if the tube is too narrow, a phenomenon so far not considered in our analysis. Then conductive loss, though not radiative, is significant.

Heat losses must be small, otherwise the flame goes out. Our analysis will, therefore, be based on $O(\theta^{-1})$ proportionality factors, but the dependence of the loss on temperature will be quite general; in particular, both conductive and radiative losses are covered. A term $\psi(T)/\theta$ is added to the right side of the temperature equation (2.4), where ψ vanishes at T_f but is otherwise an arbitrary (positive) function. In particular,

$$\begin{aligned} \psi &= k(T - T_f) \quad \text{for conductive losses} \\ \psi &= k(T^4 - T_f^4) \quad \text{for radiative losses} \end{aligned} \quad (28)$$

where k is a given constant, with $M_r^2 k$ independent of the mass-flux unit.

We are, therefore, dealing with the special case of equations (22)–(24) in which

$$\partial/\partial\tau = f = g = 0, \quad h = -\psi(T). \quad (29)$$

As already noted the continuity equation integrates to

$$\rho(V + v) = V, \quad (30)$$

with V constant, so that to leading order we are left with the unperturbed steady equations. Use can be made, therefore, of the result (12*a*) to write

$$T = T_{\mathrm{f}} + Y_{\mathrm{f}}\,\mathrm{e}^{Vn}, \quad Y = Y_{\mathrm{f}}(1 - \mathrm{e}^{\mathscr{L}Vn}) \quad \text{for } n<0, \tag{31}$$

but (12*b*) must be taken one term further. From the perturbed temperature equation we easily find

$$T = T_{\mathrm{b}} - \theta^{-1}[T_{\mathrm{b}}^2\phi_* + V^{-1}\psi(T_{\mathrm{b}})n] \quad \text{for } n>0, \tag{32}$$

a result that is needed to calculate $\partial T/\partial n$ at $0+$. (An exponentially growing term has been discarded as unmatchable with the decay to ambient temperature that occurs on a scale $O(\theta)$.)

The right side of (26) may now be calculated, giving

$$\phi_* = \Psi V^{-2} \quad \text{with } T_{\mathrm{b}}^2\Psi = \psi(T_{\mathrm{b}}) + \int_0^\infty \psi(T_{\mathrm{f}} + Y_{\mathrm{f}}\,\mathrm{e}^{-\nu})\,\mathrm{d}\nu, \tag{33}$$

so that

$$V^2 \ln V^2 + \Psi = 0 \tag{34}$$

is the equation for the flame speed. The result is graphed in Figure 1, forming a backward C. Response curves in the shape of a C are common in combustion theory. The function Ψ consists of two terms: the integral

1 Flame speed V versus heat-loss parameter Ψ. The notation $v_{\mathrm{f}}(0)$, $\Psi(0)$ refers to section 7 where S, the point at which $\Psi(0) = \frac{4}{3}v_{\mathrm{f}}^2(0)$ plays a role.

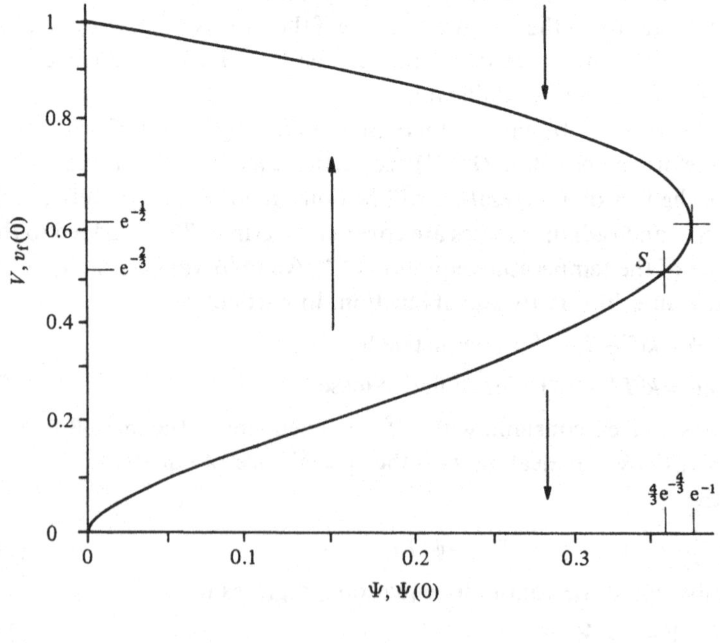

represents the total heat loss up to the flame, while the other accounts for conduction into the burnt mixture behind.

The top branch shows that heat loss decreases the flame velocity; however, if Ψ exceeds the critical value e^{-1} there is no corresponding V, i.e. no flame. The term 'extinction' is, therefore, appropriate for the point at which the curve turns back, the burning rate then being $\mathrm{e}^{-\frac{1}{2}}(=0.61)$ times the adiabatic value for the same pressure. (Note that this result is independent of all parameters; numerical integration, even though for a two-dimensional flame confined by parallel walls through which conductive heat losses occur, leads to the remarkably close value 0.64 (Aly & Hermance 1981)). The bottom branch is generally believed to be unstable (Emmons 1971) and, hence, of no interest, but we shall postpone a discussion of that point until the next section.

It has been known for over 150 years that a flame will not propagate through a wire gauze, the principle underlying the miner's safety lamp invented by Davy. This is suggested by our results: a cold wire gauze of sufficiently small mesh is an effective heat sink and corresponds to values of Ψ greater than critical.

Extinction in small tubes can also be explained. Heat loss by conduction through the wall is proportional to surface area (i.e. the radius) while the volume of the mixture is proportional to the radius squared. It follows that the heat loss per unit volume and, hence, the parameter Ψ varies inversely as the radius. There is, therefore, a definite tube radius below which steady combustion will not occur.

The loss of heat from the unburnt mixture produces a deficiency of enthalpy at the flame, so that its temperature and, hence, burning rate is reduced. We now consider the possibility of increasing the burning rate by using heat lost behind the flame to raise the temperature of the fresh mixture. Such an excess-enthalpy flame permits the combustion of mixtures that are difficult or impossible to burn under more familiar circumstances; to engineer it would be of considerable practical importance. Fuel-lean mixtures are of particular interest considering current concerns with pollution and fuel efficiency.

An experimental burner effecting this transfer by means of a porous conducting slug which envelops some of the burnt gas and the preheat zone has been constructed by Takeno & Sato (1981); a mathematical model of it may be couched in the theory of this section by introducing a counterflow to bring the flame to rest. The counterflow changes the velocity condition ahead of the flame to

$$v \to v_{\mathrm{f}} \quad \text{as } x \to -\infty \tag{35}$$

so that continuity now implies the general result

$$\rho(V+v)=V+v_\mathrm{f}\equiv W \tag{36}$$

to leading order. The term flame speed is normally reserved for W, which (as the speed of the flame sheet relative to the counterflow) is frame independent. When ρ_f is taken to be 1 (as here), W is also the dimensionless flux through the sheet, i.e. the burning rate previously designated M/M_r.

We have $V=0$, which effectively returns us to equation (30) with V replaced by v_f on the right side; the same replacement must then be made in the formulas (31). The heat-transfer device (i.e. the metal matrix) occupies the portion $0<x<a$ of a well-insulated tube and has temperature T_m. (The limiting case $a=\infty$, studied by Deshaies & Joulin (1980), misses important features of the solution.) For simplicity we shall suppose that heat transfer between the matrix and the gas phase is represented by setting

$$\psi=k(T-T_\mathrm{m}) \quad \text{for } 0<x<a \quad \text{and 0 elsewhere,} \tag{37}$$

where k is a constant, in equations (29). Since the matrix experiences no net loss or gain of heat in the steady state, T_m is a temperature intermediate between T_f and T_b defined by

$$\int_0^a (T-T_\mathrm{m})\,\mathrm{d}x=0. \tag{38}$$

It is found that the flame sheet must lie inside the device.

The formulas (25) and (26) still apply, when V is replaced by v_f, but the temperature profile (32) must be changed to

$$T=T_\mathrm{b}+\theta^{-1}[-T_\mathrm{b}^2\phi_*+kv_\mathrm{f}^{-2}(T_\mathrm{m}-T_\mathrm{b})\,\mathrm{e}^{v_\mathrm{f}(x_*-a)}(1-\mathrm{e}^{v_\mathrm{f}n})+kv_\mathrm{f}^{-1}(T_\mathrm{m}-T_\mathrm{b})n] \quad \text{for } 0<n<a-x_*. \tag{39}$$

(The derivative $\mathrm{d}T/\mathrm{d}x$ vanishes at $n=a-x_*$ because the heat flux must be continuous there and the temperature is constant downstream of the matrix.) All terms on the right side of equation (26) can now be calculated, leading to

$$T_\mathrm{b}^2v_\mathrm{f}\ln v_\mathrm{f}^2=kY_\mathrm{f}v_\mathrm{f}^{-1}(v_\mathrm{f}x_*+\mathrm{e}^{-v_\mathrm{f}x_*}-1)+kv_\mathrm{f}^{-1}(T_\mathrm{m}-T_\mathrm{b})[v_\mathrm{f}x_*-\mathrm{e}^{v_\mathrm{f}(x_*-a)}+1].$$

When T_m is eliminated by means of the condition (38), which becomes

$$T_\mathrm{b}-T_\mathrm{m}=Y_\mathrm{f}(v_\mathrm{f}x_*+\mathrm{e}^{-v_\mathrm{f}x_*}-1)/v_\mathrm{f}a,$$

we obtain

$$T_\mathrm{b}^2av_\mathrm{f}^3\ln v_\mathrm{f}^2=kY_\mathrm{f}(v_\mathrm{f}x_*+\mathrm{e}^{-v_\mathrm{f}x_*}-1)[v_\mathrm{f}a+\mathrm{e}^{v_\mathrm{f}(x_*-a)}-1-v_\mathrm{f}x_*], \tag{40}$$

an equation for x_* as a function of v_f derived by Buckmaster & Takeno (1981).

The right side of this equation vanishes when $x_*=0$ and when $x_*=a$. Since v_f takes the value 1 at either extreme, somewhere in between it must have a maximum as a function of x_*. We conclude that, for a limited range of values of v_f greater than 1, there is a corresponding x_* and, hence, a steady state. Any attempt to position the flame in too strong or too weak a flow (i.e. one with v_f outside the range) will result in blow-off or flash-back. Corresponding values of a and the maximum v_f can be obtained by reading off the ordinates of the maxima in Figure 2. (Static stability arguments (cf. section 9.1) suggest that only the smaller of the two values of x_* for a given v_f corresponds to a stable flame.)

The related problem of heat loss to inert particles in the flow has been considered by Joulin (1981). A curve similar to that in Figure 1 completed by the Ψ-axis (cf. next section) then appears as the limit of a family of Z-shaped curves depending on the thermal inertia of the particles.

5 SVFs

Flames can be unsteady for a variety of reasons. Evolution from ignition, passage down a tube of varying cross section, and instability are all examples of unsteady combustion. Here we are concerned with the one-dimensional problem of a flame evolving from initial conditions; in general, such an evolution can be expected to take place on an $O(1)$ time scale. There have been numerical integrations (with f, g, and h all 0) that show the formation of what appears to be a steady wave from a mass of hot quiescent gas on such a time scale (Zeldovich & Barenblatt 1959).

Our techniques cannot describe such phenomena, in general, since the

2 Family of curves, determined by equation (40), from which blow-off conditions can be found.

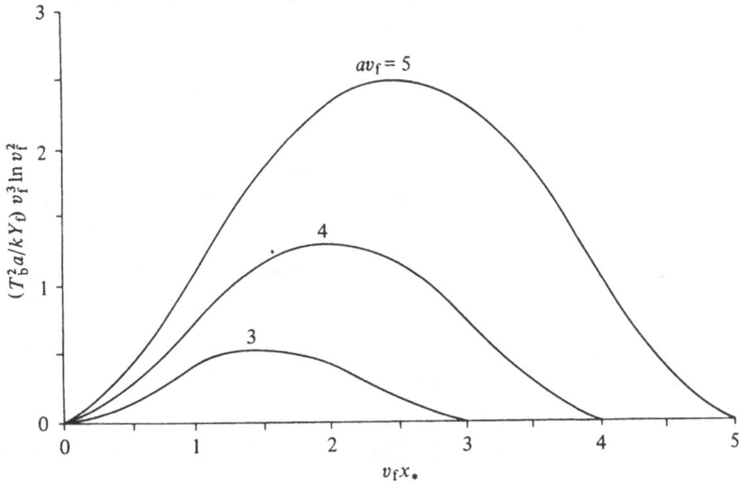

right side of (26) will not be $O(1)$. From the initial development, however, there emerges a structure, characterized by the single variable V, that undergoes a slower evolution on the time scale (27); during this phase of the unsteady process the time-derivatives are perturbations that can be incorporated into the analysis as easily as was heat loss.

To leading order the continuity equation still has the integral (30), where V is now a function of τ, so that the profiles (31) remain valid ahead of the flame sheet. To them we must add

$$\rho = T_f/(T_f + Y_f e^{Vn}), \tag{41}$$

but an expression for v will not be needed. As before, the remaining question is to calculate $\partial T/\partial n$ at $0+$, which comes from the temperature perturbation behind the flame sheet.

To begin with, consider perturbations due solely to the unsteadiness. Then $f=g=h=0$ and, hence,

$$T = T_b - \theta^{-1} T_b^2 \phi_* \quad \text{for } n>0, \tag{42}$$

so that

$$\phi_* = -bV^{-3}\dot{V} \quad \text{with } b = \frac{T_f Y_f}{T_b^2}\int_0^1 \frac{1-v^{\mathscr{L}-1}}{T_f + Y_f v}\,dv; \tag{43}$$

the dot now denotes the rate of change with respect to the time τ. The equation for $V(\tau)$ is, therefore,

$$b\dot{V} = V^3 \ln V^2. \tag{44}$$

Note that the parameter b changes from negative to positive as the Lewis number increases through 1, giving the first indication that $\mathscr{L}=1$ is not typical.

We are free to specify only the initial value of V and then the approximation to the combustion field is uniquely determined. As mentioned previously, this approach does not solve the general initial-value problem since the initial evolution takes place on the $O(1)$ time scale. But from this initial development emerges a structure, characterized by the single variable V, that changes on the $O(\theta)$ time scale.

When $\mathscr{L}>1$, b is positive and when $\mathscr{L}<1$, b is negative. The closer $\mathscr{L}$ is to 1 the more rapidly V changes, so that when $\mathscr{L}-1=O(\theta^{-1})$ or smaller there are no $O(1)$ effects on the slow-time scale and the result (44) implies

$$V=0 \quad \text{or } 1. \tag{45}$$

We infer that the initial evolution ends at one of these values, although we may expect the ultimate value to be reached more slowly in t as the Lewis number gets further from 1 on the scale of θ^{-1}. It is tempting to take a unit Lewis number in combustion problems because of the simplifications stemming from global Shvab–Zeldovich variables (section 1.7); but in view of the present atypical behavior such temptations should certainly be resisted when dealing with unsteady problems.

The values (45) are also possible steady states as $\tau \to \infty$. However, the first corresponds to $\phi_* = +\infty$, when our analysis breaks down but indicates that the flame temperature differs from its adiabatic value by much more than an $O(\theta^{-1})$ amount. A discussion of the nonuniformity, in the context of vanishingly small heat loss, has been given by Buckmaster (1976). The second is unstable when b is positive, i.e. $\mathscr{L} > 1$: according to (44), when V is less than 1 it will be driven towards 0 and when V is greater than 1 it will be driven towards ∞. This instability has a counterpart in numerical integrations for finite but large activation energy (Peters & Warnatz 1982).

It might be thought, therefore, that steady flames of Lewis number greater than 1 are of no practical interest, but laboratory experience shows otherwise. Moreover, if flames with $\mathscr{L} > 1$ are to be discarded on these grounds then so must flames with $\mathscr{L} < 1$, since they are also unstable when three-dimensional disturbances are permitted (Chapter 11). The fact that flames are stabilized by burners suggests that velocity and thermal gradients play an important role. It is not difficult to devise a one-dimensional model (section 7) that stabilizes flames in this way. It is unlikely, however, that such mechanisms will explain stable flames propagating down uniform tubes.

Lewis numbers are usually not far from 1 so that, considering the irrelevance of the slow-time scale when $\mathscr{L} - 1 = O(\theta^{-1})$, it is conceivable that predictions of a slow-time analysis are often not applicable. Alternatively, stable flames may be a phenomenon of finite activation energy about which asymptotic theory provides faulty information. Finally, stable flames may be the result of chemical kinetics more complicated than those adopted here. Whatever the explanation, activation-energy asymptotics for $\mathscr{L} \neq 1$ can provide useful qualitative insights even though the applicability of the model cannot be justified at the moment.

To determine the effect of heat losses we still let $f = g = 0$ but now take $h = -\psi(T)$. The effects of unsteadiness and heat loss are then seen to be additive, i.e.

$$\phi_* = -bV^{-3}\dot{V} + \Psi V^{-2}, \tag{46}$$

so that we end up with

$$b\dot{V} = V^3 \ln V^2 + \Psi V \tag{47}$$

as the equation governing V. To the possible steady state (34), plotted in

Figure 1, must now be added

$$V=0 \quad \text{for all } \Psi. \tag{48}$$

However, the latter corresponds to $\phi_*=+\infty$, a nonuniformity that has yet to be treated.

The question of which branch of the multivalued response in Figure 1 will be observed in practice can be answered in the context of our model. The arrows in Figure 1 give the direction in which V changes according to the differential equation (47) when b is negative (i.e. $\mathscr{L}<1$). The lower branch of (34) is, therefore, predicted to be unstable, a result in accord with general belief. Spalding & Yumlu (1959), however, claim to have observed the slow branch, which in the absence of other effects can only be reconciled with our analysis if b is positive (i.e. $\mathscr{L}>1$) so that the arrows are reversed. (The fact that they used a special stabilizing apparatus suggests that there were other effects (cf. section 7).)

6 Nonuniform ducts

We consider ducts whose cross-sectional area A varies over distances of order θ, i.e.

$$A=A(\chi) \quad \text{with } \chi=x/\theta=n/\theta-\int_0^\tau V\,\mathrm{d}\tau. \tag{49}$$

For the usual one-dimensional treatment of such slowly varying ducts, the divergences in the basic equations (1.55), (1.56), and (1.58) now give rise to extra terms

$$\left(\frac{1}{A}\frac{\mathrm{d}A}{\mathrm{d}x}\right)\rho v, \quad -\mathscr{L}^{-1}\left(\frac{1}{A}\frac{\mathrm{d}A}{\mathrm{d}x}\right)\frac{\partial Y}{\partial x}, \quad -\left(\frac{1}{A}\frac{\mathrm{d}A}{\mathrm{d}x}\right)\frac{\partial T}{\partial x},$$

which correspond to

$$f=-B(\chi_*)\rho v, \quad g=\mathscr{L}^{-1}B(\chi_*)\partial Y/\partial n, \quad h=B(\chi_*)\partial T/\partial n, \tag{50}$$

where

$$B(\chi_*)=A'(\chi_*)/A(\chi_*) \quad \text{with } \chi_*=-\int_0^\tau V\,\mathrm{d}\tau. \tag{51}$$

We have neglected n/θ in the argument of A for $n=O(1)$, i.e. B is the logarithmic derivative of the area at the location χ_* of the flame sheet.

It is easily seen that $\partial T/\partial n=0$ at $0+$, so that the extra terms

$$T_b^{-2}B(\chi_*)\int_{-\infty}^0 [(H-T_b)(1-\rho)V-\mathscr{L}^{-1}\partial Y/\partial n-\partial T/\partial n]\,\mathrm{d}n=-bB(\chi_*) \tag{52}$$

in $\phi_* V$ come from the integral. They lead to

$$b\dot{V}=V^3\ln V^2-bB(\chi_*)V^2 \tag{53}$$

as the replacement for the governing equation (44).

An immediate consequence is that a flame sheet propagating steadily ($V=1$) down a duct of constant cross section will, in general, accelerate on entering a diverging section ($A'<0$); whereas, if it enters a converging section ($A'>0$) it will slow down (Sivashinsky 1974*a*). The only exception is $b=0$ ($\mathscr{L}=1$); then $V=1$ irrespective of area changes. Again $\mathscr{L}=1$ is seen to be an exception.

The $B(\chi_*)$ makes the differential equation (53) difficult to discuss in general. However, if the area changes very slowly, more precisely $A(\chi)=\tilde{A}(\varepsilon\chi)$ where $0<\varepsilon\ll 1$, then there is an approximate solution

$$V=1+\tfrac{1}{2}\varepsilon b\tilde{B}(\varepsilon\chi_*) \quad \text{with } \tilde{B}=\tilde{A}'/\tilde{A} \tag{54}$$

in which V differs only slightly from 1. It represents a balance between the effect of area change and that of enthalpy deficiency at the flame, with temporal variations playing no role. If $\tilde{A}'$ is negative (i.e. $\tilde{B}<0$), the area of the flame sheet increases as it propagates to the left and we say the flame is being stretched. For $\mathscr{L}<1(b<0)$ such stretching accelerates the flame, but for $\mathscr{L}>1(b>0)$ there is a deceleration.

There have been attempts to discuss the effect of curvature on unsteady flames by assuming that the difference between the speed of a slightly curved flame and a plane flame is proportional to the curvature, with the constant of proportionality depending only on the mixture properties and not on the particular unsteady process. Markstein (1964, p. 22) has discussed flame stability in this way. The result (54) provides a rational justification for such an assumption, when it is realized that curvature manifests itself through fractional changes in area. In particular, it shows that the constant of proportionality changes sign as $\mathscr{L}$ passes through 1.

7 An elementary flame holder

The basic function of a flame holder is to maintain a flame in a stable rest position: the flame is brought to rest by applying a counterflow; stability is ensured by appropriate velocity and thermal gradients. For example, velocity gradients can be produced by expansion of the mixture as it leaves a tube or by the flow to the rear of a bluff body; thermal gradients arise quite naturally by conduction to the burner. These agencies can be modeled by a one-dimensional formulation of propagation along a non-uniform duct with a counterflow, incorporating heat losses that depend on location as well as temperature. For simplicity we shall consider only conductive losses through the side walls for which the linear law (28*a*) holds, with k a slowly varying function of x, as is A. The perturbation terms are now

$$f=-B(\chi_*)\rho v, \quad g=\mathscr{L}^{-1}B(\chi_*)\partial Y/\partial n, \quad h=B(\chi_*)\partial T/\partial n-k(\chi_*)(T-T_{\mathrm{f}}). \tag{55}$$

The counterflow (35) that appears in the continuity equation (36) now depends on the location χ_* of the flame sheet, as we shall see. At various places in the preceding analysis we must now replace V by W. As a consequence, the expansions (31), (32), (41), and, hence, the enthalpy result (46) with the extra term (52) still apply if V is replaced everywhere by W except in $(1-\rho)V$, which must be replaced by $W-\rho V$. Noting that

$$\psi = k(\chi_*)(T - T_f) \tag{56}$$

then gives

$$\phi_* W = -bW^{-2}\dot{W} + \Psi(\chi_*)W^{-1} - bB(\chi_*)VW^{-1} \quad \text{with } T_b^2\Psi(\chi_*) = 2Y_f k(\chi_*), \tag{57}$$

so that the governing equation becomes

$$b\dot{W} = W^3 \ln W^2 + \Psi(\chi_*)W - bB(\chi_*)VW. \tag{58}$$

Without counterflow, i.e. for $v_f = 0$ and $W = V$, we reach a combination of the two equations (47) and (53) as expected.

In the absence of heat losses ($\Psi = 0$) there is a stationary solution

$$V = 0 \quad \text{provided } W = 1, \quad \text{i.e. } v_f = 1.$$

The flame speed relative to the fresh mixture is the same as that for a uniform duct; that is, the area changes do not affect the flame speed to leading order, in striking contrast to the unsteady problem without counterflow (section 6). The difference is that the flame sheet itself does not encounter area changes here; although the preheat zone does, so that the result is not evident *a priori*. The implication is that a steady cylindrical or spherical flame of large curvature has the same speed as a plane flame. The corresponding condition is

$$v_f^2 \ln v_f^2 + \Psi(0) = 0 \tag{59}$$

when heat losses are present. Not surprisingly, this is relation (34) with V replaced by v_f and with Ψ given its value at the origin (i.e. any finite x); so that Figure 1 applies. In either case, with or without heat loss, the location of the flame is indeterminate on the x-scale.

When the flame is moving, v_f becomes a function of χ_*; for it is the mass flux of the counterflow that is held fixed, whereas the area $A(\chi_*)$ at the flame varies with its position. This area is effectively constant in the integration (36) of the equation of continuity on the x-scale, so that

$$v_f \propto A^{-1}(\chi_*). \tag{60}$$

Accordingly, we should write $v_f(0)$ in the condition (59).

To investigate the stability of the stationary flame, time-dependent solutions of equation (58) are sought in which V is small (though still large compared to θ^{-1}), so that W differs just a little from v_f. The linear equation governing small changes from the rest position is

$$\ddot{\chi}_* + (2B - b^{-1}D)v_f\dot{\chi}_* + (\Psi' - BDv_f)b^{-1}v_f\chi_* = 0, \tag{61}$$

where

$$D = 2v_f(1 + \ln v_f^2), \tag{62}$$

and B, Ψ', and v_f are to be evaluated at $\chi_* = 0$; the necessary and sufficient condition for stability is that the coefficients of both $\dot{\chi}_*$ and χ_* be positive. The stability criteria are, therefore,

$$2B > b^{-1}D \quad \text{and } \Psi' \lessgtr DBv_f \quad \text{accordingly as } b \lessgtr 0. \tag{63}$$

To see how these conditions can be met, consider a duct of uniform wall material whose cross section varies in size but not in shape. The argument used for a circular tube in section 4 then shows that the heat-loss coefficient varies inversely as a linear dimension of the cross section, i.e.

$$k \propto A^{-\frac{1}{2}}(\chi_*). \tag{64}$$

It follows that $\Psi' = -\Psi B/2$, so that the criterion (63b) may be written

$$\Psi B \lessgtr \tfrac{4}{3}v_f^2 B \quad \text{accordingly as } b \lessgtr 0, \tag{65}$$

since $Dv_f = 2(v_f^2 - \psi)$.

To interpret these conditions, note that the derivative D is positive on the upper branch in Figure 1, negative on the lower. When the Lewis number is less than 1, i.e. $b < 0$, a flame is stable only if it corresponds to a point on the curve above S and the duct converges to the left; on the lower branch the convergence must also be sufficiently rapid, as determined by the inequality (63a). When the Lewis number exceeds 1, i.e. $b > 0$, there is stability only on the lower branch and then, for points below S, the duct must converge or, for points above S, diverge sufficiently slowly, as again determined by the inequality (63a).

A well insulated duct must, therefore, converge; then the flame with v_f small is stable for $\mathscr{L} > 1$ while that with v_f close to 1 is stable for $\mathscr{L} < 1$. The results in section 5 for a uniform duct are a limiting form of these conclusions.

8 NEFs

So far in our discussion of premixed flames we have assumed that the Lewis number is fixed at an $O(1)$ value. The results in section 5–7, however, are not uniformly valid for $\mathscr{L}$ close to 1 and suggest that for $\mathscr{L} - 1$ of order θ^{-1} the time scale on which the flame evolves is not $O(\theta)$ but $O(1)$. We turn, therefore, to an important class of flames characterized by

$$\mathscr{L}^{-1}=1-l/\theta \quad \text{with } l=O(1), \tag{66}$$

which will be called near-equidiffusional. Note that the term relates to the diffusion of heat and the single reactant, not to the differential diffusion of species introduced in section 1.4.

Consider the basic equations (15)–(20) with $f=g=h=0$. Adding together the second and third eliminates the reaction term to give

$$\rho\partial H/\partial t+\rho v\partial H/\partial x-\partial^2 H/\partial x^2+\theta^{-1}l\partial^2 Y/\partial x^2=0. \tag{67}$$

Here $H\equiv T+Y$ is the enthalpy introduced in section 2.2; one possibility for its leading term is

$$H_0=T_b=H_f. \tag{68}$$

The result will hold everywhere (i.e. both inside and outside the flame sheet) at all times if it holds initially; we shall restrict attention to such initial conditions. The temperature and reactant fields are not exactly similar, as they are for $l=0$, but are approximately so.

Deviations from similarity are described by the perturbation H_1, which will be replaced by h. If equilibrium prevails behind the flame sheet, so that Y vanishes and $H=T$ there, then the choice (68) ensures that the flame temperature deviates from T_b (its adiabatic value) by at most $O(\theta^{-1})$, and, hence, that the flame speed remains $O(1)$. On either side of the flame sheet we have

$$\partial\rho/\partial t+\partial(\rho v)/\partial x=0, \quad \rho=T_f/T, \tag{69}$$

$$\rho\partial T/\partial t+\rho v\partial T/\partial x-\partial^2 T/\partial x^2=0, \tag{70}$$

$$\rho\partial h/\partial t+\rho v\partial h/\partial x-\partial^2 h/\partial x^2-l\partial^2 T/\partial x^2=0, \tag{71}$$

a closed set of equations for the variables ρ, v, T and h; here ρ, T, and v stand for the leading terms in their own expansions for θ large. There remains the task of deducing the jump conditions satisfied by these variables across the flame sheet.

For that purpose the exact equation (67) is written, in the frame moving with the flame sheet:

$$\rho\partial H/\partial t+\rho(V+v)\partial H/\partial n-\partial^2 H/\partial n^2+\theta^{-1}l\partial^2 Y/\partial n^2=0. \tag{72}$$

Using the variable $\xi=\theta n$ inside the flame sheet now yields

$$\partial^2 H_1/\partial\xi^2=0, \quad \rho_0(V+v_0)\partial H_1/\partial\xi-\partial^2 H_2/\partial\xi^2+l\partial^2 Y_1/\partial\xi^2=0, \tag{73}$$

where $\rho_0(\xi,t)$, $v_0(\xi,t)$, $H_1(\xi,t)$, $H_2(\xi,t)$, and $Y_1(\xi,t)$ are coefficients in the expansions valid in the flame sheet. As noted in section 3, V is not to be expanded. Since there are no $O(1)$ gradients of H outside the flame sheet, we may conclude from these equations first that H_1 is constant in ξ, and then that the total change across the sheet of $\partial H_2/\partial\xi$ is l times that of $\partial Y_1/\partial\xi$. For the expansion functions outside the flame sheet we find, therefore,

$$\delta(h)=0, \quad \delta(\partial h/\partial n)=l\delta(\partial Y/\partial n)=-l\delta(\partial T/\partial n) \tag{74}$$

where $\delta(\ )$ denotes the jump across the flame sheet of the quantity concerned.

Behind the flame sheet there are no $O(1)$ temperature gradients, so that to complete the formulation it is only necessary to calculate $\partial T/\partial n$ ahead. For this purpose, note that the eigenvalue (11) is still correct provided $\mathscr{L}$ is replaced by 1, J_s by the value of $-\partial Y/\partial n=\partial T/\partial n$ just ahead of the flame sheet, T_b by $T_b-\theta^{-1}T_b^2\phi_*$, and M by M_r, the reference mass flux defined by (14) with J_s replaced by Y_f. (The first corresponds to the leading-order specification (66); the second recognizes that the mass-flux fraction is not conserved for $O(1)$ time-derivatives; the third replacement corresponds to the change (13); and the fourth is due to the definition (19).) In short

$$\partial T/\partial n|_{0-}=Y_f\,e^{-\frac{1}{2}\phi_*} \tag{75}$$

which, together with the conditions (74) and

$$\delta(T)=0, \tag{76}$$

completes the formulation. It was first presented by Matkowsky & Sivashinsky (1979*b*) in the context of their constant-density approximation (cf. section 8.6). (The problem as formulated so far can be reduced to its constant-density form by a von Mises transformation (cf. Stewartson 1964, p. 30)).

NEFs are more difficult to treat than SVFs because their nonlinearity is intrinsic. A constant-density approximation is, therefore, often adopted: we shall resort to such a device in our discussion of multi-dimensional flames (section 8.6); Buckmaster & Nachman (1981) have done so in treating general duct flow. In considering propagation in a duct of slowly varying cross section, however, no such simplification is required. Here we present that analysis as a complement to section 6.

To account for area changes, terms

$$\left(\frac{1}{A}\frac{dA}{dx}\right)\rho v, \quad -\left(\frac{1}{A}\frac{dA}{dx}\right)\frac{\partial T}{\partial x}, \quad -\left(\frac{1}{A}\frac{dA}{dx}\right)\left(\frac{\partial h}{\partial x}+l\frac{\partial T}{\partial x}\right)$$

must be added to the left sides of equations (69*a*), (70), and (71). If the slow variations are characterized by a small parameter ε, i.e.

$$A=A(\chi) \quad \text{with } \chi=\varepsilon x, \tag{77}$$

then it is appropriate to seek a solution depending on n and the slow time

$$\tau = \varepsilon t. \tag{78}$$

In particular, the flame speed V is a function of τ; its determination is the central goal of the analysis. With error $O(\varepsilon^2)$ the governing equations are, therefore,

$$\partial[\rho(V+v)]/\partial n + \varepsilon(\partial\rho/\partial\tau + B\rho v) = 0, \quad \rho = T_f/T, \tag{79}$$

$$\rho(V+v)\partial T/\partial n - \partial^2 T/\partial n^2 + \varepsilon(\rho\partial T/\partial\tau - B\partial T/\partial n) = 0, \tag{80}$$

$$\rho(V+v)\partial h/\partial n - \partial^2 h/\partial n^2 - l\partial^2 T/\partial n^2 + \varepsilon(\rho\partial h/\partial\tau - B\partial h/\partial n - lB\partial T/\partial n) = 0, \tag{81}$$

where the area effect may be evaluated at the flame sheet, i.e.

$$B = A'(\chi_*)/A(\chi_*) \quad \text{with } \chi_* = -\int_0^\tau V \, d\tau \tag{82}$$

(cf. section 6).

In the limit $\varepsilon \to 0$ these equations are compatible with the conditions (74)–(76) provided $V=1$; the solution is then

$$\rho = T_f/T, \quad v = T/T_f - 1, \tag{83}$$

$$T = \begin{cases} T_f + Y_f e^n \\ T_b = H_f \end{cases}, \quad h = \begin{cases} -lY_f n \, e^n \\ 0 \end{cases} \quad \text{for } n \lessgtr 0. \tag{84}$$

This can be recognized as the steady plane deflagration wave treated in section 2.

The procedure for evaluating the effect of ε is similar to that used in section 3 to establish the result (26). Integrating the equation for T from $n = -\infty$ to $0-$, using the continuity equation, and applying the flame-sheet conditions yields

$$\begin{aligned} V - e^{-\frac{1}{2}\phi_*} &= \frac{\varepsilon}{Y_f}\int_{-\infty}^{0-}\left\{T_b\frac{\partial\rho}{\partial\tau} + B\left[\frac{\partial T}{\partial n} + \rho v(T_b - T)\right]\right\} dn \\ &= \varepsilon\tilde{b}B \quad \text{with } \tilde{b} = (T_b/Y_f)\ln(T_b/T_f) > 0. \end{aligned} \tag{85}$$

The integral has been evaluated by means of the formulas (83), (84), a step that introduces an error $o(\varepsilon)$. Since ϕ_* is $O(\varepsilon)$ according to the result (87) that follows, we may also replace $e^{\frac{1}{2}\phi_*}$ with $1+\frac{1}{2}\phi_*$, whence the relation

$$V = 1 - \tfrac{1}{2}\phi_* + \varepsilon\tilde{b}B \tag{86}$$

is seen to hold between the speed and temperature of the flame. A similar integration of the equation for h (noting that $\partial h/\partial n|_{0+}$ is $o(\varepsilon)$) yields

$$\phi_* = -2\varepsilon\hat{b}lB \quad \text{with } \hat{b} = -\frac{T_f}{2T_b^2}\int_0^1 \frac{\ln m}{m + T_f/Y_f} dm > 0, \tag{87}$$

so that we finally have

$$V = 1 + \varepsilon(\tilde{b} + \hat{b}l)B, \tag{88}$$

to be compared with the result (54) for SVFs.

If A' is negative (i.e. $B < 0$), the flame is being stretched (section 6). For $l = 0$ such stretching slows the flame; and

$$l < -\tilde{b}/\hat{b} \tag{89}$$

is required for the stretch to accelerate the flame. Here we have a refinement of the result (54), where the effect of area change was reversed as $\mathscr{L}$ passed through 1. The concept of flame stretch plays a role in the discussion of multidimensional flames (section 8.4), with the critical value of l appearing in the stability criteria for a plane flame (section 11.4) as its constant-density approximation $-2T_b^2/Y_f$ for $T_f \to \infty$ or $Y_f \to 0$.

4

STEADY BURNING OF A LINEAR CONDENSATE

1 Responses

In sections 2.3–2.5 several methods of supplying a reactant at a station $x=0$ were considered. In each case four relations between the five parameters M, D, and the three supply values (2.24) are prescribed; the flame eigenvalue (2.22) then determines them. For the switch-on reaction the supply values are prescribed along with D; the eigenvalue then gives M. For the burner M, D, T_s, and J_s are prescribed and the eigenvalue gives T_b, i.e. T_s+Y_s. (The third example – the vaporizing liquid – is discussed at the end of section 7.)

The switch-on reaction provides the simplest example of a response. With the supply values, i.e. the switch-on temperature T_s and the upstream state Y_f and T_f held fixed, the flame speed, i.e. M is determined as a function of pressure, i.e. D. The response curve $DM^{-2}=$constant is a parabola, where the constant is determined by $T_b=T_f+Y_f$ and $J_s=Y_f$. Note that while x_* is fixed, the location of the flame varies along the parabola because M is involved in the length unit.

For the burner, where T_s and J_s are fixed, the flame temperature T_b becomes a function of both M and D. The corresponding surface can be described by its section $MT_b^{-2}\,e^{\theta/2T_b}=\sqrt{2D}/J_s\theta=$constant (cf. equation (2.26*b*)), which is a parabola flattened along the T_b-axis near the origin. Not all of the curve, however, represents the response of T_b to M for fixed D; T_b must lie between T_s and T_s+J_s if x_* is to lie in the range $(0, \infty)$. These limits correspond, respectively, to all and none of the heat released at the flame being conducted back to the supply. When $T_b=T_s$ the flame has reached the surface and M may then be decreased indefinitely without changing T_b (see section 2.6). When $T_b=T_s+J_s$ the flame has become remote and M may then be increased indefinitely. The resulting response is similar to that in Figure 1, with M and D replaced by T_b and M. In practice, the remote flame signals an effective extinction (section 2). Now x_* varies along the central part of the response curve, producing the behavior in the location of the flame discussed in section 2.5.

A more complicated response, again of M to D, occurs when gasification at the surface of a liquid or solid, lying in $x<0$, supplies the reactant. We shall suppose that $J_s=1$ and calculate $T_b(=T_s+Y_s)$ from an overall enthalpy balance. Under adiabatic conditions T_b is thereby determined; otherwise, it is expressed in terms of T_s and M. A further relation comes from specifying the nature of the gasification: in pyrolysis, the surface temperature T_s determines the production rate M; in evaporation, it determines the partial pressure $\bar{p}_c Y_s$, i.e. DY_s for $v=1$. The fourth relation is the prescription of D; then the flame eigenvalue is an equation for M, the T_b in it being known (implicitly) as a function of M. Our task is to determine how M varies with D when the remote temperature of the condensate and the background temperature are held fixed.

Motivation for discussing the combustion of condensates comes from the use of solids and liquids as fuels and propellants. These are often in the form of small particles or drops that are sprayed into the combustion chamber. Chapters 6 and 7 stem from an interest in the burning of single drops; an understanding of that phenomenon is the first step in treating spray combustion. Propellants of a different kind are used in rocket motors and projectiles. Often fuel and oxidant, normally in solid form (although liquid mixtures, such as organic and inorganic molten salts with isopropyl ammonium nitrate, have been used experimentally in guns) are packed together. The packing may be loose, as in the case of gunpowder, or tight so as to ensure that gasification takes place only at the surface. Such tightly packed propellants may be macroscopically homogeneous (called double-based) or heterogeneous (called composite), as for oxidant particles embedded in a fuel matrix. Examples of fuels are rubber, polystyrene, and polymethylemethacrylate; the oxidant is often ammonium perchlorate. Homogeneous propellants permit plane burning (the subject of Chapters 4 and 5): a semi-infinite cylinder of propellant forms a linear condensate that is allowed to gasify at its end, the liberated reactants burning in a premixed plane flame. In practice, transverse effects must also be taken into account, leading to so-called erosive burning. Plane burning can also occur in the interior of solids; this model was used for earlier propellants, such as nitrocellulose–nitroglycerine. For some materials the product is also solid (these are discussed in Chapter 12).

Surface gasification of a solid is a complex process, not yet completely understood. We shall introduce two extremes: irreversible disintegration, and equilibrium; both processes being completed in an infinitely thin surface layer. Irreversible disintegration is known as (solid) pyrolysis. Equilibrium is, strictly speaking, unattainable, since the vapor must be continuously produced for burning; it is exemplified by ordinary evaporation of a liquid, and that is the term used even when discussing a solid. Equilibrium is a good approximation for liquid propellants, and, hence, should apply to solid propellants that melt prior to gasification. In practice, surface gasification probably lies somewhere between irreversible disintegration and equilibrium; the results for such an intermediate case have not been worked out, but there would be no great difficulty in doing so.

Fuels and oxidants have often been investigated separately in an attempt to understand the behavior of the propellants that they constitute. (The connection, however, is not completely clear.) In particular, the decomposition of ammonium perchlorate has been studied extensively; some of the results are quoted in sections 2 and 5.

A glance at the report of Williams (1969) and the references therein should convince the reader that the use of activation-energy asymptotics clarifies the discussion.

2 Experimental results; extinction

If a solid or liquid contained in the left half of a long, straight, uniform tube gasifies into a combustible mixture at its surface, then a steady state may be attained in which the surface recedes and is followed at a fixed distance by a flame separating unburnt gas from burnt. An observer moving with the surface sees a constant flux M of unburnt gas towards the flame and an equal flux of burnt gas away. The problem of determining the combustion field is of the type considered in section 2.3 (while resolving the cold-boundary difficulty) provided the combustible mixture can be treated as a single reactant. The evaporating liquid treated in that section constitutes one of two main questions and it will have to be reconsidered. The other main question concerns the pyrolyzing solid.

Since our aim is to determine the response of M to D, one might think that equation (2.22) has the answer: the response is parabolic. The condensate, however, has two important effects. First, it fixes the position of the flame, which must be somewhere between its surface and infinity. Use of the formula (2.22) is thereby restricted; when the flame approaches the surface or recedes to infinity the appropriate replacement from section 2.6 must be made. Secondly, it can make T_b dependent on M or D and, hence, completely change the response predicted by the original formula.

These two effects constitute our subject for discussion.

Experiments that approximate the situation described have been conducted for solids that gasify by surface pyrolysis (Johnson & Nachbar 1962; Nir 1973). Three basic phenomena are observed: (i) the burning rate increases with ambient pressure, but very little beyond a certain pressure; (ii) there is a minimum pressure beyond which gaseous deflagration will not take place (extinction); and (iii) preheating the solid lowers this minimum. All three effects are exhibited by the theory, even when absorption or emission of radiant energy by the surface and heat loss or gain through the wall of the tube are neglected. These departures from adiabatic deflagration will be considered later; here we merely note the partial analysis given by Johnson & Nachbar (1962) based on distributed heat loss from the condensed phase.

The analysis follows Buckmaster, Kapila, & Ludford (1976); it reveals that extinction, by which is meant the disappearance of gaseous deflagration and not the cessation of pyrolysis, can be of two kinds. In true extinction, the flame suddenly disappears as the pressure is lowered. The pyrolysis continues, however, at a lower temperature, the products passing away unburnt in what is known as flameless combustion. Effective extinction, followed by flameless combustion without a sudden drop in surface temperature, occurs when the flame recedes to infinity so that the products of continuing pyrolysis effectively pass away unburnt. The disappearance of a flame is generally taken to be true extinction, the concept of effective extinction being mainly unknown.

We shall demonstrate that extinction is always effective for adiabatic deflagration, and may be so for radiative loss or gain (when a separate phenomenon of ignition is possible). We shall also show that effective extinction is changed to true extinction by heat losses through the tube wall (Kapila & Ludford 1977).

Apparently no comparable experiments have been reported on evaporating liquids. (See, however, Adams & Wiseman (1954).) A more realistic treatment than that presented in section 2.4 reveals striking effects (detailed later), which deserve experimental verification.

3 Solid pyrolysis under adiabatic conditions

The combustible mixture will be treated as a single reactant with Lewis number 1. The object is to determine the combustion field, in general, and M, in particular, given the remote temperature $\hat{T}_f$ of the condensate (i.e. far to the left) and the pressure level in the gas phase, i.e. D. For that purpose (see section 2.4) we need only determine (in addition to M) the values T_s, Y_s and $-T_s'=Y_s'$, which the solid in $x<0$ presents to the gas phase in $x>0$.

To determine the eigenvalue $\mathscr{D}=DM^{-2}$ it is only necessary to know the mass flux fraction $J_s=Y_s-Y_s'$ and the flame temperature $T_b=T_s+Y_s$. If no product of the gaseous reaction or inert is produced or absorbed at the surface of the condensate, then $J_s=1$ and, according to equation (2.22),

$$DM^{-2}=\theta^2\,e^{\theta/T_b}/2T_b^4, \tag{1}$$

a relation determining M once T_b is known. The formula has a mixed character: θ and T_b have no dimensions, while M and $D^{\frac{1}{2}}$ have those of mass flux. Several such equations will appear in the sequel. Note that the presence of the condensate (or whatever else is creating the stream of reactant) is felt solely through T_b, a fact first recognized by Williams (1973). Note also that of the equations determining the burning rate here, and in subsequent sections concerned with solid pyrolysis, only (1) contains D and θ; so that in discussing the solution in the limit $\theta\to\infty$ we may suppose that all quantities are $O(1)$, except for D.

It is possible to obtain $T_b=T_s+Y_s$ without calculating surface values. In the absence of radiation to or from the surface and of losses elsewhere, the flux of enthalpy at $x=+\infty$ must equal that at $x=-\infty$ (since the kinetic energy is negligible). Reverting to dimensional quantities shows that

$$(h^0-Q)+c_p(T_b-T^0)=\hat{h}^0+\hat{c}_p(\hat{T}_f-T^0),$$

where h^0-Q is the heat of formation of the product of the deflagration and the circumflex denotes values for the condensate. Solving for T_b and writing the result in its dimensionless form yields

$$T_b=\kappa\hat{T}_f+(1-\kappa)T^0+q+1 \quad \text{with } \kappa=\hat{c}_p/c_p,\ q=(\hat{h}^0-h^0)/Q. \tag{2}$$

Here q, the heat of gasification, is the difference between the heat of formation of the condensate and that of the gaseous reactant, both at the standard temperature T^0, expressed in units of the heat of combustion Q. Equation (2a) shows that T_b (which we shall suppose to be positive) is independent of M and is a linear function of the remote temperature $\hat{T}_f$ of the condensate. Figure 1 sketches the curves (1) for two values of $\hat{T}_f$.

A third condition must be added to $J_s=1$ and $T_s+Y_s=T_b$ to determine all of the surface values; for a solid this can take the form of a pyrolysis law. The rate of gasification is supposed to depend only on the surface temperature and not, for example, on the pressure. We shall take

$$M=kT_s\,e^{-\hat{\theta}/T_s}, \tag{3}$$

where k and $\hat{\theta}$ are constants. Note that the pyrolysis occurs at any surface temperature, so that the term extinction (or ignition) must refer to the gaseous deflagration, as we have supposed. The law should be viewed as determining the surface temperature T_s that produces the required flux M into the gas phase. Various modifications have been suggested, all of which lead to similar results. It is only essential that M should be a monotonically increasing function of T_s. Apart from its simplicity, the law adopted here has the advantage of being firmly founded theoretically in gas kinetics and reactive solids. Its role in Figure 1 is merely to make T_s a parameter on the curves.

The gas phase is now completely determined, without reference to the conductivity of the solid. Only the enthalpy of the solid enters into the result (1), and the remaining details of the gas phase come from adding the pyrolysis law. The role of the conductivity is to fix the distribution of temperature between $\hat{T}_f$ and T_s so that the total heat flux at each point of the moving solid is constant.

Whatever the value of θ, the surface temperature must lie within certain bounds: $T_b = T_s + Y_s$ and $0 < Y_s < 1$ imply

$$T_b - 1 < T_s < T_b. \tag{4}$$

For finite θ it is not immediately clear that a solution is then guaranteed; but in the limit $\theta \to \infty$ such is the case. Only the location (2.14) of the flame need be checked. If we note that $T_s' = 1 + T_s - T_b$, a result expressing heat-flux balance in the gas phase, then

$$x_* = -\ln(1 + T_s - T_b) \tag{5}$$

1 Solid pyrolysis under adiabatic conditions: ---- parabola (1); —— response curve. On horizontals through P_s and P_∞, flame sheet is at surface and infinity, respectively.

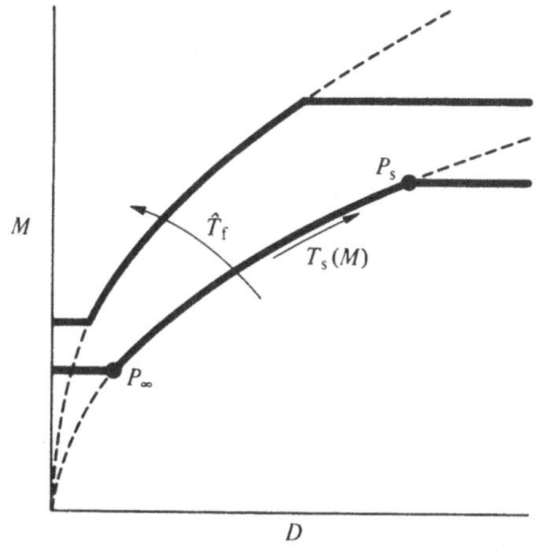

and the inequalities (4) ensure $0<x_*<\infty$, as supposed. It is a characteristic feature of activation-energy asymptotics that such questions can be settled in a simple way.

Figure 1 shows the corresponding end points P_s and P_∞, so designated because at P_s the flame has moved back to the surface and at P_∞ has receded to infinity. These two possibilities were dealt with before (section 2.6), requiring reconsiderations of the structure. It was found that, with conditions at the surface fixed, the remote flame allows D to be decreased to zero while the surface flame allows it to be increased indefinitely, both without change in M; that is just what the pyrolysis law demands. Thus, the M, D-curve is completed by horizontal lines, one stretching from P_s to $D=\infty$ and the other from P_∞ to $D=0$. Of course, if T_b is less than 1, the left inequality (4) is ineffective, i.e. P_∞ does not exist and the curve extends to the origin. The following remarks will apply when $T_b>1$; otherwise they must be slightly modified.

The response curve now shows that: (i) M increases with D, but not beyond P_s; and (ii) there is a minimum D, corresponding to P_∞, before which there is effective extinction. The phenomenon (iii) discussed in section 2 concerns the change in D, for $T_s=T_b-1$, as T_b increases linearly with $\hat{T}_f$. We find

$$\frac{\mathrm{d}D}{\mathrm{d}\hat{T}_f}=-\frac{\theta^3M^2}{2T_b^6}\,\mathrm{e}^{\theta/T_b}\frac{\mathrm{d}T_b}{\mathrm{d}\hat{T}_f}<0 \tag{6}$$

with relative error $O(\theta^{-1})$, since $\mathrm{d}M/\mathrm{d}T_b=\mathrm{d}M/\mathrm{d}T_s$ is $O(1)$. The effect (iii) is clearly present and is caused by the exponential factor, which decreases as T_b increases, irrespective of how M changes.

4 Radiation from the surface

It is of interest to see how the picture is modified by radiation between the surface of the solid and the surroundings. (Distributed exchange will be treated later.) If the background temperature is T_g, there is a heat loss (or gain, if negative)

$$\psi_s=k_s(T_s^4-T_g^4), \tag{7}$$

where k_s is a positive constant that is not always small in practice. Experiments have been reported by Levy & Friedman (1962), showing qualitative agreement with the results we shall obtain. As for mathematical treatments, only Spalding (1960) has treated radiative exchange with surroundings at nonzero temperature ($T_g\neq0$) but in an approximate fashion that precluded heat gain ($T_g>T_s$). As a consequence, he did not uncover the most striking results.

The loss changes the overall enthalpy balance to

$$T_b = T_a - \psi_s/M, \tag{8}$$

where T_a is the flame temperature under adiabatic conditions, given by the formula (2); but nothing else changes. The actual flame temperature is now a function of M, both directly and through T_s in ψ_s. Inserted in the response formula (1), it provides a D that is no longer proportional to M^2. Note that the pyrolysis law now affects the shape of the M, D-curve.

Consider first $T_g = 0$. To determine the new response curve it is helpful to sketch the graph of T_b versus T_s for fixed $\hat{T}_f$; that has been done in Figure 2. There is a maximum at E, corresponding to $T_s = \frac{1}{3}\hat{\theta}$, from which the descent is monotonic to $-\infty$ as $T_s \to 0$ or ∞. The shape of the M, D-curve is obtained by rotation through 90° since the asymptotically correct result

$$\frac{1}{D}\frac{dD}{dM} = -\left(\frac{\theta}{T_b^2}\frac{dT_s}{dM}\right)\frac{dT_b}{dT_s} \tag{9}$$

shows that all slopes have their signs changed. Figure 3 gives sketches of the resulting C-shaped curve, with the point on the extreme left corresponding to E.

As in the absence of radiation, not all of the curve is acceptable. In Figure 2 only the portions lying between the lines $T_b = T_s$ and $T_b = T_s + 1$ satisfy the inequalities (4), and these portions depend on $\hat{T}_f$. (An increase in $\hat{T}_f$ merely translates the curve upwards.) If $\hat{T}_f$ is too small, there are no

2 Solid pyrolysis with radiative loss (only) from surface: T_b versus T_s according to equation (8) with $T_g = 0$.

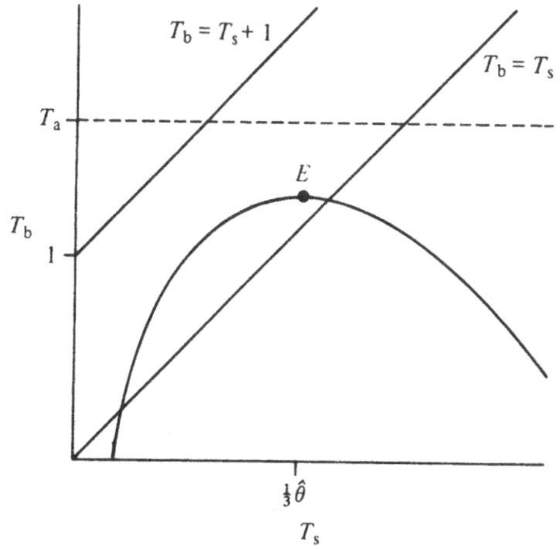

portions; otherwise five possibilities arise depending on whether the curve intersects one or both of the lines with E either between them or not. These lead to the five parts of Figure 3, where progress is from (*a*) to (*b* or *b*′) to (*c*) to (*d*) as $\hat{T}_f$ increases. In each case the end points of the portion are marked to correspond with the position of the flame, and horizontal completions are included. The complete responses are shown by unbroken lines.

It is widely believed that if M decreases as D increases the combustion is

3 Solid pyrolysis with radiative loss (only) from surface: ---- curve (1) with T_b a function of M; —— response curve; —— presumably unstable. On horizontals through P_s and P_∞, flame sheet is at surface and infinity, respectively.

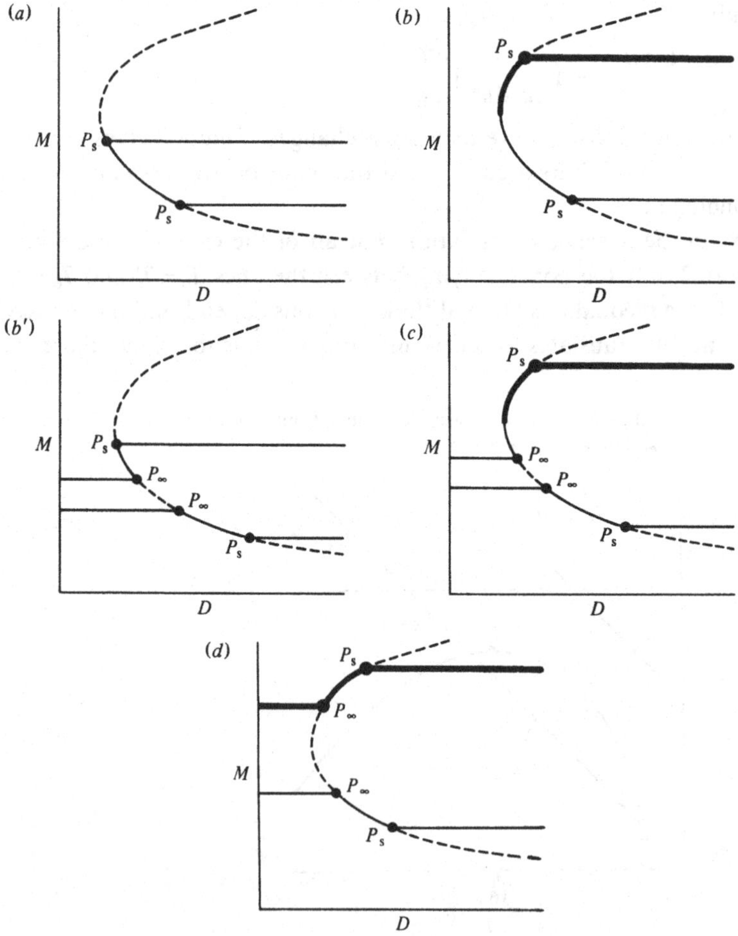

unstable (Emmons 1971); we shall adopt the idea here. (Surface burning, corresponding to a horizontal extension from such a point, will also be assumed unstable.) The whole of Figures 3(*a*) and 3(*b'*) are then discarded, the preheating being too weak for stable combustion. The surviving parts of Figures 3(*b*), 3(*c*), and 3(*d*) are drawn heavily, as the responses to be expected in an experiment. Clearly, they exhibit the phenomena (i) and (ii) mentioned in section 2.

The minimum $\hat{T}_f$ for stable burning can be found by setting $T_b = T_s = \frac{1}{3}\hat{\theta}$ in (8) and solving for $\hat{T}_f$. As the preheating increases, the true extinction of Figures 3(*b*) and 3(*c*) will occur until $\hat{T}_f$ reaches the value given by (8) for $T_b - 1 = T_s = \frac{1}{3}\hat{\theta}$. For all larger values of $\hat{T}_f$ the effective extinction of Figure 3(*d*) will occur, as under adiabatic conditions. The extinction value of D is given explicitly by the formula (1), where M and T_b are to be calculated from (3) and (8) by setting $T_s = \frac{1}{3}\hat{\theta}$ for true extinction and $T_s = T_b - 1$ for effective extinction. The third effect, i.e. that the extinction value of D decreases as $\hat{T}_f$ increases, then follows from the derivative (6) since T_b still increases with $\hat{T}_f$.

5 Background radiation

If $T_g \neq 0$ in the radiation term (7) several responses can occur. These are not associated with T_s large (where the results of section 4 apply) but with T_s small, which not only allows T_g to change the sign of ψ_s, but also enhances the latter's effect in (8) through the smallness of M. The left side of the curve in Figure 2 is thereby bent upwards to give Figure 4. Whether the result is a monotonic curve (Figure 4(*a*)) or one with a minimum $\check{E}$ and a maximum $\hat{E}$ (Figure 4(*b*)) depends on the size of T_g; the polynomial factor $3T_s^5 - \hat{\theta}T_s^4 + T_g^4 T_s + \hat{\theta}T_g^4$ in $\mathrm{d}T_b/\mathrm{d}T_s$ has just two positive zeros (both lying between T_g and $\frac{1}{3}\hat{\theta}$) if T_g is less than $0.168\,\hat{\theta}(<\hat{\theta}/3)$ and no positive zero otherwise. Since the curve must cross the strip described by (4) there is always deflagration (albeit unstable) for some range of D, in contrast to section 4 where $\hat{T}_f$ had to be large enough.

The monotonic curve for $T_g > 0.168\,\hat{\theta}$ leads to a response that is qualitatively the same as for adiabatic conditions (Figure 1). The strong background radiation tends to compensate for losses from the surface.

The minimum–maximum curve for $T_g < 0.168\,\hat{\theta}$ is cut by 45° lines in at most three points. Such curves present 30 possibilities depending on how the lines $T_b = T_s$, $T_s + 1$ are cut (times and order) and on whether $\check{E}$ and $\hat{E}$ lie above, between, or below them. Three of the possibilities can be ruled out because this curve has only one inflexion point; others lead either to unacceptable responses or to responses that have already appeared for $T_g=0$. The remainder fall into five groups, according to similarity of response; one member (the simplest) from each of the five will be presented here. Figure 5 shows these responses, which all have the shape of an S (before modification). We meet here, for the first time, an ignition–extinction response that is ubiquitous in combustion and chemical reactors. The upper turning point corresponds to extinction, as for the C-response in section 4. There it also corresponds to ignition, while

4 Solid pyrolysis with radiative exchange between surface and background: T_b versus T_s according to equation (8) with $T_g \neq 0$: (*a*) $T_g > 0.168\hat{\theta}$; (*b*) $T_g < 0.168\hat{\theta}$.

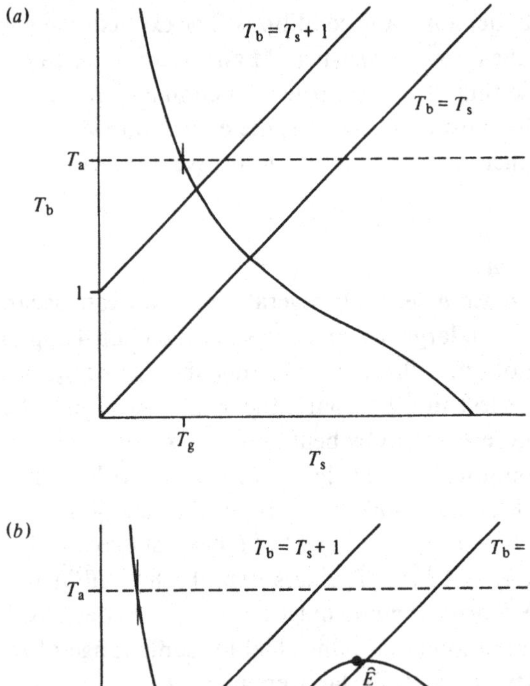

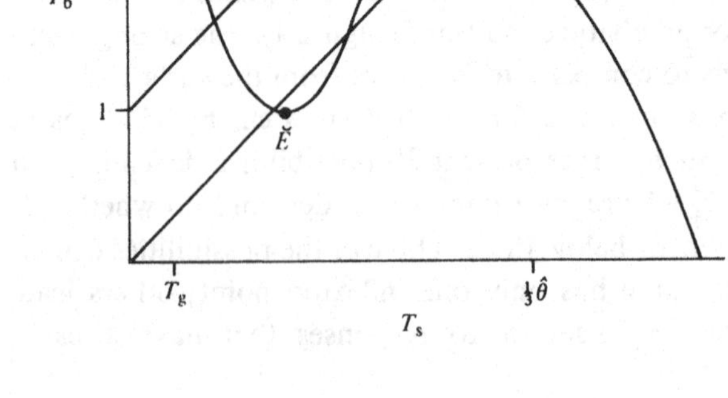

here that role is reserved for the lower turning point. (Note that following extinction there is now weak burning on the lower branch, rather than flameless combustion.)

The modified responses are these S-curves with parts deleted by the limitations (4) and the stability requirement of positive slope, but completed by horizontal lines. The first four of these responses have two

5 Solid pyrolysis with radiative exchange between surface and background: ---- curve (1) with T_b a function of M; —— response curve; —— presumably unstable. On horizontals through P_s and P_∞, flame sheet is at surface and infinity, respectively. (*a*), (*b*), (*c*), (*d*), (*e*) are representative members of five groups of responses.

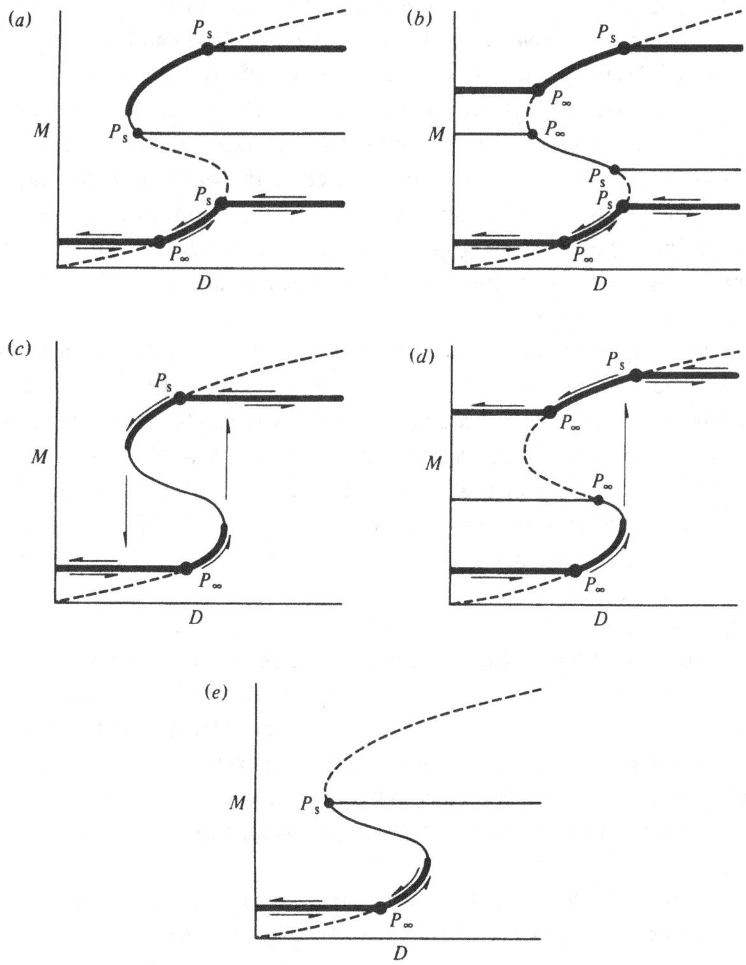

branches, each ending on the left with either true or effective extinction. In (*a*) and (*b*) both branches have a plateau to the right; we shall see that the plateau on the upper branch in each of (*c*) and (*d*) is reached for D large enough. The three phenomena in section 2 are, therefore, exhibited once more. The sole exception is (*e*), which has no upper branch.

We shall now construct the probable sequence of events, first as D increases from 0 to ∞ and then as D decreases from ∞ to 0. When there is only one effective extinction on the left, as in (*a*), (*c*), and (*e*), the lower branch is followed as D increases. The same is true for (*b*) and (*d*) if, as we shall assume, weak burning is preferred to strong when the latter is not already established. In (*a*) and (*b*) the lower branch will be followed all the way through surface burning to infinity, but in (*c*) and (*d*) there will be a jump to the upper branch, i.e. ignition, as the point furthest to the right is approached on the lower branch. As D decreases in (*c*) and (*d*) the upper branch will be followed, for (*d*) all the way to effective extinction but for (*c*) with a jump to the lower branch as true extinction is approached on the upper branch. In (*a*) and (*b*) we invoke the weaker burning assumption to see that the lower branch will be followed all the way to effective extinction. In (*e*) there is true extinction as D increases and ignition as it decreases. After extinction and before ignition there is no steady burning if we persist in believing that the surface burning that emanates from the unstable branch is itself unstable.

This description is based on several unproved principles: no part of an M, D-curve with negative slope can be attained physically because the combustion is unstable; a branch will be followed as far as it can be as D increases or decreases; and when a choice remains weaker burning will occur. Whether the other branches in (*a*) and (*b*) can ever be attained is an interesting open question; but supposing they cannot be, then (*c*), (*d*), and (*e*) are the only really new responses. Even then, only (*e*) fails to exhibit the three basic phenomena.

Data of Johnson & Nachbar (1962) and Guirao & Williams (1971) for ammonium perchlorate, however, can be interpreted as lying on the strong burning branch of Figure 5(*a*), see Buckmaster, Kapila, & Ludford (1976), suggesting that the weak burning branch is not applicable. That would be the case if pyrolysis ceased below a certain temperature, so that the law (3) only applied above and M was zero below. The whole lower branch could then be eliminated, leaving only the upper branch for the combustion to follow.

It should be noted that the bending back of the M, D-curve from a C-shape into an S-shape occurs for any nonzero $T_g < 0.168\,\hat{\theta}$. For $T_g > 0.168\,\hat{\theta}$, the S is pulled out into a monotonic curve. Even weak background radiation can, therefore, result in quite different responses. Far from being negligible, it can be the dominant effect under suitable conditions. The apparent contradiction as $T_g \to 0$ is due to the nonuniformity of the limit. However small T_g may be there are smaller values of T_s, for which background radiation changes the heat loss into a heat gain. Moreover, the effect of the gain on the enthalpy of the reactant is magnified by the smallness of the mass flux at such surface temperatures. It is unlikely, however, that such a limiting behavior could be observed, because of pyrolysis cut-off at low temperatures.

In practice, the most important features of the response are the ignition and extinction values of D. In any particular case, these are easily determined by finding the points $\breve{E}, \hat{E}$, or the appropriate intersection of the T_s, T_b-curve with $T_b = T_s + 1$.

6 True nature of effective extinction

The object of this section is to determine how the previous results are modified when $O(\theta^{-1})$ heat is lost or gained (by radiation or lateral conduction) throughout the gas phase. For simplicity we shall start by supposing that the solid phase is perfectly insulated. The main conclusion is that all effective extinctions are changed into true extinctions by heat loss, the remote flame only being maintainable as the limit for vanishingly small heat gain.

For each point of the relevant response curve determined previously, we calculate the new D (due to heat exchange) for the same value of M. The general effect of heat loss can be seen from the formula (1). If the flame temperature is reduced, M can only be maintained by increasing D: to maintain the rate of reaction at a lower temperature the concentration of reactant must be increased, i.e. the pressure raised. The pyrolysis law (3) ensures that T_s is unaffected, from which it follows that the temperature distribution in the solid and T_s' are also unaffected (the latter because of an energy balance at the surface); Y_s and Y_s' do change by $O(\theta^{-1})$ amounts, but the combination $Y_s - Y_s'$ does not.

The problem then is to calculate the perturbation of the flame temperature, which Kapila & Ludford (1977) did by straightforward matching. We shall follow section 3.3, however, in calculating it directly from the change in enthalpy from the surface up to (and including) the flame sheet. Rather than modifying equation (3.26), we shall derive the result *ab initio* for these circumstances.

The governing equations are

$$\mathfrak{L}(Y,1) = -\mathfrak{L}(T,1) + \theta^{-1}\psi(T) = \mathscr{D}Y\,\mathrm{e}^{-\theta/T}, \tag{10}$$

where ψ is one of the functions (3.28) with T_f replaced by T_g. Integration of the first equation between 0 and x_*+0 immediately yields the flame-temperature perturbation

$$T_b^2\phi_* = -\theta\frac{\mathrm{d}T}{\mathrm{d}x}\bigg|_{x_*+0} + \int_0^{x_*}\psi(T)\,\mathrm{d}x, \tag{11}$$

since the surface values T_s, T_s', and $Y_s - Y_s'$ are not perturbed. Both terms on the right side are to be evaluated to leading order and, for that purpose, the approximations

$$T = \begin{cases} T_s + T_s'(\mathrm{e}^x - 1) & \text{for } 0<x<x_*, \\ T_b - \theta^{-1}[T_b^2\phi_* + \psi(T_b)(x - x_*)] & \text{for } x > x_* \end{cases} \tag{12}$$

are used to obtain

$$\phi_* = \Psi \quad \text{with } T_b^2\Psi = \psi(T_b) + \int_0^{x_*}\psi[T_s + T_s'(\mathrm{e}^x - 1)]\,\mathrm{d}x, \tag{13}$$

a result that should be compared with the formula (3.33). The two terms in Ψ represent heat exchanges of the unburnt mixture (between the surface and the flame) with the burnt mixture and surroundings, respectively. (The burnt mixture later exchanges the same heat with its surroundings.) It follows that

$$D = D^\circ(M)\,\mathrm{e}^\Psi, \tag{14}$$

where $D^\circ(M)$ is given by the eigenvalue (1).

The last three sections have been an investigation of $D^\circ(M)$, which involves the dependence (8) of T_b on M both directly and through the pyrolysis law. The exponent Ψ depends directly and indirectly on M, with the additional involvement of the function ψ. It is, therefore, a complicated matter to describe the dependence of D on M in detail, especially when ψ is left arbitrary. However, if ψ is not too large, the general shape of the response will be maintained, except near P_∞ where $x_*\to\infty$ in Ψ. What happens there depends on whether the surface is hotter or cooler than the background.

When the background temperature is low enough, namely

$$T_g < T_s(M_\infty), \tag{15}$$

the temperature in the gas phase for M above M_∞ (its value at P_∞) is everywhere higher than T_g, so that ψ is positive. Consequently, Ψ is positive and tends to $+\infty$ as $M \to M_\infty$ because x_* does so. The response lies to the right of that in Figure 1, 3, or 5 and bends around as $M \to M_\infty$ to form a C. We conclude that heat loss in the gas phase changes an effective extinction into a true extinction.

For higher background temperatures, namely,

$$T_g > T_s(M_\infty), \tag{16}$$

the temperature in the gas phase near the surface is lower than T_g for at least a range of values above M_∞. It is easily seen that, as $M \to M_\infty$, the resulting heat gain eventually overwhelms any heat losses further away, so that Ψ becomes negative (if it is not already so) and tends to $-\infty$. The response, therefore, lies ultimately to the left of that in Figures 1, 3, or 5 and ends at $D=0$. Even then the effective extinction may be preceded by a true extinction: if $T_s(M_\infty)$ is sufficiently close to T_g, the function Ψ will only become negative very close to M_∞ and before that will be increasing rapidly (because of x_*), so that an S is formed. For smaller values of $T_s(M_\infty)$ the S straightens out into a monotonic response. Thus, the remote flame corresponding to the horizontal line through P_∞, is seen to be the limit for vanishingly small heat gain.

Kapila & Ludford (1977) have given details of this picture for the linear law (3.28*a*) when the ambient temperature is the same as that for the remote part of the solid ($T_g = \hat{T}_f$) and there is no radiative exchange with the surface. They also consider less than perfect insulation of the solid; then T_g is necessarily the same as $\hat{T}_f$ since otherwise the solid exchanges an infinite amount of heat. The termination points are unaltered even though the curves are somewhat distorted. The reason is that heat exchange in the solid, which adds a term of Ψ, is bounded as $M \to M_\infty$ and, therefore, cannot affect the unboundedness of Ψ.

Distributed heat loss from the solid phase alone was considered by Johnson & Nachbar (1962) as a way of modifying the C-shaped responses they obtained analytically when there is radiative loss from the surface. Under their assumptions effective extinction cannot occur.

7 Liquid evaporation under adiabatic conditions

With one important exception, the analysis follows that in section 3, the circumflex now referring to the liquid. The latent heat of evaporation

$$L = -q + (1-\kappa)(T_s - T^0) \tag{17}$$

is positive, however, at all temperatures T_s of interest, corresponding to $q<0$ and $\kappa<1$; so that we are dealing with a strictly endothermic process (in contrast to pyrolysis, which is usually taken to be exothermic). The exception to section 3 is the use of the pyrolysis law (3), which must be replaced by the Clausius–Clapeyron relation

$$Y_s \bar{p}_c = kT_s^\beta \, e^{-\hat{\theta}/T_s} \quad (\beta > 0) \tag{18}$$

in calculating boundary values. Here $\beta = \gamma(1-\kappa)/(\gamma - 1)$, where κ has the definition (2*b*) and γ is the ratio of specific heats in the gas. The new law relates the partial pressure of the vapor at the surface to the temperature there, and is applicable below the critical point whenever the evaporation is rapid enough for thermodynamic equilibrium to be achieved. (As the law is derived, the mole fraction stands in place of Y_s; but approximation by the intermediate mass (1.49) and subsequent absorption of masses into k results in the formula given.) The critical pressure is invariably very large so that we shall allow $\bar{p}_c$ to become indefinitely large, as is needed for the Damköhler numbers involved in the asymptotic theory. Equation (18), with Y_s replaced by $T_b - T_s$ from the Shvab–Zeldovich relation, should be viewed as determining T_s for given D (i.e. $\bar{p}_c$) so that once more the surface temperature is a parameter on the parabola (1). For the pyrolyzing solid the surface values depended on M; here they depend on D.

Suppose $T_b > 1$ (otherwise the discussion is modified in an obvious way). The point P_∞, corresponding to the left end of the range (4), then exists and is marked in Figure 6; the point P_s, corresponding to the right end,

6 **Liquid evaporation under adiabatic conditions: ---- parabola (1); —— response curve; —— presumably unstable. Vertical through P_∞ represents complete evaporation with flame at infinity. Modification for large D is due to flame approaching surface, reached at P_s.**

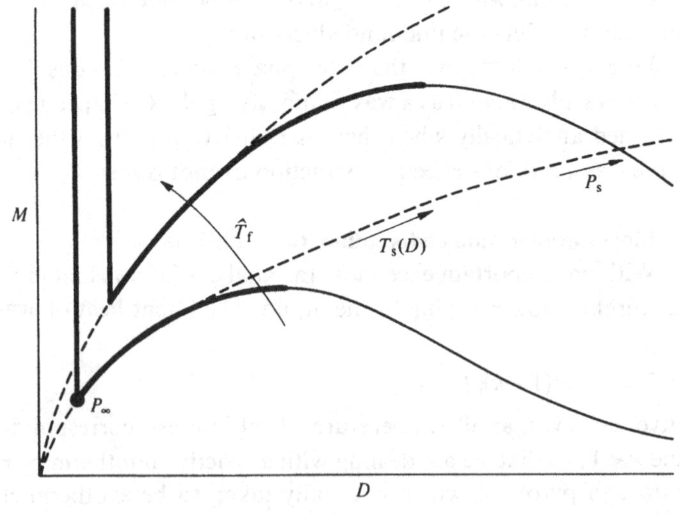

lies at infinity (according to equation (18)) because Y_s vanishes when $T_s = T_b$. As long as T_s is away from the ends of its range the response is similar to that for a pyrolyzing solid. Near the ends, however, the two responses differ markedly, due entirely to the difference between the two laws (3) and (18).

As $T_s \to T_b - 1$, the flame sheet recedes to infinity and, since $Y_s \to 1$, the pressure $\bar{p}_c$ (and hence D) tends to a finite nonzero value. The analysis of remote flames, given in section 2.6, shows that M can then increase indefinitely, with conditions at the surface fixed, without change in D; and that is just what the Clausius–Clapeyron law demands. The liquid, which is at its saturation temperature for the ambient pressure, must evaporate completely before the pressure can be lowered. Combustion plays no role because heat transfer from the gas to the liquid has ceased; the entire heat needed for evaporation is supplied from the remote end of the liquid, which, since the evaporation is endothermic ($L > 0$), is now hotter than the surface (cf. equations (2), with $T_b = T_s + 1$, and (17)). The process is represented in Figure 6 by the vertical line through P_∞; before it occurs there will, of course, be effective extinction.

As $T_s \to T_b$ the flame moves to the surface and, since $Y_s(= T_b - T_s)$ tends to zero, the pressure $\bar{p}_c$ (and hence D) tends to infinity like $(T_b - T_s)^{-1}$. When the temperature difference becomes $O(\theta^{-1})$ the eigenvalue (1) has to be replaced, the surface-flame result (2.30) showing that DM^{-2} now behaves like $(T_b - T_s)^{-2}$. Hence, M behaves like $(T_b - T_s)^{\frac{1}{2}}$ or $D^{-\frac{1}{2}}$, and we conclude that the M, D-curve bends down from the parabola to asymptote the D-axis. The later portion of the curve, however, does not provide an acceptable response since its slope is negative.

As in the case of the pyrolyzing solid, preheating the liquid (and, therefore, increasing T_b) lowers the pressure needed to sustain a given burning rate. The effect of preheating on P_∞ follows from the relation (18) on setting $Y_s = 1$ and $T_s = T_b - 1$, and is the opposite of that for the solid pyrolysis: the pressure $\bar{p}_c$ (hence D) increases with T_b (hence $\hat{T}_f$). A hotter liquid requires a higher pressure to produce the sudden complete evaporation.

The evaporating liquid was used as an example in section 2.4; the treatment there can now be seen as an incomplete answer to a somewhat different question. There the temperature was supposed to be uniform throughout the liquid phase, i.e. the remote temperature was maintained at the surface value. As a consequence T_b, instead of remaining constant, varied with the surface temperature ($\hat{T}_f = T_s$ in the determination (2)). The M, D-relation (1), which was not treated thoroughly there, is no longer parabolic since T_b now depends on D through the Clausius–Clapeyron relation. (We shall not go into details since they are of limited interest.) Finally, the formula (17) shows that the requirement (4) is simply $0<L<1$, as found before.

8 Radiative exchange at the surface; distributed heat exchange

Janssen (1981) has shown that, when radiative transfer is allowed, dramatic changes occur in the response of an evaporating liquid; these were overlooked by Kapila & Ludford (1977). The exchange does not have to be strong so that, to simplify the discussion, we set $k_s=\theta^{-1}k_1$ in the radiation law (7). The enthalpy balance (8) then shows that T_b is within $O(\theta^{-1})$ of T_a whenever M is at least $O(1)$.

To leading order, the Clausius–Clapeyron law (18) becomes

$$\bar{p}_c=kT_s^{\beta}\,e^{-\hat{\theta}/T_s}/(T_a-T_s) \tag{19}$$

when Y_s is replaced by T_b-T_s from the Shvab–Zeldovich relation. Except for T_s-T_a small, $\bar{p}_c$ must be $O(1)$ and then, if M is assumed to be $O(1)$ also, equation (1) only balances when

$$D=\tilde{D}\bar{p}_c\theta^2\,e^{\theta/T_a}\quad\text{with } \tilde{D}=O(1); \tag{20}$$

it becomes

$$2T_a^4\tilde{D}\bar{p}_c=M^2\exp[k_1(T_s^4-T_g^4)/MT_a^2] \tag{21}$$

when the expression (8) for T_b is used. We have taken D proportional to $\bar{p}_c$, i.e. $\nu=1$, since that is the reaction order assumed in deriving the eigenvalue (1).

The main response of M to $\bar{p}_c$ is given by equations (19) and (21); T_s is again a parameter on the corresponding M, D-curve subject to the limitation (4), which becomes

$$T_a-1<T_s<T_a. \tag{22}$$

For $k_1=0$ we recover the parabolic central portion of Figure 6; as there the end points of the range (22) require separate treatment for $k_1\neq0$. The form of the central part of the M, D-curve, which depends on the values of T_a and T_g, can be inferred from a simple observation: for each $T_s<T_g$, equation (21) gives a unique value for M, but for $T_s>T_g$ there are either two values (possibly coincident) or none. Three cases arise accordingly as T_g is less than T_a-1, between T_a-1 and T_a, or greater than T_a.

For $T_g>T_a$, the limitation (22) ensures that T_s is always less than T_g. There is a unique M for each D (i.e. T_s); the result (Figure 7*a*) is a curve similar to that for $k_1=0$ (Figure 6), including the ends.

For $T_a-1<T_g<T_a$, there is a single value of M for $T_s<T_g$ and either

two values or none for $T_s > T_g$. There are two values of M over the whole range $T_g < T_s < T_a$ if k_1 is less than a certain critical value (Figure 7*b*); otherwise, a central portion of the range yields no value of M (Figure 7*c*). The critical value of k_1 is that for which

$$8T_a^8 \tilde{D} k T_s^\beta \, e^{-\tilde{\theta}/T_s} = e^2 k_1^2 (T_s^4 - T_g^4)^2 (T_a - T_s), \tag{23}$$

considered as an equation for T_s, has a double root. The ends of the upper curve(s) are completed as in Figure 6; the right end of the lower curve(s) is also modified as shown.

For $T_g < T_a - 1$, no portion of the range (22) corresponds to a single-

7 Liquid evaporation with radiative exchange between surface and background: ---- curve (19, 21) with parameter T_s; —— response curve; —— presumably unstable. Vertical through P_∞ represents complete evaporation with flame at infinity. Modifications for large D are due to flame approaching surface, reached at P_s. (*a*) $T_g > T_a$; (*b*) or (*c*) $T_a - 1 < T_g < T_a$ with k_1 small or large; (*d*), (*e*) or (*f*) $T_g < T_a - 1$ with k_1 small, moderate or large.

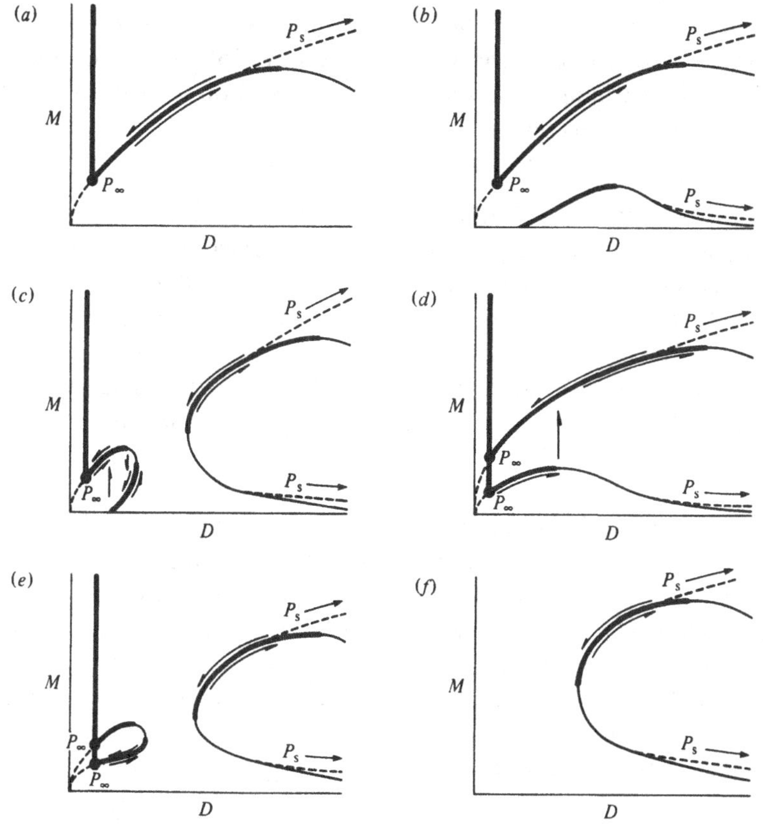

valued M. When k_1 is less than the critical value just defined, M is double valued over the entire range (22) as before (Figure 7d). When k_1 is greater, a gap again appears, which is centrally located (Figure 7e) if k_1 is less than a second critical value but extends to the left end (Figure 7f) if it is greater. The new critical value is given by equation (23) with $T_s = T_a - 1$. The ends of the upper curve(s) are completed as before; there is also sudden complete evaporation at the left end of the lower curve(s) and modification at the right end.

Half-arrows in Figure 7 show the probable sequence of events in each case as D increases from 0 to ∞ or decreases from ∞ to 0. As in section 5, we assume that only those parts of the response with a positive slope are stable and that weak burning is preferred to strong when the latter is not already established.

Finally, we come to the effect of distributed heat exchange on the response. The analysis in section 6 makes no use of the pyrolysis law, which only serves to locate T_s on the response curve. It is, therefore, equally valid for the evaporating liquid, with the Clausius–Clapeyron law playing the same role. In fact, complete analogy is obtained when there is no radiative exchange at the surface: T_s is a function of D alone and the response may be written

$$M = M^0(D)\,\mathrm{e}^{-\Psi} \quad \text{with } M^0 = T_b^2\theta^{-1}\,\mathrm{e}^{-\theta/2T_b}\sqrt{2D} \tag{24}$$

since T_b is a constant. (Otherwise, T_b is a function of both M and D.)

Arguments similar to those in section 6 determine the shape of the response as $D \to D_\infty$, showing that it depends on whether $T_s(D_\infty) = T_a - 1$ is greater or less than T_g. When it is less the curve eventually goes above that in Figure 6 and is asymptotic to $D = D_\infty$. The sudden evaporation is, therefore, replaced by an increasingly rapid one as the flame recedes to infinity. This behavior was reported by Kapila & Ludford (1977) who, by taking $\hat{T}_b = T_g$, ensured $T_a - 1 < T_g$ because of the endothermic evaporation. For $T_a - 1 > T_g$, however, the curve continues to dip down until the D-axis is approached, when it bends back to the right (because of the M^{-2} in the exponential factor) into a C shape, exhibiting true extinction. One possibility is that an oval is formed, the part beyond the crest closing with the lower part of the C. Such ovals have been reported by Spalding (1960) and are sketched by Williams (1965, p. 212); they were supposedly valid for heat loss by surface radiation alone. We now see that, while they are obtained under those conditions (Figure 7), they may also occur when there is solely distributed heat loss.

5

UNSTEADY BURNING OF A LINEAR CONDENSATE

1 The problem

Chapter 4 was concerned with the steady combustion of the gases produced by vaporization of a linear condensate at its surface through pyrolysis or evaporation. The results were characterized by response curves of burning rate versus pressure (represented by the Damköhler number). If the applied pressure varies in time, then so also must the burning rate; the nature of the dependence is examined in this chapter.

The effect of variations in pressure on solid pyrolysis has received considerable attention because of its relevance to the stability of solid-propellant rocket motors. Acoustic waves bouncing around the combustion chamber will impinge on the propellant surface and thereby generate fluctuations in the burning rate. These fluctuations will affect the reflected wave which, it is argued, might have a larger amplitude than the incident wave. If so, the transfer of energy (provided it is greater than losses through dissipation and other mechanisms) implies instability.

Our discussion will focus on the response of a burning condensate (solid or liquid) to an impinging acoustic wave. Mathematically we must deal with the disturbance of a steady field containing large gradients; consequently, a frontal attack on the governing equations is not feasible. Six regions can be distinguished: condensate, preheat zone, flame, burnt gas, entropy zone, and far field; without rational approximation, the discussion soon degenerates into either a numerical or *ad hoc* analysis (or both). There are three parameters other than θ^{-1} whose smallness can be exploited, namely the Mach number, the ratio of characteristic time in the gas to that in the condensate, and the ratio of densities. A low Mach number enables one to use the combustion approximation (with its spatial constancy of pressure) in the preheat zone, flame sheet, and burnt gas; a small time ratio ensures that the gas phase responds much more rapidly than the condensate, so that the three combustion regions are quasi-steady if the acoustic frequency is sufficiently small (a restriction of apparently little practical significance); and a small density ratio implies that fluctuations in the location of the surface of the condensate may be neglected. How small the acoustic frequency must be is determined by the activation-energy asymptotics needed for an analytical description of the combustion regions: the resulting changes in flame temperature (due entirely to fluctuations in the condensate) must be $O(\theta^{-1})$.

Only adiabatic burning will be considered; no work on heat loss has so far been reported. The first task is to determine the response of the burning rate to a general (small) change in pressure level of the combustion regions, without regard to its cause. Section 2 determines the response for solid pyrolysis and section 3 for liquid evaporation. The incident acoustic wave excites such a response, and the response induces a reflected wave, through the intermediary of an entropy zone that converts the isentropic state in the far field into an isothermal one in the near field (section 4). Conditions for the reflected wave to be stronger than the incident wave are developed in section 5.

The analysis also enables us to discuss stability characteristics, at least on the time scale adopted, by holding the pressure level fixed (i.e. eliminating the incident acoustic wave). The stability analysis of Chapter 3 is thereby extended to certain anchored flames and continues in section 6 to flat burners, where the combustion field need no longer be taken quasi-steady but still varies on a long-time scale. Only in this last section, where the combustion field is truly unsteady, does the Lewis number play a crucial role; so that in earlier sections we assign $\mathscr{L}=1$.

In view of the many small parameters involved we shall eschew writing formal expansions. Instead, the symbol $\ll$ will be used to indicate that only the leading term in the corresponding parameter is being considered.

2 Solid pyrolysis

The unsteadiness will be treated as the perturbation of a steady state (between P_s and P_∞ in Figure 4.1) with Damköhler number D_r. Then, as in section 3.3, the burning rate M_r in the steady state will be taken as the mass flux on which units are based; according to the relation (4.1), we have

$$M_r=\sqrt{2D_r}T_b^2\,e^{-\theta/2T_b}/\theta \qquad (1)$$

when no product of the gaseous reaction or inert is produced or absorbed at the surface of the condensate (i.e. $J_s = 1$). The flame temperature T_b is to be calculated from the formula (4.2).

The smallness of the Mach number ensures that the combustion approximation is valid. If, in addition, the fluctuations in the applied pressure are not too rapid, then the combustion is quasi-steady and the analysis of chapter 2 is applicable. More precisely, if ω_d is a (dimensional) characteristic frequency of the fluctuations (e.g. the frequency of an impinging acoustic wave) then the time ω_d^{-1} should be long compared to the response time $\rho_c\lambda/c_p M_r^2$ of the gas phase, determined by diffusion. In short, we require

$$\omega_d \ll c_p M_r^2/\rho_c\lambda, \tag{2}$$

a restriction of apparently little practical consequence since the diffusion time is always very small. The same formula (1), with M_r and D_r replaced by M and D, therefore, holds for the burning rate, where now both D and T_b are functions of time, the former prescribed and the latter to be determined from an analysis of heat transfer in the condensate and the nature of the gasification.

This conclusion, which is not restricted to solid pyrolysis, was reached by Denison & Baum (1961) using *ad hoc* arguments that Williams (1973) later replaced with rational asymptotic analysis. None of these authors noted, however, that (in general) a further restriction of ω_d is required if a perturbation theory is to be consistent. In our formulation the restriction comes from allowing the flame temperature to be perturbed only by $O(\theta^{-1})$ amounts, so that the change in burning rate is $O(1)$. More precisely,

$$M = D^{\frac{1}{2}} e^{-\frac{1}{2}\phi_*}, \tag{3}$$

where $-\theta^{-1}T_b^2\phi_*$ is the perturbation (T_b now being the unperturbed flame temperature), M and D being measured in units of M_r and D_r.

This perturbation is due solely to fluctuations of temperature in the condensate, and it may be determined by calculating the total change in enthalpy they produce up to the surface. In general, slow variations on the time scale $\hat{\rho}\hat{\lambda}/\hat{c}_p M_r^2$ of the condensate are needed if the result is to be $O(\theta^{-1})$; i.e., to within an $O(1)$ multiplicative factor,

$$\omega_d = \theta^{-1}\hat{c}_p M_r^2/\hat{\rho}\hat{\lambda} + \ldots \tag{4}$$

Note that, since the condensate is much denser and more conductive than the gas, i.e.

$$\rho_c/\hat{\rho} \ll 1, \; \lambda/\hat{\lambda} \ll 1, \tag{5}$$

the change in enthalpy in the gas phase is negligible; in response to changing conditions the gas (unlike the condensate) has no inertia. Thus, the restriction (4) implies

$$\omega_d \ll \theta^{-1} c_p M_r^2 / \rho_c \lambda, \tag{6}$$

and time derivatives in the gas are small even on the slow scale of section 3.3. (Note that the condition (2) is satisfied *a fortiori*.) In special circumstances there is an alternative to slow variation, which we shall cover later.

To calculate the change in enthalpy we consider the dimensionless temperature equation

$$\partial T/\partial t + M \partial T/\partial x - \partial^2 T/\partial x^2 = 0 \tag{7}$$

in the condensate. Here $\hat{\rho}\hat{\lambda}/\hat{c}_p M_r^2$ and $\hat{\lambda}/\hat{c}_p M_r$ have been taken as units of time and length, while T is still referred to Q/c_p. Since temporal variations occur on the long-time scale $\tau = t/\theta$, the term $\partial T/\partial t = \theta^{-1} \partial T/\partial \tau$, which represents local storage of enthalpy, is seen to be a perturbation. If it is neglected, the leading approximation

$$T = \hat{T}_f + (T_s - \hat{T}_f)\, e^{Mx} \tag{8}$$

is obtained, where M varies with τ. The analysis so far is valid whatever kind of gasification occurs at the surface of the condensate; for solid pyrolysis, T_s is connected to M by the law (4.3), which now reads

$$M = (k/M_r) T_s\, e^{-\hat{\theta}/T_s}. \tag{9}$$

From the leading approximation we can calculate the neglected perturbation term, which represents a heat source or sink in equation (7), and hence the perturbation in flame temperature. Thus, when account is taken of this rate of change of enthalpy stored in the solid, an overall balance of enthalpy flux (as used in section 4.3) leads to

$$M\phi_* = (\kappa/T_b^2) \int_{-\infty}^{0} (\partial T/\partial \tau)\, dx = (p+q) M^{-2}\, dM/d\tau, \tag{10}$$

where κ is the ratio of specific heats (4.2*b*) and

$$p = \kappa T_s^2 / T_b^2 (\hat{\theta} + T_s) > 0, \quad q = \kappa (\hat{T}_f - T_s)/T_b^2; \tag{11}$$

so that, from the result (3), the connection between variations in M and D is

$$(p+q)\, dM/d\tau = M^3 \ln(D/M^2). \tag{12}$$

The steady state $M = D = 1$ clearly satisfies the last equation; and we are mainly concerned with relating the small disturbances of M and D caused by an acoustic wave. The equation may, however, also be used to investigate the inherent stability of the steady state, i.e. the way M changes when D is held fixed at 1. We see that when M is disturbed from 1, either up or down, it is driven further away whenever

$$p + q < 0, \quad \text{i.e. } \hat{T}_f < \hat{\theta} T_s / (\hat{\theta} + T_s). \tag{13}$$

The condition is difficult to interpret because T_s itself depends on M; but certainly there is linear instability if it is satisfied by the value of T_s corresponding to $M=1$ in the pyrolysis law (9). This instability was predicted by Denison & Baum (1961) but has apparently never been observed. Their condition, after extraneous terms are eliminated by letting $\theta \to \infty$, is actually

$$\hat{T}_f < T_s(\hat{\theta} - T_s)/\hat{\theta} \tag{14}$$

because their pyrolysis law lacks the factor T_s. Figure 1 shows the stability boundaries for a factor T_s^α with $\alpha = 0, \frac{1}{2}$, and 1.

A possible reason why the instability has not been observed is revealed by the result (14): if T_s is greater than $\hat{\theta}$ the inequality cannot be satisfied. To make the reason convincing a general pyrolysis law $M(T_s)$ should be

1 Stability boundaries for solid pyrolysis when conditions in the solid are slowly varying.

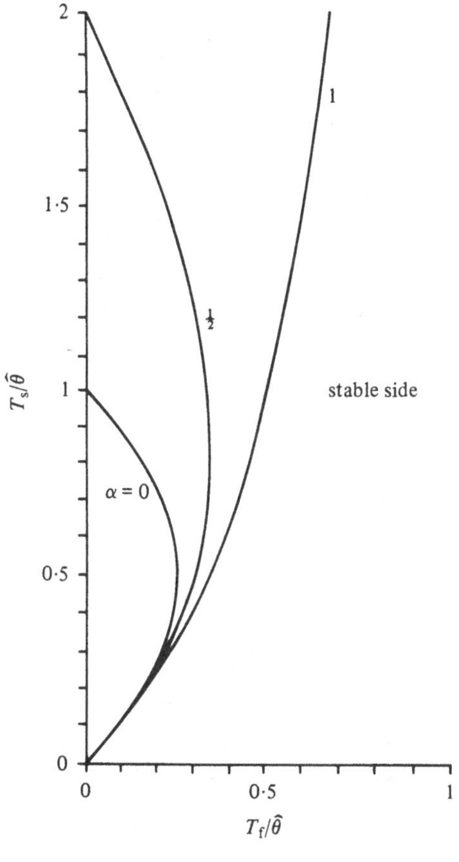

used, in which case the right side of the inequality becomes $T_s - M\,\mathrm{d}T_s/\mathrm{d}M$. It is conceivable that $\mathrm{d}T_s/\mathrm{d}M$ is greater than T_s/M in the operating range of all propellants. The procedure for checking a given $\hat{T}_f$ would be to compute the burning rate (1) for the assumed D_r and then determine T_s and $\mathrm{d}T_s/\mathrm{d}M$ for $M=1$ from the pyrolysis law.

For oscillations about the steady state forced by an acoustic wave, we write

$$M=1+m\,\mathrm{e}^{\mathrm{i}\tilde{\omega}\tau},\quad D=1+d\,\mathrm{e}^{\mathrm{i}\tilde{\omega}\tau} \tag{15}$$

and linearize. Then

$$m=cd \quad \text{with } c^{-1}=2+\mathrm{i}\tilde{\omega}(p+q), \tag{16}$$

where T_s now has its value for $M=1$. This relation determines the reflected wave in the far field, as we shall see in section 4. (The instability result (13) is obtained by requiring c^{-1} to vanish.)

Alternatively changes in flame temperature are $O(\theta^{-1})$ when the temperature in the solid fluctuates by that amount everywhere. The fluctuations can then be on the scale of t, rather than τ, so that we are faced with solving the full heat equation (7), with M an arbitrary function of t. A general analysis would, therefore, be quite complicated; accordingly just the case of interest will be considered, namely small departures from the steady state.

With $\omega=\theta^{-1}\tilde{\omega}$ a dimensionless frequency, we write

$$M=1+m\,\mathrm{e}^{\mathrm{i}\omega t},\quad D=1+d\,\mathrm{e}^{\mathrm{i}\omega t} \quad \text{with } m, d \ll 1. \tag{17}$$

The temperature in the solid then has a steady component, given by the distribution (8) with $M=1$ and T_s the corresponding surface temperature, and a fluctuating component

$$(T_b^2 m/\kappa)[(-\mathrm{i}q/\omega)\,\mathrm{e}^{x+\mathrm{i}\omega t}+(p+\mathrm{i}q/\omega)\,\mathrm{e}^{\Omega x+\mathrm{i}\omega t}] \quad \text{with } \Omega=\tfrac{1}{2}(1+\sqrt{1+4\mathrm{i}\omega}). \tag{18}$$

The two contributions to this last result are due to fluctuations in M within the solid and in T_s at its surface, respectively. We now find

$$\theta^{-1}\phi_*=(\kappa/T_b^2)\int_{-\infty}^{0}(\partial T/\partial t)\,\mathrm{d}x=(\Omega-1)(p+q/\Omega)m\,\mathrm{e}^{\mathrm{i}\omega t}, \tag{19}$$

in place of the result (10), so that

$$m=cd \quad \text{with } c^{-1}=2+\theta(\Omega-1)(p+q/\Omega), \tag{20}$$

which reduces to the previous response (16) for ω small but is quite different otherwise, as we shall see in section 4.

The inherent stability of the steady state is determined by setting

$c^{-1}=0$. Then

$$P\Omega^2+(Q-P+2)\Omega-Q=0 \quad \text{with } P=\theta p, Q=\theta q; \tag{21}$$

for instability the parameter values P and Q must make the real part of $i\omega=\Omega(\Omega-1)$ positive for an Ω whose real part is positive. Figure 2 shows the region of instability in the P, Q-plane; its boundary asymptotes $P+Q=-6$, in accord with the previous result (13) for P and Q large.

For consistency, we must have

$$p, q=O(\theta^{-1}), \tag{22}$$

which certainly requires both T_s and $\hat{T}_f$ to be of the same order if $\hat{\theta}/T_s$ is $O(1)$. Such small temperatures do not mean that the condensate is cold, but only that the corresponding enthalpies are small compared to the heat of combustion, which is being used as the unit. The requirement (4.4) for the steady state does mean, however, that T_b, as calculated from the formula (4.2), must be no greater than 1 (i.e. the pyrolysis must be endothermic).

2 Stability boundary for solid pyrolysis when the heat of combustion is large. The curve H is the hyperbola $P(2-Q)=(2+Q)^2$, with asymptote A given by $P+Q=-6$.

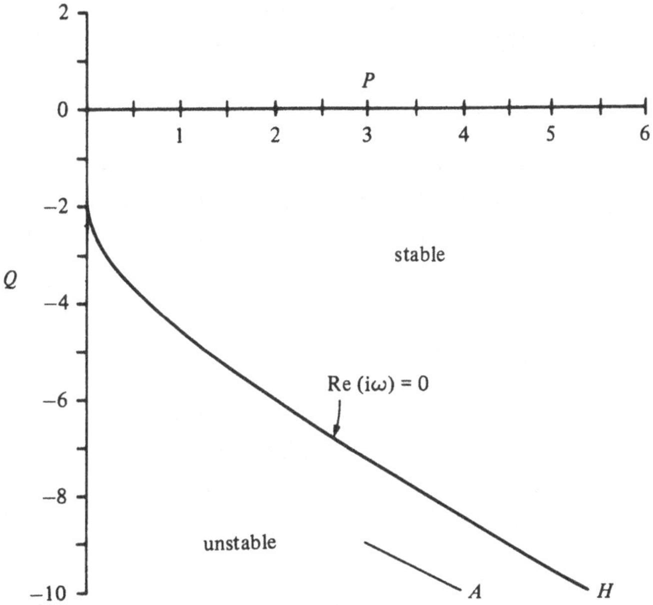

3 Liquid evaporation

If attention is again focused on adiabatic combustion, our analysis follows that in the last section until the pyrolysis law is reached, when the Clausius–Clapeyron relation (4.18) is used instead, i.e. we write

$$D = kT_s^{\beta}(T_b - T_s)^{-1}\,e^{-\hat{\theta}/T_s}. \tag{23}$$

Here we have taken the Damköhler number proportional to $\bar{p}_c$ in accordance with the formula (1.60*c*) for a first-order reaction (otherwise a power of D is required); and various parameters have been absorbed into k.

Since T_s is now a function of D, the latter enters into the temperature distribution (8) in the liquid, so that the formula (10) for the change in flame temperature becomes

$$M\phi_* = qM^{-2}\,dM/d\tau + pM^{-1}D^{-1}\,dD/d\tau, \tag{24}$$

where now

$$p = \kappa T_s^2(T_b - T_s)/T_b^2[T_s^2 + (\beta T_s + \hat{\theta})(T_b - T_s)] > 0. \tag{25}$$

The equation connecting the variations in M and D is, therefore,

$$qM^{-1}\,dM/d\tau + pD^{-1}\,dD/d\tau = M^2\ln(D/M^2). \tag{26}$$

In particular, there is inherent instability ($D \equiv 1$) when q is negative, i.e. for

$$\hat{T}_f < T_s. \tag{27}$$

Note that this instability (which has not been reported before) is independent of the precise form of the gasification law at the surface: it occurs whenever the surface temperature is determined by the surface pressure.

For general small departures (15) from the steady state we find

$$m = cd \quad \text{with } c = (1 - i\tilde{\omega}p)/(2 + i\tilde{\omega}q). \tag{28}$$

This result will be needed when we come to the acoustic response in section 5. When p and q are $O(\theta^{-1})$, higher frequencies ω can be considered, as for solid pyrolysis. The fluctuating component in the condensate now becomes

$$(T_b^2/\kappa)[(-iqm/\omega)\,e^{x+i\omega t} + (pd + iqm/\omega)\,e^{\Omega x + i\omega t}],$$

where p has its new definition (25), so that

$$m = cd \quad \text{with } c = [1 + \theta(1-\Omega)p]/[2 - \theta(1-\Omega)q/\Omega]. \tag{29}$$

When D is held fixed at 1, i.e. $d = 0$, burning fluctuations only occur if the denominator of c vanishes, so that $\Omega = Q/(2+Q)$. Instability requires $\Omega > 1$, i.e.

$$Q < -2 \quad \text{with } Q = \theta q, \tag{30}$$

a condition in accord with the earlier result (27) for Q large.

4 Response to an impinging acoustic wave

Analysis of the combustion zone provides, among other things, a complete description of the state of the burnt gas behind the flame sheet in terms of the time-dependent, but spatially uniform, pressure there. Conditions are essentially isothermal with the velocity v_∞ determined by the instantaneous value of M.

The varying pressure is generated by some source far from the flame; we shall confine ourselves to acoustic waves traveling normal to the surface of the condensate, which immediately raises a problem. The far field is isentropic, whereas the near field is isothermal, so that the two cannot be matched. In other words, it is not simply a matter of evaluating the pressure in the far field as the near field (represented by a single point) is approached. An 'entropy' layer is needed to transform conditions from isentropic to isothermal in a distance small on the scale of the far field, but large on that of the near field.

In the chemistry-free region beyond the flame sheet, the equations with which we have to deal are those of a compressible, heat-conducting and viscous fluid, i.e. the dimensional form of equations (1.55), (1.57), and (1.58) with pressure and viscous terms restored in the energy balance, with the reaction term deleted there, and with the perfect-gas law reinstated. Linearized about the steady state ρ_∞, v_∞, $\bar{p}_c$, and T_∞ they read

$$\partial\rho/\partial t + v_\infty \partial\rho/\partial x + \rho_\infty \partial v/\partial x = 0, \tag{31}$$

$$\rho_\infty(\partial v/\partial t + v_\infty \partial v/\partial x) = -\partial p/\partial x + (4\kappa/3)\partial^2 v/\partial x^2, \tag{32}$$

$$\rho_\infty c_p(\partial T/\partial t + v_\infty \partial T/\partial x) - \lambda \partial^2 T/\partial x^2 = \partial p/\partial t + v_\infty \partial p/\partial x. \tag{33}$$

It is natural to take ρ_∞, $a_\infty = \sqrt{\gamma R T_\infty/m}$, $\bar{p}_c = \rho_\infty a_\infty^2/\gamma$, and T_∞ as units of density, velocity, pressure, and temperature in such a combustion-free region. Length will be referred to distances l, to be specified later, and time to l/a_∞. Then, for disturbances proportional to $e^{i\omega t}$, the dimensionless equations governing the perturbations are

$$i\omega\rho + m_\infty \partial\rho/\partial x + \partial v/\partial x = 0, \quad i\omega v + m_\infty \partial v/\partial x = -\gamma^{-1}\partial p/\partial x + \kappa_l \partial^2 v/\partial x^2, \tag{34}$$

$$i\omega T + m_\infty \partial T/\partial x - \lambda_l \partial^2 T/\partial x^2 = (\gamma - 1)(i\omega p + m_\infty \partial p/\partial x)/\gamma, \quad p = T + \rho, \tag{35}$$

where

$$m_\infty = v_\infty/a_\infty, \quad \kappa_l = 4\kappa/3\rho_\infty a_\infty l, \quad \lambda_l = \lambda/c_p\rho_\infty a_\infty l. \tag{36}$$

Solution of these equations determines the dimensional density, velocity, pressure, and temperature as

$$\rho_\infty(1 + \rho\, e^{i\omega t}), \quad v_\infty + a_\infty v\, e^{i\omega t}, \quad \bar{p}_c(1 + p\, e^{i\omega t}), \quad T_\infty(1 + T e^{i\omega t}), \tag{37}$$

respectively.

These equations, being linear with constant coefficients, can be solved exactly and the solution used to trace the transition from isothermal to isentropic conditions. However, to see how the burning-rate fluctuations are caused by the incident acoustic wave across the entropy layer and then how they determine the reflected wave across that layer, it is easier to use expansions appropriate to the various regions.

To describe the acoustic waves, l is chosen to be the wavelength, defined in terms of the frequency by

$$l_a = a_\infty/\omega_d \gg l_c/m_\infty \gg l_c, \tag{38}$$

where $l_c = \lambda/c_p\rho_\infty v_\infty$ is the unit of length used in the combustion zone. We have used the restriction (2), which now reads

$$\omega_d \ll v_\infty/l_c, \tag{39}$$

and the fact that the Mach number m_∞ is small, to obtain these extreme inequalities. It follows that ω is 1, but that κ_l and λ_l are small, so that discarding terms that are $o(m_\infty)$ gives

$$p = A\,e^{i(1+m_\infty)x_a} + B\,e^{-i(1-m_\infty)x_a}, \quad T = (\gamma - 1)p/\gamma, \tag{40}$$

$$\rho = p/\gamma, \quad v = [-A\,e^{i(1+m_\infty)x_a} + B\,e^{-i(1-m_\infty)x_a}]/\gamma \tag{41}$$

where x_a is distance measured in units of l_a. Here A and B are the amplitudes of the pressure in the incident and reflected waves, respectively. To leading order, A and B are equal so that $v \to 0$ as the surface is approached ($x_a \to 0$). Corrections to B are $O(m_\infty)$ and it is for this reason that terms of this magnitude are retained in the acoustic zone. Dissipation will change the amplitude by a like amount over a distance

$$L_a = m_\infty l_a/\lambda_a \gg l_a/m_\infty \gg l_a \quad (\lambda_a \text{ is } \lambda_l \quad \text{for } l = l_a), \tag{42}$$

so that for the theory to be applicable L_a must be large compared to an overall dimension. Since the latter is normally comparable to l_a, that provides no restriction in practice. The goal is to determine B when A is given, so as to calculate the reflection coefficient $|B/A|$.

On the acoustic scale the whole combustion field is represented by a single point, which we may take to be $x_a = 0$. The temperature disturbance there is effectively zero since the fluctuations only generate $O(\theta^{-1})$ changes in the flame temperature, and this clearly contradicts the isentropic requirement (40*b*) of the acoustic waves when there is a pressure disturbance. A layer is needed to change the entropy from its constant value in the acoustic field as $x_a \to 0$. In the absence of dissipation, entropy is convected with the fluid; to describe these waves, we choose for l the wavelength

$$l_e = v_\infty/\omega_d = m_\infty l_a \ll l_a. \tag{43}$$

Now, the frequency ω equals m_∞ and, hence, is small. We also find that κ_l and λ_l are still small, their dissipative effect having an $O(1)$ impact over a distance

$$L_e = m_\infty l_e/\lambda_e \gg l_e \quad (\lambda_e \text{ is } \lambda_l \quad \text{for } l = l_e). \tag{44}$$

If L_e is to be small compared to l_a, we must have

$$\omega_d \gg m_\infty v_\infty / l_c, \tag{45}$$

which now places a lower bound on the frequency for the theory to apply. If the restriction is not introduced, entropy effects extend into the far field and our simple picture is destroyed. The same results must then be obtained from the full solution of equations (34) and (35).

We now consider the structure of the entropy layer, which the momentum balance shows to be isobaric, as might have been expected. The complete solution may, therefore, be written

$$T = (1 - 1/\gamma)p(1 - e^{-ix_e - X_e}), \quad \rho = (p/\gamma)[1 + (\gamma - 1)\, e^{-ix_e - X_e}],$$
$$v = m_\infty[-ipx_e/\gamma + (c-1)p] \tag{46}$$

to leading order, where x_e and X_e are distances measured in units of l_e and L_e, respectively. Here the constants of integration have been determined so that

$$T = 0, \quad \rho + v/m_\infty = cp \text{ at } x_e = X_e = 0, \tag{47}$$

as required by the analysis of the combustion regions: to $O(1)$ accuracy in θ the temperature is unaffected and $m = cd$ (in the notation of sections 2 and 3), with D proportional to pressure. Note that $v \to \infty$ as $x_e \to \infty$, a fact that makes a careful analysis of the entropy layer necessary, as we shall see shortly.

As $X_e \to \infty$ the solution (46) becomes isentropic, i.e. the relations (40*b*) and (41*a*) are satisfied. A formal matching is obtained by taking m_∞ as the vanishingly small parameter, which is suggested by the transformation $x_a = m_\infty x_e$. When applied to v the matching shows that

$$B = A(1 + m_\infty a) \quad \text{with } a = 2\gamma(c-1) \tag{48}$$

correct to first order in m_∞. The reflected wave is thereby determined in terms of the incident wave. (The use of m_∞ as a small parameter makes the inequalities (39) and (45) become $O(m_\infty^3) \ll \omega \ll O(m_\infty^2)$.)

Usually such reflection problems are discussed in terms of the acoustic admittance. In this context Williams (1962) equates the value of v/p as $x_a \to 0$ to its value in the combustion region, which happens to give the correct result (48). It would be more logical to equate it to the value at the edge of the entropy layer, but that is impossible since $v \to \infty$ as $x_e \to \infty$. (The difficulty is not a creature of our analysis: the continuity equation

(31) requires v to become unbounded whenever ρ does not vanish at the edge of the layer.) Only proper examination of the entropy layer resolves the dilemma.

5 Amplification

Since A is the amplitude of the incident wave and B is that of the reflected wave, there is amplification for small values of m_∞ if, and only if,

$$\mathrm{Re}(a)>0, \quad \text{i.e. } \mathrm{Re}(c)>1. \tag{49}$$

We shall now examine the two cases of solid pyrolysis in section 2 and the two cases of liquid evaporation in section 3 for this instability.

3 Amplification condition of an acoustic wave for solid pyrolysis with large heat of combustion, drawn for $Q=-(P+2)$: (*a*) small P; (*b*) large P.

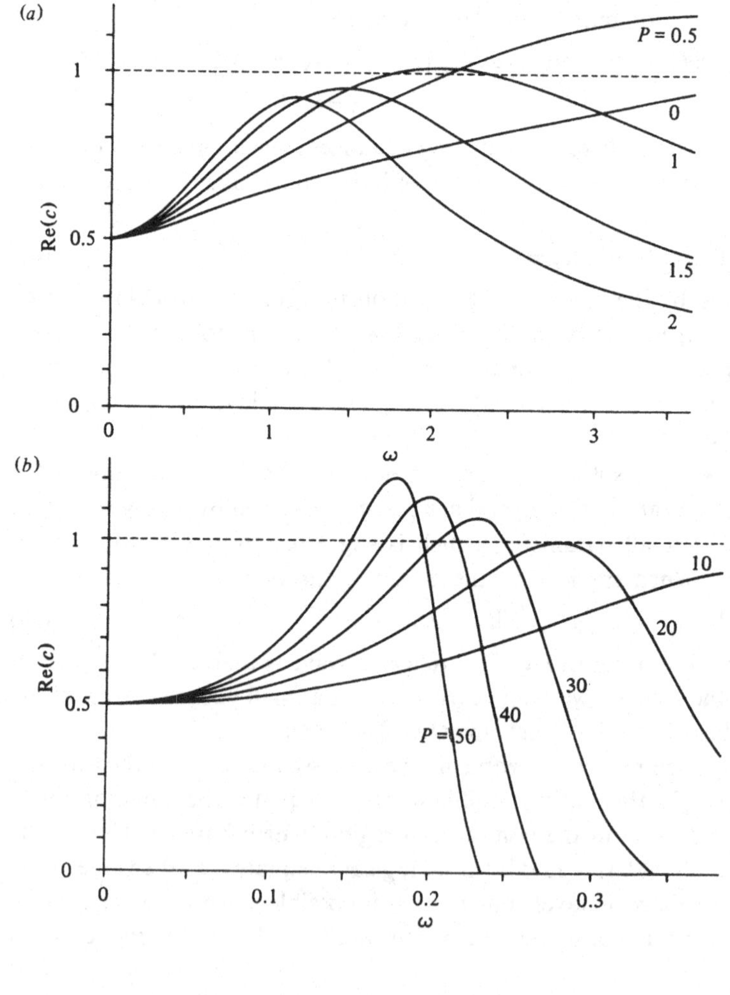

The response (16) for solid pyrolysis shows that Re(c) is never greater than $\frac{1}{2}$, its value for $\tilde{\omega}=0$, so that there is no instability. The conclusion is changed by giving D a pressure dependence p^n with $n>2$ since then c is replaced by nc, whose maximum real part exceeds 1; but in practice n is never that large.

On the other hand, the response (20) shows that

$$\mathrm{Re}(c)=\tfrac{1}{2}+\tfrac{1}{8}[2P+4-(Q+P+2)^2]\omega^2+O(\omega^4) \qquad (50)$$

for small ω, which increases initially with ω for a band of values around

$$Q=-(P+2), \qquad (51)$$

suggesting that the inequality (49) may be satisfied for some values of ω when P and Q are chosen appropriately. This is confirmed by graphs of Re(c) versus ω for several values of P in Figure 3, where the value (51) has been taken for Q to ensure the fastest rate of initial increase. Experiments provide values of the acoustic admittance for different frequencies, and for some propellants there is rough agreement with these predictions when the frequencies are high enough for the slope to be negative; the imaginary part of c is not well predicted, however (Culick 1968). The model analyzed here is suggestive, but no doubt important physics has been omitted.

Turning now to the evaporating liquid, we find that the response (28) gives

$$\mathrm{Re}(c)=1-[2+\tilde{\omega}^2 q(p+q)]/(4+\tilde{\omega}^2 q^2). \qquad (52)$$

4 Amplification condition of an acoustic wave for liquid evaporation with large heat of combustion, drawn for $P=1$ and Q near -2.

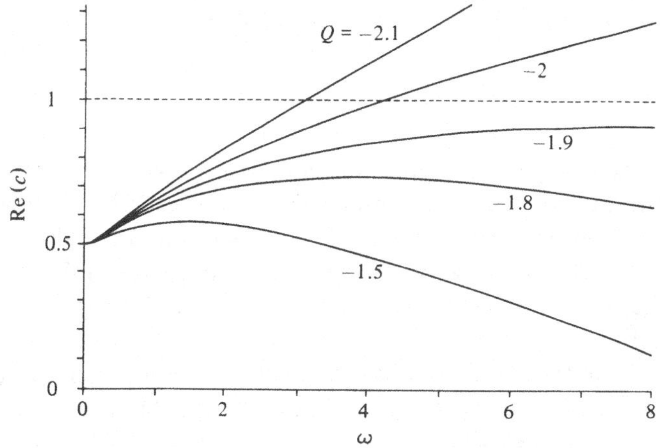

For instability q must be negative but greater than $-p$; then all frequencies

$$\tilde{\omega} > \sqrt{-2/q(p+q)} \tag{53}$$

are amplified. On the other hand, Re(c) for the response (29) behaves like $-P\sqrt{\omega}/\sqrt{2}(2+Q)$ as $\omega \to \infty$, so that 1 is certainly exceeded for $Q < -2$ whatever the value of P. Figure 4 shows graphs for $P=1$ and various values about $Q=-2$.

The absence of amplification at all frequencies $\tilde{\omega}$ for a pair of values p and q does not necessarily mean that stable burning will take place. There may be an inherent instability. For solid pyrolysis, parameter values for which $p+q$ is negative are ruled out even though there is no amplification of acoustic waves. For liquid evaporation there is stable burning only when q is positive. Similar remarks apply to the cases governed by P and Q.

6 Stability of anchored flames

The flames considered in this chapter are anchored to the condensate in the sense that they are controlled by the manner in which the reactant is liberated at the surface (which in turn is affected by heat from the flame). The stability of certain anchored flames has, therefore, been covered already, in particular under conditions of slow variations in the condensate with the gas phase quasi-steady. We shall now consider an anchoring for which there can be slow variations in the gas phase, and test its stability. Such variations only exist when $\mathscr{L} \neq 1$.

The discussion revolves around the flame-temperature expression corresponding to the result (3.26), derived in the same way except that the basic equations (3.15), (3.16), (3.17), and (3.20) are integrated directly, now over the interval $0 \leqslant x \leqslant x_* + 0$. We find

$$T_* = J_s + E_s + M_s^{-1} \int_0^{x_*} [(T_* - Y)\partial\rho/\partial t - \rho\,\partial Y/\partial t]\,dx, \tag{54}$$

where

$$J_s = Y_s - \mathscr{L}^{-1} M_s^{-1} Y_s', \quad E_s = T_s - M_s^{-1} T_s', \tag{55}$$

and M_s is the mass flux at the supply, which in general will be different from the burning rate M, i.e. the flux through the flame sheet. The formula shows that variations in T_* can conveniently be divided into those due to changes in the supply and those stemming from unsteadiness in the gas phase. The latter are independent of the former since they also depend on the initial conditions.

Slow variations in the gas phase will ensure $O(\theta^{-1})$ perturbations of T_* if the supply restricts changes in $J_s + E_s$ to the same order. The flat burner of section 2.5 can do this without necessarily controlling T_s itself; and, with that in mind, we shall suppose that the composition of the fresh mixture, more precisely J_s, its injection rate M_s, and the pressure level of the gas phase, represented by D, are held constant. For slow variations the burning rate M will equal M_s to leading order because the flame velocity is $O(\theta^{-1})$. Moreover, the dimensional mass flux is still governed by (3.11), with $T_b = T_*$; so that fixing M_s and D ensures that T_* is fixed, at a value $O(\theta^{-1})$ away from T_{*0}, the temperature determining M_r (the unit of mass flux). Thus

$$T_* = T_{*0} - \theta^{-1} T_{*0}^2 \phi_* + \ldots, \tag{56}$$

where M and ϕ_* are related by (3) when D in that equation is set equal to 1. One further condition is required to determine the combustion field to leading order; in section 2.5 the temperature T_s was assumed to be known. But fixing it here would prevent slow variations: the $O(1)$ terms in the relation (54) would require T_s' and hence Y_s and Y_s', also to be fixed (to leading order). There is no experimental difficulty in holding T_s constant; indeed it has much to commend it for the unsteady problem, given the large thermal capacity of the burner relative to that of the gas. But it leads to a combustion field that cannot vary on the τ-scale.

In steady problems Clarke & McIntosh (1980) have used the condition

$$L \equiv M^{-1} T_s' = k(T_s - T_f), \tag{57}$$

where T_f is the temperature of the fresh mixture entering the rear of the porous plug, whose so-called conductance is denoted by k. (The condition has also been used by Carrier, Fendell, & Bush (1978) in a modified form.) If $k = 1$, all the heat conducted back to the plug is used to raise the temperature of the mixture from T_f to T_s. Otherwise, there is heat loss to, or gain from, the plug. We shall suppose the same relation holds for unsteady conditions and write

$$k = 1 + \theta^{-1} k_1 \quad \text{with } k_1 > 0 \tag{58}$$

to ensure a small heat loss.

The slowly varying combustion field is completely determined (to leading order) by L. Thus, according to the formulas in section 3.2 (with allowance for the mass-flux unit), we have

$$-\mathscr{L}^{-1} M^{-1} Y_s' = J_s^{1-\mathscr{L}} L^{\mathscr{L}} \quad \text{and hence } Y_s = J_s - J_s^{1-\mathscr{L}} L^{\mathscr{L}};$$
$$\text{while } T_s = T_f + L. \tag{59}$$

These all hold to leading order so that the same accuracy is obtained in the formulas

$$Y = J_s - J_s^{1-\mathscr{L}} L^{\mathscr{L}} e^{\mathscr{L} M x}, \quad T = T_f + L e^{Mx} \quad \text{for } 0 < x < x_*$$
$$\text{with } x_* = M^{-1} \ln(J_s/L). \tag{60}$$

The basic relation (54) yields $T_{*0}=J_s+T_f$, the adiabatic flame temperature, so that M_r is the mass flux associated with the adiabatic flame. In addition,

$$\phi_*=(k_1L-M^{-2}b\,\mathrm{d}L/\mathrm{d}\tau)/T_{*0}^2 \quad \text{with } b=T_f[1-(L/J_s)^{\mathscr{L}-1}]/T_s \tag{61}$$

so that the steady state is given by

$$L=\phi_* T_{*0}^2/k_1. \tag{62}$$

L must be positive so that $M<1$ and $L\to 0$, $x_*\to\infty$ as $M\to 1$. Since x_* cannot be negative we have $L/J_s<1$, so that $b \lesseqgtr 0$ accordingly as $\mathscr{L} \lesseqgtr 1$. Stability is determined by the sign of b; the criterion is

$$\mathscr{L}<1(b<0) \text{ stable}, \quad \mathscr{L}>1(b>0) \text{ unstable}, \tag{63}$$

which is the same as for the unbounded flame (section 3.5).

In this chapter we have dealt exclusively with anchored flames that permit slow variations. The question of stability on the t-scale, not only for the anchored flames considered, but also for others, has been left untouched. (The relation (54) shows that the perturbations would have to be at most $O(\theta^{-1})$.) Recent results of Margolis (1980), the first so far, for the porous plug with T_s fixed, to which the slow-variation analysis has nothing to contribute, suggest that much remains to be uncovered. It is no more than a suggestion because Margolis's analysis does not lie within the framework of our self-consistent asymptotic treatment (cf. section 11.8).

6

SPHERICAL DIFFUSION FLAMES

1 Diffusion flames

Earlier chapters have been concerned with flames for which the reactants are supplied already mixed. When two reactants are initially separate and diffuse into each other to form a combustible mixture, the term diffusion flame applies. A Bunsen burner with its air hole closed supports a diffusion flame between the gas supplied through the tube and the surrounding oxygen-rich atmosphere. A candle supports a vapor diffusion flame, so called because the fuel is produced by liquefaction and subsequent evaporation of the wax caused by the heat of the flame.

Premixed and diffusion flames share certain features, but there are important differences. For example, there is no unlimited plane flame with fuel supplied far upstream, oxidant far downstream. The only bounded solutions of the chemistry-free equations behind the flame sheet are constants, so that the oxidant fraction would be constant there, with no mechanism to generate the necessary flux towards the reaction zone. Supplying the fuel at a finite point upstream does not change this picture. Moreover, since cylindrical flames are geometrically attenuated versions of plane flames (cf. section 7.2), there is no cylindrical diffusion flame either.

To gain insight into the nature of diffusion flames we seek a simple one-dimensional configuration that has some physical reality. One possibility is to introduce the oxidant at a finite point in the plane flame; details of that choice have been worked out by Lu (1981). To be realistic, however, it must be treated as the limit of a chambered diffusion flame, in which the fuel supply is also at a finite point. This problem has recently been considered by Matalon, Ludford, & Buckmaster (1979) but is too complicated for our purposes.

Two choices have been popular in the literature: the counterflow flame, and the spherical flame. In the counterflow flame (Figure 1) opposing jets of fuel and oxidant collide, supporting a one-dimensional flame in the neighborhood of the stagnation point (Liñán 1974, the seminal work in the application of activation–energy asymptotics to diffusion flames). The configuration has the advantage of being relatively easy to set up experimentally but suffers analytically from the compromises that have to be made with the fluid mechanics involved. For that reason, and because of its mathematical similarity to the other choice, we shall not consider it further.

The spherical diffusion flame, because it occurs in the burning of fuel drops (Figure 2), is of great technological interest. It is then a vapor flame, the liquid fuel being vaporized at its surface by heat conducted back from the surrounding flame where the vapor burns with the ambient oxidant.

1 Counterflow diffusion flame.

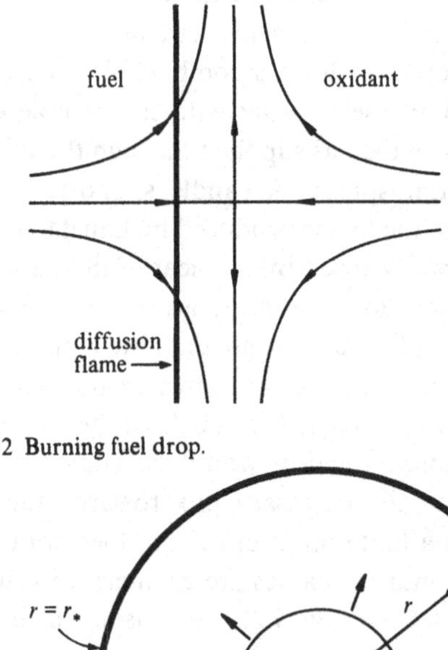

2 Burning fuel drop.

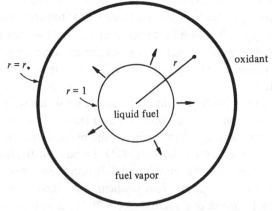

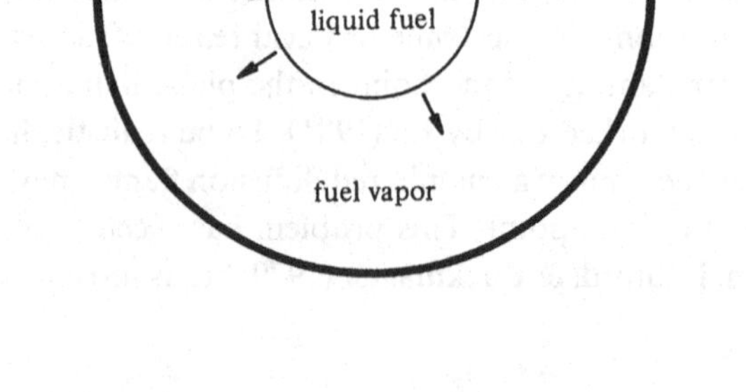

Although the size of the drop continually decreases as fuel is consumed, the quasi-steady approximation (Williams 1965, p. 48) neglects this effect. Such steady burning can be produced experimentally by forcing the fuel through a porous sphere in a gravity-free environment; by supplying the fuel as a gas, rather than as an evaporating liquid, the whole range of spherical flames can be examined.

If such phenomena are studied for the express purpose of solving the practical problems from which they derive, then we have to be concerned with the validity of the modeling. Thus, in several studies of the fuel-drop problem, the validity of the quasi-steady approximation has been both questioned and examined. Such practical considerations are not our main concern, since we regard the models primarily as mathematical idealizations whose study can provide some insight into the nature of diffusion flames. The experimentalist may then feel challenged to devise real flames that model the mathematics, as in the porous-sphere experiment mentioned previously. While such an experiment does not appear to have been carried out in a gravity-free environment to date, fuel-drop studies have been (Knight & Williams 1980). In that connection, however, gravity plays an even smaller role as the size of the drop decreases, so that it is negligible in sufficiently fine sprays. Potter & Riley (1980) have made a start on a theory which accounts for gravity-induced convection for large drops.

We have to deal with the spherically symmetric form of the equations (1.55)–(1.58), in particular

$$\frac{1}{\theta}\frac{\partial\rho}{\partial\tau}+\frac{1}{r^2}\frac{\partial}{\partial r}(r^2\rho v)=0,\quad \rho\left(\frac{1}{\theta}\frac{\partial T}{\partial\tau}+v\frac{\partial T}{\partial r}\right)-\frac{1}{r^2}\frac{\partial}{\partial r}\left(r^2\frac{\partial T}{\partial r}\right)=\Omega, \tag{1}$$

$$\rho\left(\frac{1}{\theta}\frac{\partial Y_i}{\partial\tau}+v\frac{\partial Y_i}{\partial r}\right)-\frac{1}{\mathscr{L}_i}\frac{1}{r^2}\frac{\partial}{\partial r}\left(r^2\frac{\partial Y_i}{\partial r}\right)=\alpha_i\Omega \qquad (i=1,2), \tag{2}$$

where the radial velocity v is the only component, $\rho=1/T$, $\tau=t/\theta$ and (if the radius, a, of the supply sphere is used to define units)

$$\Omega=DY_1^{\nu_1}Y_2^{\nu_2}\,\mathrm{e}^{-\theta/T} \tag{3}$$

with D given by the formula (1.61). The equations for the product and inert species, which have not been written, are complicated by the implicit assumption that $\mathscr{L}_1$ is not necessarily equal to $\mathscr{L}_2$ (section 1.4). The momentum equation, which has also been omitted, plays no role in this spherical combustion field other than determining the pressure. Throughout this chapter we shall take, for the sake of simplicity,

$$\nu_1=\nu_2=1,\quad m_1=m_2\quad \text{so that } \alpha_1=\alpha_2=-\tfrac{1}{2}, \tag{4}$$

(corresponding to a bimolecular reaction) even though retaining general values would also cover the corresponding analysis in Chapter 7 (where $\nu_1=1$, $\nu_2=0$), showing why there are so many similarities. Finally, unsteady terms have been retained with a view to stability considerations on the long-time τ. To avoid subscripts we shall change the notation to

$$Y_1=X,\quad Y_2=Y,\quad J_1=I,\quad J_2=J,\quad \mathscr{L}_1=\mathscr{K},\quad \mathscr{L}_2=\mathscr{L}. \tag{5}$$

Seven boundary conditions must be appended to the system (1), (2) to form a well-posed problem. We shall suppose that the two mass-flux fractions

$$I_s=X_s-M^{-1}X_s'/\mathscr{K},\quad J_s=Y_s-M^{-1}Y_s'/\mathscr{L} \tag{6}$$

are given, where $M(\tau)$ is the total flux from the supply sphere. A pure diffusion flame requires $I_s=0$, $J_s>0$, i.e. the sphere is a source of fuel, but neither a source nor sink of oxidant; for simplicity we shall take

$$J_s=1,\quad I_s=0 \tag{7}$$

making the sphere a source of fuel only. In addition,

$$L=M^{-1}T_s'>0 \tag{8}$$

(the heat conducted to the sphere per unit mass of fuel supplied) will be assigned. For the fuel drop or the liquid-saturated porous sphere, L is the latent heat of evaporation, thereby motivating the restriction to positive values of L. Negative values have been considered by Janssen (1982). The surface temperature T_s is also given, a choice difficult to justify in the cases mentioned, unless we abandon the specification of L; a more realistic condition will be considered in section 7. The remaining three conditions are imposed at infinity, where T_∞, X_∞, and

$$Y_\infty=0 \tag{9}$$

are given; the last condition ensures that all of the fuel has originated at the supply. Note that T_∞ may be assigned, a matter discussed later within the context of the premixed flame (section 7.2).

For the slow variations being considered here $M=M(\tau)$ and allowable initial conditions may be characterized by a single parameter $M(0)$ just as for the premixed flame (section 3.5). The steady state corresponds to particular values of M determined by D (and the other parameters). For certain parameter values (see Figure 3) there is no steady solution for any value of θ (Janssen 1982).

2 Steady combustion for $\mathscr{K}=\mathscr{L}=1$; D-asymptotics

The qualitative nature of a steady solution is not affected by the Lewis numbers; only in the stability question do they play an important role. To simplify an already complicated discussion we shall assume for the present that both Lewis numbers are unity.

In the steady state the continuity equation can be integrated to give

$$r^2\rho v = M(\text{constant}) \quad \text{or } \rho v = M/r^2. \tag{10}$$

The mass flux M appears explicitly for the non-dimensionalization used here; it is not concealed in the length unit, as in previous chapters. Note that $4\pi M$ is the rate at which fuel is supplied, in units of $\lambda a/c_p$, but it is the burning rate only if the total outflow of fuel at infinity is zero. Nevertheless, M is loosely called the burning rate, and we shall adopt the term also. We expect M to be determined as a function of D, and the calculation of this function is the fundamental goal of the analysis.

The Shvab–Zeldovich variables $H_i = T - Y_i/\alpha_i$, which are conditional enthalpies (section 1.7), satisfy

$$\mathfrak{L}(H_i) \equiv \left[\frac{1}{r^2}\frac{\mathrm{d}}{\mathrm{d}r}\left(r^2\frac{\mathrm{d}}{\mathrm{d}r}\right) - \frac{M}{r^2}\frac{\mathrm{d}}{\mathrm{d}r}\right]H_i = 0 \quad \text{for } 1 < r < \infty. \tag{11}$$

Since all solutions are bounded at infinity, the analysis immediately differs from that for the plane (premixed) flame in having extra integration constants. Correspondingly, and as anticipated, the basic parameters T_s, L, I_s, J_s, T_∞ and X_∞ may be assigned to give, in the case (7),

$$G \equiv T + 2X = (T_s - L)(1 - \mathrm{e}^{-M/r}) + (T_\infty + 2X_\infty)\,\mathrm{e}^{-M/r}, \tag{12}$$

$$H \equiv T + 2Y = T_\infty + (T_a - T_\infty)(1 - \mathrm{e}^{-M/r}), \tag{13}$$

where

$$T_a = T_s - L + 2 \tag{14}$$

is known as the adiabatic flame temperature.

When L is the latent heat of evaporation, the concept of adiabatic flame temperature is involved as in premixed combustion (section 2.5). If the liquid fuel at temperature T_s were allowed to react adiabatically with an equal amount of oxidant at temperature T_a, the temperature of the product would also be T_a, with heat (represented by L) being required to evaporate the fuel before reaction. Such an adiabatic process occurs for a spherical diffusion flame consuming all the fuel when no heat is gained from, or lost to, the ambient atmosphere, a condition that can be ensured by making T_∞ equal to T_a.

The problem now reduces to one for T alone, namely

$$\mathfrak{L}(T) = -DXY\,\mathrm{e}^{-\theta/T} \quad \text{for } 1 < r < \infty, \tag{15}$$

with T_s, L and T_∞ prescribed. Since there are three boundary conditions for this second-order differential equation, M is determined as a function of D as part of the solution. Until the advent of activation-energy asymptotics, analytical discussion was preoccupied with Damköhler-number asymptotics (i.e. $D \to 0$ or ∞). Since knowledge of these limits is of some importance, they will be briefly discussed. (See Figure 3.)

Frozen combustion occurs when the chemical reaction is very weak, i.e. $D \to 0$. The heat conducted back to the supply sphere then comes from the far field (provided $T_\infty > T_s$) and the limit solution is

$$T = (T_s - L) + L\,e^{M_w(1-1/r)} \quad \text{with } M = M_w \equiv \ln[1 + (T_\infty - T_s)/L]. \tag{16}$$

For Y to be positive M_w must be positive, so that

$$T_s < T_\infty \tag{17}$$

is required according to the restriction (8).

The result (16) is not uniformly valid, since for large values of r the neglected reaction term decays more slowly than the convection term. Defining

$$\varepsilon^2 = \tfrac{1}{2} D X_\infty e^{-\theta/T_\infty} \tag{18}$$

3 Parameter plane for spherical diffusion flame, showing form of response in the various regions. The sign of $\gamma = (T_a - T_\infty)/(1 + X_\infty)$ determines whether the environment is heated ($\gamma > 0$) or cooled ($\gamma < 0$).

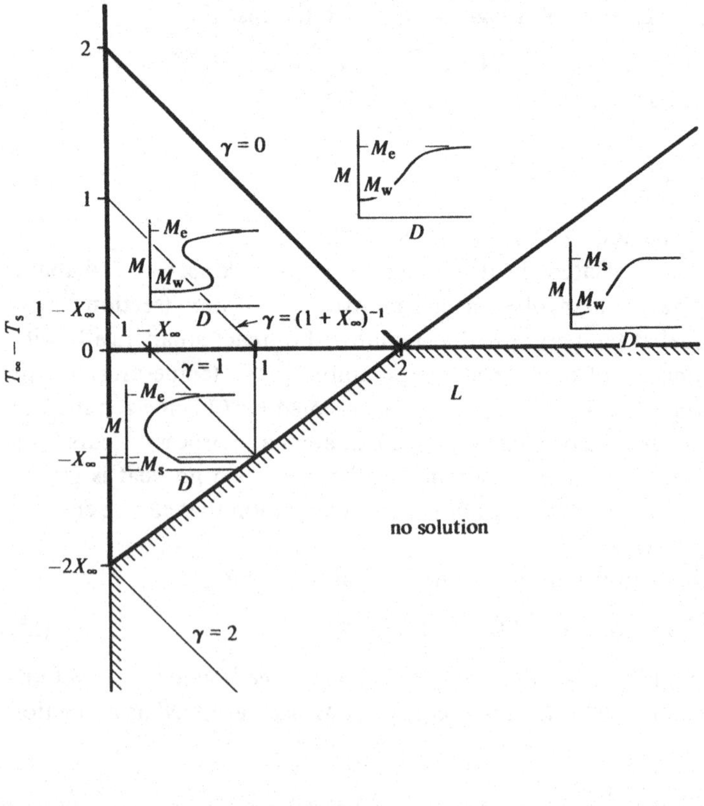

leads to the expansion

$$T = T_\infty - \varepsilon M_w[T_\infty - T_s + L - 2(1 - e^{-R})]/R + \dots \quad (19)$$

in the variable $R = \varepsilon r$. The temperature gradient far from the supply is thereby corrected, so that the heat flow from the far field can be calculated as

$$\lim_{R \to \infty} (4\pi r^2 \, dT/dr) = 4\pi M_w(T_\infty - T_a). \quad (20)$$

For $T_a > T_\infty$, a finite amount of heat is supplied to the environment, which is usually the goal of combustion. The frozen limit is not an extinguished state, but rather one in which all the reaction takes place at essentially constant temperature far from the supply.

The equilibrium limits are obtained as $D \to \infty$. They are of great practical interest since under standard atmospheric conditions Damköhler numbers are often large. Clearly the limit $D \to \infty$ is singular, since the highest derivative in equation (15) is multiplied by a vanishingly small parameter D^{-1}. We may anticipate thin regions where the second derivative is very large; when located in the interior of the combustion field, they are called Burke–Schumann flame sheets (Kassoy & Williams 1968). They should not be confused with the flame sheets that occur in activation-energy asymptotics. The thin regions can also be boundary layers but we shall consider the other, more common, possibility first.

Outside the flame sheet the limit equation is simply

$$XY = 0, \quad (21)$$

which represents chemical equilibrium for the irreversible reaction. The combustion field is divided into regions where either X or Y vanishes. Since fuel is supplied at the sphere and oxidant at infinity it is natural to put an oxidant-free region contiguous with the sphere, separated by a flame sheet at $r = r_*$ (say) from a fuel-free region extending to infinity. The Shvab–Zeldovich relations (12), (13) then imply

$$T = \begin{cases} (T_s - L)(1 - e^{-M_e/r}) + (T_\infty + 2X_\infty)\, e^{-M_e/r} \\ T_\infty + (T_a - T_\infty)(1 - e^{-M_e/r}) \end{cases} \quad \text{for } r \lessgtr r_*, \quad (22)$$

where we must have

$$M = M_e \equiv \ln[1 + (T_\infty - T_s + 2X_\infty)/L] \quad (23)$$

for the temperature at the sphere to be T_s. Continuity of temperature across the flame sheet (justified by an examination of its structure) shows that it must be located at

$$r_* = M_e/\ln(1 + X_\infty), \quad (24)$$

so that the flame temperature is

$$T_* = T_\infty + \gamma X_\infty \quad \text{with } \gamma = (T_a - T_\infty)/(1 + X_\infty). \tag{25}$$

These equations are the essence of the so-called Burke–Schumann solution (Burke & Schumann 1928), although it is still necessary to show that there is a structure linking the two sides of the flame sheet. We shall be content with setting up a well-posed problem for this structure.

It is clear from the oxidant species equation (2*a*) that, for D large but not infinite, there can be no algebraic perturbation of $X=0$ in $r<r_*$; similarly there is none of $Y=0$ in $r>r_*$. These conclusions follow from the choice (4*a*), the situation being more complicated for other choices (Kassoy & Williams 1968). The temperature is, therefore, given to all orders by the formulas (22) with M_e unperturbed from its value (23). Within the flame sheet we set

$$T = T_0 + D_*^{-\frac{1}{2}} T_1 + \dots, \quad \xi = D_*^{\frac{1}{2}}(r - r_*) \text{ with } T_0 = T_*, \quad D_* = D\,\mathrm{e}^{-\theta/T_*} \tag{26}$$

to obtain

$$\mathrm{d}^2 T_1/\mathrm{d}\xi^2 = -\tfrac{1}{4}[T_1 + \gamma M_e \xi/r_*^2][T_1 + (\gamma - 2) M_e \xi/r_*^2]. \tag{27}$$

Here we see for the first time a difference in the structure of curved flame sheets, one that carries over to activation-energy flame sheets: ξ appears explicitly in the structure equation, due to variation in Shvab–Zeldovich variables (i.e. conditional enthalpies). Matching with the expansions outside requires that

$$T_1 = (2-\gamma) M_e \xi/r_*^2 + o(1) \quad \text{as } \xi \to -\infty,$$
$$T_1 = -\gamma M_e \xi/r_*^2 + o(1) \text{ as } \xi \to +\infty. \tag{28}$$

Note that any solution of the equation (27) having the correct slopes as $\xi \to \pm\infty$ would have to satisfy the boundary conditions (28), so that the problem is not overdetermined. The integration can only be carried out numerically, which leaves no doubt that there is a structure to smooth out the discontinuous gradient obtained from the outer expansions. Uniqueness and existence have recently been proved by Holmes (1982).

The Burke–Schumann solution does not hold for all parameter values. An underlying assumption is that a flame sheet forms in the interior of the combustion field, so that for consistency r_* must be greater than 1, i.e.

$$T_\infty - T_s > (L-2) X_\infty. \tag{29}$$

Then M_e is automatically positive, as expected. As equality is approached the flame sheet moves to the surface $r=1$ and our analysis breaks down.

Although the resulting surface flame, identified by Buckmaster (1975*a*), does not play as important a role in practice as the Burke–Schumann flame, a brief description of the solution will be given here.

When the flame sheet is a boundary layer the oxidant-free region is absent and only the fuel-free region (22*b*) remains. Since the temperature must tend to T_s as $r \to 1$, M_e must be replaced by

$$M_s \equiv \ln[1 + (T_\infty - T_s)/(L-2)], \tag{30}$$

a result quite different from the Burke–Schumann value (23), but to which it is equal at the limit of the inequality (29). The formula (30) is that for weak burning with L replaced by $L-2$, where the 2 can be identified as the heat released in the flame.

In addition, the flame-sheet structure is different. The Shvab–Zeldovich relation (12) shows that

$$X_* = [(T_s - T_\infty) - (2-L)X_\infty]/(T_\infty - T_a) \tag{31}$$

no longer vanishes. As a consequence, the flame sheet now has thickness $O(D_*^{-\frac{1}{2}})$ and the structure is governed by a linear equation that can easily be integrated. We find the perturbation result

$$M = M_s[1 + 2D_*^{-\frac{1}{2}}/(2-L) + \ldots] \tag{32}$$

and the structure

$$T = T_s + D_*^{-\frac{1}{2}} M_s[2(1 - \mathrm{e}^{-\xi}) + (L-2)\xi] + \ldots, \tag{33}$$

where now

$$\xi = D_*^{\frac{1}{2}}(r-1), \quad D_* = \tfrac{1}{2} X_* D\,\mathrm{e}^{-\theta/T_*} \quad \text{with } T_* = T_s. \tag{34}$$

Restrictions on the parameters come from M_s and X_*. The (small) mass fraction of fuel is positive only for $M_s > 0$, so that we must have

$$T_\infty \gtrless T_s \quad \text{accordingly as } L \gtrless 2; \tag{35}$$

the positivity of X_* imposes the additional restriction

$$T_\infty - T_s \lesseqgtr (L-2)X_\infty. \tag{36}$$

Two angular regions are thereby determined in the parameter plane (Figure 3); one coincides with the region where the Burke–Schumann solution exists but the weak-burning solution does not, and the other with the region where the reverse is true.

3 The nearly adiabatic flame for $\mathscr{K}$, $\mathscr{L} \neq 1$ and its stability

Having determined the possible ends of the M, D-response curve, we now consider its shape in between using activation-energy asymptotics. A general discussion is still quite complicated, requiring extensive numerical calculations. For the moment we shall restrict attention to the special case considered by Buckmaster (1975*b*) in which

$$T_a - T_\infty = k/\theta \quad \text{with } k = O(1). \tag{37}$$

A great deal of insight into the nature of diffusion flames can thereby be obtained with a minimum of computation, and the results are not without practical significance. The most important aspects of the general case (for $\theta \to \infty$) will be dealt with later.

To obtain different values of k it is convenient to vary T_s keeping T_∞ and L fixed, with

$$L < 2 \tag{38}$$

to ensure a solution; clearly the frozen and Burke–Schumann limits exist. Restriction to small values of $T_a - T_\infty$ means that in both limits the combustion field is nearly adiabatic (in the sense that the heat flux at infinity is small); in fact, that will be true along the whole of the response curve, as we shall see. The unsteady formulation is reinstated, with Lewis numbers again arbitrary. Global Shvab–Zeldovich relations do not exist and it is necessary to reexamine the complete system (1) and (2).

For slow variations, the only information needed from such a system in section 3.5 was the change in enthalpy up to the flame sheet. Here the changes in both enthalpies (12) and (13) are needed, but the procedure is the same except for integrating from the supply (as in section 5.6) and taking account of the areal factor r^2. We find

$$M(T_a - 2 - G_*) = -r^2\left(\frac{\partial T}{\partial r} + 2\mathscr{K}^{-1}\frac{\partial X}{\partial r}\right)\bigg|_{r_*+0} + \theta^{-1}\int_1^{r_*} r^2 \frac{\partial}{\partial \tau}[\rho(2X - G_*)]\,\mathrm{d}r, \tag{39}$$

$$M(T_a - H_*) = -r^2\left(\frac{\partial T}{\partial r} + 2\mathscr{L}^{-1}\frac{\partial Y}{\partial r}\right)\bigg|_{r_*+0} + \theta^{-1}\int_1^{r_*} r^2 \frac{\partial}{\partial \tau}[\rho(2Y - H_*)]\,\mathrm{d}r, \tag{40}$$

to $O(\theta^{-1})$, where the integrals are to be evaluated to $O(1)$ only. The first integral will not be evaluated since the corresponding equation is only needed to $O(1)$.

In determining leading terms, we may concentrate on the steady version of equations (1*b*) and (2), where ρv is the function (10) with M now dependent on τ. For this near-adiabatic situation the temperature beyond the flame sheet must be constant at T_∞. The combustion field is, therefore, similar to that for the plane deflagration wave treated in Chapter 2. The temperature increases from T_s at the supply to T_∞ at the flame, beyond which it stays constant. The reaction is frozen for $1<r<r_*$, while there is equilibrium with

$$Y=0 \quad \text{for } r_*<r<\infty. \tag{41}$$

In fact, the latter holds to all orders.

With this picture in mind the solution in the frozen region is easily derived. We find, to leading order,

$$T=T_s-L+L\,e^{M(1-1/r)}, \quad X=X_*\,e^{\mathscr{K}M(1/r_*-1/r)}, \quad Y=1-e^{\mathscr{L}M(1/r_*-1/r)}, \tag{42}$$

where

$$r_*=M/[M+\ln(L/2)]. \tag{43}$$

These satisfy the supply conditions and give $T=T_\infty$ and $Y=0$ at $r=r_*$. The condition (38), which means that the heat conducted back into the supply sphere is less than that released in the flame, ensures that r_* is greater than 1. Similarly (though to one more term in temperature) in the equilibrium region, where chemistry-free equations are still satisfied,

$$T=T_\infty+\theta^{-1}T_{1*}(1-e^{-M/r})/(1-e^{-M/r_*}), \tag{44}$$

$$X=[X_*(1-e^{-\mathscr{K}M/r})+X_\infty(e^{-\mathscr{K}M/r}-e^{-\mathscr{K}M/r_*})]/(1-e^{-\mathscr{K}M/r_*}). \tag{45}$$

The unknown constants X_* and T_{1*} can be found (in terms of M) from the basic equations (39) and (40) of slow variations, which yield

$$T_{1*}=(1-e^{-M/r_*})(k+bM\,dM/d\tau), \quad X_*=(1+X_\infty)\,e^{-\mathscr{K}M/r_*}-1 \tag{46}$$

with

$$b=\int_0^{\ln(2/L)}\frac{L\,e^l-2(L\,e^l/2)^{\mathscr{L}}}{T_s-L+L\,e^l}\frac{d}{dl}\left[\frac{l}{(l-M)^4}\right]dl. \tag{47}$$

The chemical reaction determines the burning rate at a given temperature, so that we may look to the structure of the flame sheet for another relation between M and the perturbation T_{1*} of the flame temperature. Writing

$$\xi=\theta(r-r_*) \quad \text{and } T_0=T_\infty, \quad X_0=X_*, \quad Y_0=0 \tag{48}$$

leads to the equations

$$T_\infty^2\partial^2\phi/\partial\xi^2=(2/\mathscr{L})\partial^2 Y_1/\partial\xi^2=\tilde{D}Y_1\,e^{-\phi}$$
$$\text{with } \phi=-T_1/T_\infty^2, \quad \tilde{D}=DX_*\theta^{-2}\,e^{-\theta/T_\infty}. \tag{49}$$

The structure problem is completed by the boundary conditions

$$\phi=\begin{cases}-2M\xi/r_*^2T_\infty^2+\dots\\ \phi_*+\dots\end{cases}, \quad Y_1=\begin{cases}-\mathscr{L}M\xi/r_*^2+\dots\\ o(1)\end{cases} \quad \text{as } \xi\to\mp\infty, \tag{50}$$

which come from matching on the two sides of the flame sheet. Clearly there is a local Shvab–Zeldovich relation

$$Y_1 = \tfrac{1}{2}\mathscr{L}T_\infty^2(\phi - \phi_*) \tag{51}$$

enabling the equation for ϕ to be integrated to give

$$\phi_* = \ln(D\mathscr{L}X_* r_*^4 T_\infty^4 / 4M^2\theta^2\,\mathrm{e}^{\theta/T_\infty}). \tag{52}$$

Elimination of $T_{1*} = -T_\infty^2\phi_*$ between this and the result (46*a*) finally yields an equation for M, namely

$$bM\,\mathrm{d}M/\mathrm{d}\tau = (1 - \mathrm{e}^{-M/r_*})^{-1} T_\infty^2 \ln(4M^2\theta^2\,\mathrm{e}^{\theta/T_\infty}/D\mathscr{L}X_* r_*^4 T_\infty^4) - k \tag{53}$$

where r_* and X_* are given by equations (43) and (46*b*).

We shall first describe the steady state, determined by setting the right side of this last equation equal to zero, and then discuss its stability. Different values of k lead to different steady-state responses, i.e. curves of M versus D with all other parameters fixed (Figure 4), all joining the frozen limit (16*b*) to the Burke–Schumann limit (23). In between, a curve is monotonic or S-shaped depending on the value of k. When there is a gain of heat from infinity ($k<0$) or when the loss is sufficiently small, the response is monotonic. But for a sufficiently large loss the burning rate is triple valued over some range of Damköhler number. Conditions under

4 Steady-state responses when the combustion is nearly adiabatic ($T_a - T_\infty$ small), as measured by k in the distinguished limit (37). Drawn for $\mathscr{K} = 1$, $L = 0.2$, $T_\infty = 0.2$, $X_\infty = 1$. At left ends $r_* \to \infty$; at right ends $X_* \to 0$.

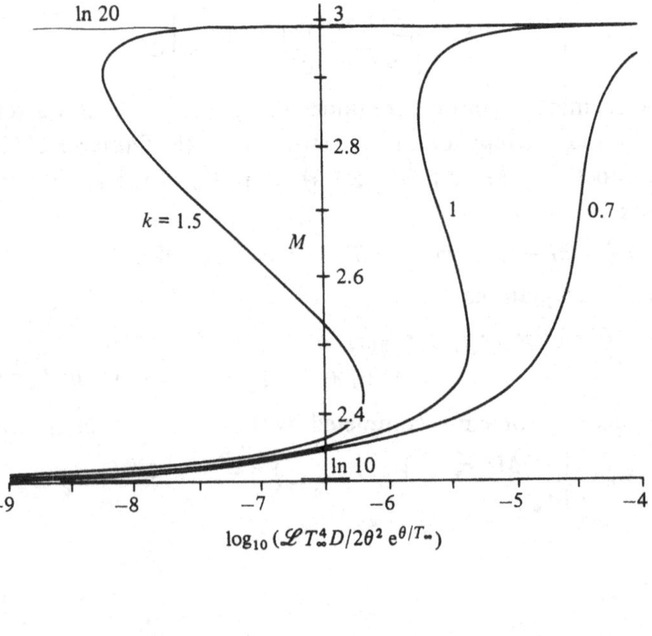

which heat is delivered to the environment are important, so that the S-shaped response is of special interest.

Explicit formulas can be given for the turning points on the S-shaped response when k is large. At the lower point

$$M \sim \ln(2/L) + 4T_\infty^2/k, \quad D \sim 4(4/\mathrm{e})^4 T_\infty^4 \theta^2 \,\mathrm{e}^{\theta/T_\infty}/[\ln(2/L)]^2 \mathscr{L} X_\infty k^4, \tag{54}$$

where the leading term is the limiting value of M_w, while at the upper point

$$M \sim M_e - T_\infty^2 L \,\mathrm{e}^{M_e}/2k,$$

$$D \sim 8\,\mathrm{e}[M_e + \ln(L/2)]^4 \theta^2 \,\mathrm{e}^{\theta/T_\infty} k \times \exp\{-k[1-(1+X_\infty)^{-\mathscr{K}}]/T_\infty^2\}/L\mathscr{K}\mathscr{L} M_e^2 T_\infty^6 \,\mathrm{e}^{M_e}, \tag{55}$$

where

$$M_e = \mathscr{K}^{-1} \ln(1+X_\infty) + \ln(2/L) \tag{56}$$

is the Burke–Schumann value (23) corrected for $\mathscr{K} \neq 1$.

Conclusions about the instability of the steady state can easily be drawn from the full equation (53), which holds in the strip $M_s < M < M_e$ where $X_* > 0$ and $1 < r_* < \infty$. The right side is negative to the right of the steady-state response curve and positive to the left (Figure 5), while $b \lessgtr 0$ accordingly as $\mathscr{L} \lessgtr 1$. Consider first $\mathscr{L} < 1$. If the burning state lies off the steady-state curve, then M will change in the direction of the arrows in Figure 5 so that the middle branch of the S is unstable. Now suppose that $\mathscr{L} > 1$; the arrows must be reversed so that then the upper and lower branches of the S-curve, and the whole of the monotonic curve, are seen to be unstable.

5 Stability of an S-curve in Figure 4 for $\mathscr{L} < 1$.

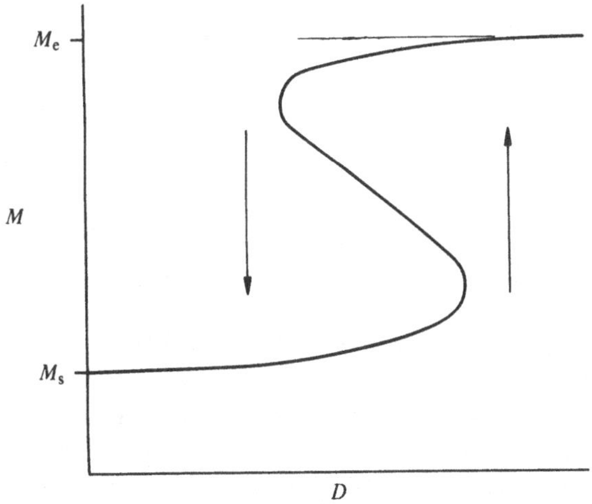

The result should be compared with general beliefs and expectations. On the grounds that a flame would be expected to burn more vigorously when the pressure is increased, the middle branch is assumed to be unstable. Our failure to uncover the instability for $\mathscr{L}>1$ is readily explained by the special class of disturbances considered. Evidence is presented in section 8 that a different class leads to the anticipated result. The instability of the upper and lower branches for $\mathscr{L}>1$ is analogous to that for premixed plane flames subjected to plane disturbances (section 3.5). The observation of stable burning with heavy hydrocarbons (when $\mathscr{L}>1$ is appropriate) is, therefore, an embarrassment.

Note again that equation (53) does not apply when $b=O(\theta^{-1})$ since the slow-time scale is not then appropriate. No conclusion can be drawn from it for $\mathscr{L}$ close to 1, and it could well be that stable flames are associated with this restriction. Such a possibility is examined in Chapter 11 in the context of premixed flames.

If it is assumed that the stability results deduced here for $\mathscr{L}<1$ are valid for arbitrary disturbances, i.e. the middle branch is unstable and the other two are stable, then the turning points of the S-curve represent ignition and extinction. The asymptotic results (54) and (55) then give ignition and extinction conditions, respectively.

4 General ignition and extinction analyses for $\mathscr{K}=\mathscr{L}=1$

We proceed now to the general steady state (Law 1975), i.e. arbitrary values of T_a-T_∞, with emphasis in this section on the top and bottom branches of the S-responses. As noted in section 2, the qualitative nature of the solution is not affected by the Lewis numbers so that we shall restrict attention to $\mathscr{K}=\mathscr{L}=1$, for which the analysis is complicated enough.

We must deal with the Shvab–Zeldovich relations (12), (13), and the temperature equation (15). The condition for an S-response is

$$T_s<T_\infty<T_a, \tag{57}$$

the left inequality ensuring a weak-burning limit and the right being suggested by the discussion in section 3 where such a response was only obtained for sufficiently large $k=\theta(T_a-T_\infty)>0$.

The lower branch will be considered first, the main goal being to locate the ignition point, which is a fundamental characteristic of the burning response. The value of D is, in general, characterized by the flame temperature T_*, so that a key question is the value of T_* in the vicinity of the point. Section 3 suggests that the burning will take place far from the supply on the whole lower branch, i.e. with M close to its frozen value M_w: the ignition point (54) has this property – the measure of closeness is k^{-1} rather than the θ^{-1} that we must expect now. The conclusion is, therefore, that $T_* = T_\infty$.

At any finite value of r the combustion field is frozen to all orders, so that we may write

$$T_0 = T_s - L + L\,e^{M_w(1-1/r)}, \quad T_1 = M_1 L(1-1/r)\,e^{M_w(1-1/r)} \tag{58}$$

where M_1 is the perturbation of M_w. The corresponding perturbation of the temperature in the reaction zone significantly affects the combustion field through the Arrhenius factor. The correct values of T and T' are taken at $r=1$ but the condition $T \to T_\infty$ as $r \to \infty$ is violated, showing that a new coordinate is required, which turns out to be

$$R = r/\theta. \tag{59}$$

Setting $T_0 = T_\infty$ then leads to the structure equation

$$\frac{1}{R^2}\frac{d}{dR}\left(R^2\frac{d\phi}{dR}\right) = D_w\left[\phi + \frac{(T_a - T_\infty)M_w}{T_\infty^2 R}\right]e^{-\phi}$$
$$\text{with } \phi = -T_1/T_\infty^2, \quad D_w = \tfrac{1}{2}DX_\infty\theta^2\,e^{-\theta/T_\infty} \tag{60}$$

subject to the boundary conditions

$$\phi = L\,e^{M_w}(-M_1 + M_w/R)/T_\infty^2 + \ldots \quad \text{as } R \to 0, \quad \phi = o(1) \quad \text{as } R \to \infty, \tag{61}$$

the first of which comes from matching with the frozen expansion. With all parameters specified, including D_w, the perturbation T_1 can be found (numerically) without using the M_1-term; thus, M_1 is determined as a function of the Damköhler number.

By suitably scaling M_1 and D_w the relation between them can be made to depend on the single parameter

$$\beta = (T_a - T_\infty)/L\,e^{M_w} \tag{62}$$

(Law 1975). It is found (numerically) that the graph of M_1 versus D_w displays a turning point if, and only if, β is positive, in agreement with section 3. Curves for various values of $\beta > 0$ are shown in Figure 6; note that they are independent of X_∞.

We turn now to the upper branch. The burning rate at the extinction point (55) of the nearly adiabatic flame lies close to the Burke–Schumann value when k is large. This suggests that the general extinction curve can be uncovered by an analysis that considers $O(\theta^{-1})$ perturbations of the burning rate from its Burke–Schumann value.

The flame sheet has the location (24), while on both sides of it the reaction is frozen to all orders. Thus,

$$T_0 = T_s - L + L\,e^{M_e(1-1/r)}, \quad T_1 = M_1 L(1-1/r)\,e^{M_e(1-1/r)} \quad \text{for } r<r_*$$
$$\text{and } T_0 = T_a + (T_\infty - T_a)\,e^{-M_e/r} \quad \text{for } r>r_*, \tag{63}$$

where M_1 is now the perturbation of M_e. Only M_1 is unknown and for that we must turn to the structure of the flame sheet. With the same stretched coordinate (48*a*) and with

$$T_0 = T_* \quad \text{and } X_0 = Y_0 = 0, \tag{64}$$

where T_* is the flame temperature (25), the structure equation becomes

$$d^2\phi/d\xi^2 = D_e[\phi + \gamma(M_1/r_* - M_e\xi/r_*^2)/T_*^2]$$
$$\times[\phi + (\gamma-2)(M_1/r_* - M_e\xi/r_*^2)/T_*^2]\,e^{-\phi}$$
$$\text{with } \phi = -T_1/T_*^2, \quad D_e = \tfrac{1}{4}DT_*^2\theta^{-3}\,e^{-\theta/T_*}, \tag{65}$$

and γ taking the value (25*b*); it is subject to the boundary conditions

$$\phi = (\gamma-2)[M_e\xi/r_*^2 + (1-1/r_*)M_1]/T_*^2 + \ldots \quad \text{as } \xi \to -\infty,$$
$$d\phi/d\xi = \gamma M_e/r_*^2 T_*^2 + \ldots \quad \text{as } \xi \to +\infty, \tag{66}$$

which come from matching. This should be compared with the problem (27) and (28); the only (but crucial) differences are that here the Arrhenius

6 Steady-state response near ignition for unit Lewis numbers, as determined by the problem (60, 61). When $\beta = (T_a - T_\infty)/Le^{M_w}$ is negative (corresponding to heat being supplied by the environment) there is no turning point.

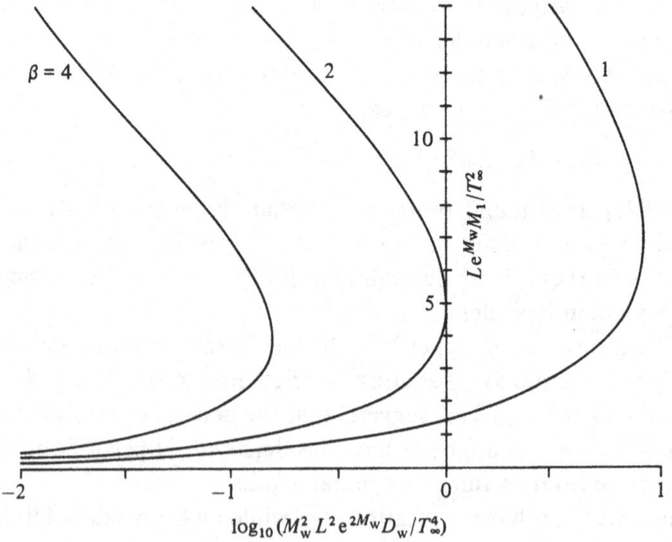

factor is not approximated by a constant, since the perturbations are as large as $O(\theta^{-1})$, and M_1 is not zero, because the combustion field is frozen between the supply and the flame, not in equilibrium. Note that the temperature only achieves its maximum in the flame sheet, as it must, if γ lies between 0 and 2, i.e.

$$0 < T_a - T_\infty < 2(1 + X_\infty). \tag{67}$$

The problem determines M_1 and, by suitable scaling, the relation between it and D_e can be made to depend on the single parameter γ (Law 1975). Thus,

$$\tilde{D}_e = r_*^4 T_*^6 D_e / M_e^2, \quad \tilde{M}_1 = (1 + \tilde{\gamma}) M_1 / T_*^2,$$
$$\tilde{\phi} = \phi + \tilde{\gamma}\tilde{\xi}, \quad \tilde{\xi} = (M_e \xi / r_*^2 - M_1 / r_*) / T_*^2 \quad \text{with } \tilde{\gamma} = 1 - \gamma \tag{68}$$

lead to the canonical problem

$$d^2\tilde{\phi}/d\tilde{\xi}^2 = \tilde{D}_e(\tilde{\phi}^2 - \tilde{\xi}^2)\, e^{-\tilde{\phi} + \tilde{\gamma}\tilde{\xi}}, \tag{69}$$

$$\tilde{\phi} = -\tilde{\xi} - \tilde{M}_1 + \ldots \quad \text{as } \tilde{\xi} \to -\infty, \quad d\tilde{\phi}/d\tilde{\xi} = 1 + \ldots \quad \text{as } \tilde{\xi} \to +\infty. \tag{70}$$

It is found (again numerically) that the graphs of M_1 versus D_e display the extinction phenomenon. Curves for various values of γ are shown in Figure 7. Note that the actual Damköhler number for extinction is

7 Steady-state response near extinction for unit Lewis numbers, as determined by the problem (69, 70). When $\gamma = (T_a - T_\infty)/(1 + X_\infty)$ is negative (corresponding to heat being supplied by the environment) there is no turning point.

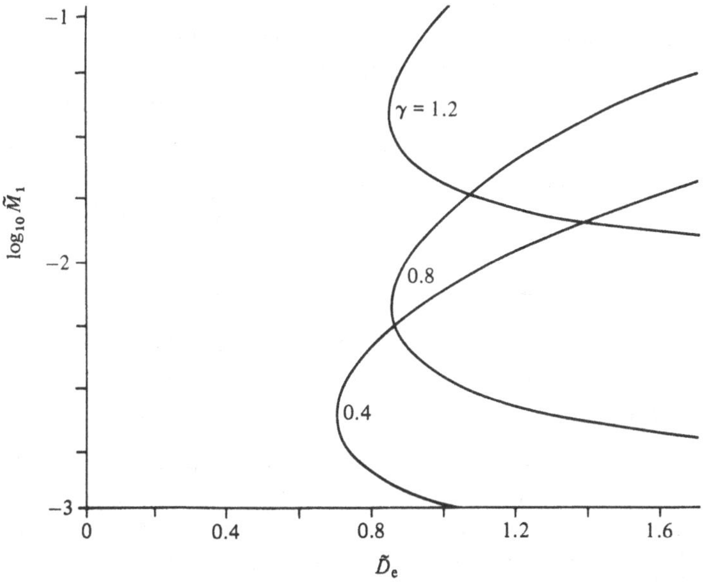

$O(\theta^3 e^{\theta/T_*})$ as compared to $O(\theta^{-2} e^{\theta/T_\infty})$ for ignition, so that (since $T_* > T_\infty$) the extinction point lies to the left of the ignition point, as it should.

5 Remarks on the middle branch of the S-response

It is widely believed that the middle branch, where M is a decreasing function of D, is not of practical interest because the burning states there are unstable. (Some aspects of this question are considered in section 8; here we sketch the nature of the steady-state solution.) Our discussion is justified not only because of the significant difference from what we have heretofore considered, but also because of the need for an appropriate mathematical description on which to base a stability analysis.

It is clear from the ignition and extinction analyses that the flame temperature on the middle branch must span the range T_∞ to $T_\infty + \gamma X_\infty$ and, since it is the maximum temperature at both ends, we may expect it to be so in between. On either side of the flame sheet the reaction is frozen, just as in the extinction analysis. The key difference is that here X_* and Y_* do not both vanish; we shall first assume that neither does.

It is more convenient now to specify M and determine

$$D = D^0[1 + O(\theta^{-1})] \tag{71}$$

as part of the solution. The leading terms (63) are, therefore, replaced by

$$T_0 = \begin{cases} T_s - L + L\,e^{M(1-1/r)} \\ T_\infty + L\,e^M(1 - e^{-M/r}) \end{cases} \quad \text{for } r \lessgtr r_* \tag{72}$$

if we anticipate a property of the flame structure, namely that the heat generated by the reaction is conducted equally to the two sides. Continuity of temperature to leading order at the flame sheet now determines

$$r_* = M/\ln[2/(1 + e^{M_w - M})] \quad \text{and} \quad T_* = \tfrac{1}{2}(T_s + T_\infty - L + L\,e^M), \tag{73}$$

the latter covering its range as M increases from M_w to M_e.

The structure is investigated, in the usual way, in terms of the variable (48*a*) and we find

$$d^2\phi/d\xi^2 = D_m\,e^{-\phi} \quad \text{with } D_m = D^0 X_* Y_* \theta^{-1} e^{-\theta/T_*}/T_*^2, \tag{74}$$

where X_* and Y_* are determined by the Shvab–Zeldovich relations (12) and (13). Integration yields

$$(d\phi/d\xi)^2 = C - 2D_m\,e^{-\phi}, \tag{75}$$

where C is a constant; from which it follows that the temperature gradients on the two sides of the flame sheet are of equal magnitude but of opposite sign, as anticipated earlier. Matching shows that $C = [(L\,e^M M/r_*^2)\,e^{-M/r_*}]^2$, but we shall not need this result.

Another integration gives a relation between the maximum value of T_1 and its location; to determine D_m it is necessary to consider the second perturbation. No useful purpose is served by presenting these rather complicated algebraic details. They are described by Liñán (1974) for the counterflow flame and by Law (1975), in adapting Liñán's analysis to the spherical flame. Another reason for not pursuing the determination of D_m is that

$$\ln D^0 = \theta/T_* + \ln\theta + \ln(D_m/X_*Y_*) \tag{76}$$

is only affected to $O(1)$ by it, T_* being a known function (73*b*) of M. The expected property $dD^0/dM < 0$ as $\theta \to \infty$ follows, irrespective of the value of D_m, since $dT_*/dM > 0$.

The solution derived here is valid only as long as $0 < X_*, Y_* < 1$ and $r_* > 1$. These inequalities are satisfied automatically with the exception of $Y_* > 0$, which, combined with the requirement that M be greater than M_w, implies

$$\ln[(2 + T_\infty - T_a)/L] < M < \ln[(T_a - T_\infty)/L]. \tag{77}$$

Thus the condition necessary for a range of validity is

$$T_\infty < T_a - 1, \quad \text{i.e. } \gamma > (1 + X_\infty)^{-1}; \tag{78}$$

the range will extend up to M_e if

$$T_\infty < T_a - (1 + X_\infty), \quad \text{i.e. } \gamma > 1. \tag{79}$$

The corresponding regions are shown in Figure 3.

When the inequality (79) is satisfied the solution predicts $X_* \to 0$ and $Y_* \not\to 0$ as $M \to M_e$. A different analysis, involving a flame structure with $X_* = O(\theta^{-1})$, is needed to complete the middle branch (cf. section 6). The transition is characterized by $o(1)$ changes in M and by Y_* decreasing to zero so as to merge finally with the extinction branch.

For $T_a - (1 + X_\infty) < T_\infty < T_a - 1$, corresponding to $(1 + X_\infty)^{-1} < \gamma < 1$, the mass fraction Y_* vanishes before M reaches M_e. The remaining values of M involve a flame structure with $Y_* = O(\theta^{-1})$, resulting in an X_* which tends to zero (as required) when $M \to M_e$. For $T_\infty > T_a - 1$, i.e. $\gamma < (1 + X_\infty)^{-1}$, the new structure is needed over the whole range of M.

6 Other responses

The last two sections have been concerned with the S-shaped response that occurs when the combustion field supplies heat to the ambient atmosphere. When

$$T_\infty > T_a \tag{80}$$

the flux is in the opposite direction and we may expect the response to be monotonic (if only on the basis of the analysis in section 3). Fuel drops, to which the theory will be adapted in section 7, are usually burnt to provide heat to their surroundings; for such an application this discussion is not pertinent.

The temperature is a monotonically increasing function of r so that, as before, there is a frozen region up to the flame, but an equilibrium region beyond, where

$$Y=0 \quad \text{for } r>r_* \tag{81}$$

must hold to all orders. A similar situation arises in a quite different context, that of a dissociating gas flowing through a Laval nozzle, where equilibrium is suddenly converted to frozen chemistry as the temperature drops below a critical value. This problem was treated by Blythe (1964) in a singularly early application of activation-energy asymptotics.

The expression (72*a*) is still correct, but now the expansion (72*b*) must be replaced by

$$T_0=T_\infty+(T_a-T_\infty)(1-e^{-M/r}) \quad \text{for } r>r_*. \tag{82}$$

Continuity at the flame sheet then determines

$$r_*=M/\ln[\tfrac{1}{2}(T_a-T_\infty+L\,e^M)], \quad T_*=T_a+2(T_\infty-T_a)/(L\,e^M-T_\infty+T_a). \tag{83}$$

The monotonicity of the response is now clear without going into details of the structure, by extracting just

$$\ln D^0=\theta/T_*+2\ln\theta+O(1) \tag{84}$$

in analogy with the expansion (76). Since $dT_*/dM<0$ under the condition (80), the leading term D^0 in the Damköhler number is an increasing function, as required.

As M increases from its frozen value, r_* decreases from ∞; there are two alternatives for M. If the inequality (29) is satisfied (see Figure 3), r_* reaches the value (24) as M tends to M_e; the combustion field is then identical to the Burke–Schumann limit; in particular, the oxidant no longer penetrates through the flame. The solution joins the frozen limit to the Burke–Schumann limit; correct descriptions of the approaches to the latter follow those in section 5. If

$$0<T_\infty-T_s<(L-2)X_\infty \tag{85}$$

(see Figure 3), then M_s is smaller than M_e and as M reaches it the flame settles down on the surface of the supply sphere. Our solution now joins the frozen limit to Buckmaster's limit (we shall not describe the approaches here).

The discussion so far in this section can be justified by the fuel drop, albeit under conditions of little practical interest. The same is true for

$$T_\infty < T_s, \tag{86}$$

though the inequality can no longer hold over the whole response of the drop (see the next section). For this reason, no analysis for the remaining triangle in Figure 3 has been published even though it is not difficult to see its outlines.

While there is no frozen limit now, it turns out that both equilibrium limits exist, provided

$$T_\infty - T_s > (L-2)X_\infty. \tag{87}$$

Otherwise, no limits exist and we must conclude that there is no solution in the corresponding region of the parameter plane, marked in Figure 3. Janssen (1982) has shown that no solution exists there whatever the value of θ.

It is reasonable to suppose that, when the conditions (86) and (87) are satisfied, the response takes the form of a C joining Buckmaster's limit to the Burke–Schumann limit (which is always higher). The analysis sketched in section 5 will still apply; only the conditions under which the basic solution must be modified will change: $r_* > 1$ is no longer automatic and the basic solution applies over at least part of the range $M_s < M < M_e$ only when $L < 1$ and $T_\infty - T_s > -X_\infty$. The necessary modifications have been treated by Janssen (1982).

The turning point of the C-shaped curve represents both ignition and extinction conditions. The burning rate there again lies within $O(\theta^{-1})$ of the Burke–Schumann value M_e, so that the extinction analysis of section 4 locates these conditions.

7 The burning fuel drop

In our discussion of the spherical diffusion flame we have supposed the seven parameters $T_s, L, I_s, J_s, T_\infty, X_\infty$, and Y_∞ to be given; for simplicity we have taken the particular values (7) and (9). If L is the latent heat of the evaporation of the fuel (taken to be constant for the temperature range of interest) then these are, in fact, prescribed in fuel-drop burning with the exception of T_s, in place of which the Clausius–Clapeyron relation

$$Y_s = (T_s/T_b)^\beta \, e^{\hat{\theta}/T_b - \hat{\theta}/T_s} \tag{88}$$

is substituted. Here T_b is the boiling temperature corresponding to the ambient pressure; i.e., in the notation of equation (4.18),

$$T_b^\beta \, e^{-\hat{\theta}/T_b} = \bar{p}_c/k. \tag{89}$$

Now Y_s may be calculated in terms of the existing parameters and M by means of the Shvab–Zeldovich relation (13), so that it may be eliminated to give

$$2(T_s/T_b)^\beta \, e^{\hat{\theta}/T_b-\hat{\theta}/T_s}=2-L+(T_\infty-T_s+L-2)\,e^{-M} \tag{90}$$

as the equation determining T_s in terms of M. The new feature, therefore, is that T_s is no longer fixed but must be calculated at each point of the response curve. Once T_s is found, the parameter plane of Figure 3 may be used to find the appropriate structure for determining D.

The properties of the resulting responses have been determined by Normandia & Ludford (1980), assuming that D is varied through the radius of the drop and not the pressure $\bar{p}_c$ appearing in equation (90) via T_b. (Janssen (1981) has recently considered pressure variations, with completely different results.) These properties are not obvious and it helps to characterize the corresponding loci in the parameter plane. First note that there is always a weak-burning limit, whatever the values of L, T_∞, X_∞, $\hat{\theta}$, and T_b. That is, the equation (90) invariably has a solution, $T_s=T_w$ (say), when M is set equal to M_w; in fact, T_w is less than T_∞, as we should expect from the absence of weak burning for $T_s>T_\infty$. Next note that there is always a Burke–Schumann limit $T_s=T_e$ (say), for $T_\infty>(L-2)X_\infty$; that $T_e<T_\infty+(2-L)X_\infty$, corresponding to a point above the equilibria divider; and that $T_e \lesseqgtr T_w$ accordingly as the point is above or below the adiabatic line. These results follow from setting $M=M_e$ in equation (90). Finally, note that when $M=M_s$ the equation has the solution $T_s=0$. While this is not relevant when the inequality (38) holds, it provides a termination point for $T_\infty<(L-2)X_\infty$.

The possible loci are illustrated in Figure 8, where dashes indicate that

8 Application to the burning fuel drop with surface equilibrium: loci of T_s in the parameter plane as the radius increases with pressure fixed. T_l is the value of T_s for which $Y_s=(2+L)/(2-L/X_\infty)$.

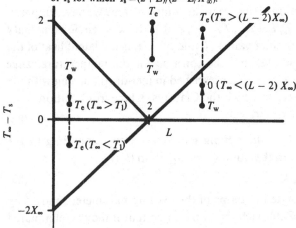

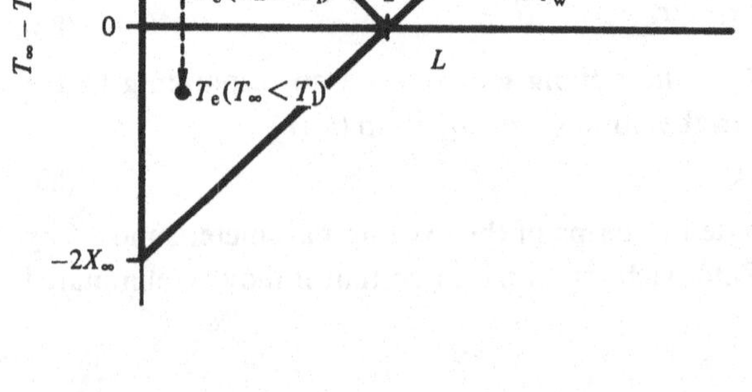

the corresponding lines may or may not be crossed (depending on T_b, β, and $\hat{\theta}$ in one case). The arrows show the direction of increasing M and follow from the property

$$\mathrm{d}T_s/\mathrm{d}M \gtrless 0 \quad \text{accordingly as } T_s \gtrless T_\infty + L - 2 \tag{91}$$

of equation (90). (The condition $T_s \leqslant T_b$, i.e. $Y_s \leqslant 1$, is always satisfied; the boiling temperature plays no role.) Our knowledge of the response curves for fixed points in the parameter plane now suggests that the location of the point T_w determines whether the fuel-drop response is S-shaped or monotonic. Proof lies in the inequalities $\mathrm{d}T_*/\mathrm{d}M \gtrless 0$, which continue to hold for the functions (73*b*) and (83*b*) when T_s varies with M according to the equation (90).

In short, qualitative results for fuel-drop burning can be deduced from the analysis of general spherical flames given in previous sections. Quantitative results, however, such as the graphs in Figures 6 and 7 cannot be carried over: the variations of T_s, while they are only $O(\theta^{-1})$ on the upper and lower branches, will modify the ignition and extinction values (Normandia & Ludford 1980).

These remarks apply to steady burning; the unsteady analysis of section 3 is not directly applicable since no account is taken of conditions in the interior of the drop. Some of the heat flux at the surface now accounts for changes in the interior temperature, so that there is a connection between it and the surface temperature. As a consequence, L is the sum of the latent heat of vaporization and a complicated integral term depending on T_s. Such difficulties do not arise for the counterflow flame.

8 Further results on stability

For combustion problems, such as those considered here and in earlier chapters, at least three time scales arise. Two have already been identified: $t = O(1)$, for which time derivatives in the main body of the combustion field are comparable to space derivatives; and the slow-time $t = O(\theta)$, encountered in both premixed and diffusion flames when the Lewis number is different from 1. On either time scale the flame-sheet structure is quasi-steady. The third time scale, $t = O(\theta^{-2})$, arises when time derivatives inside the flame sheet are comparable to the spatial derivatives there. Such rapid changes do not affect the combustion field on either side to leading order. All three scales can be expected to play a role in stability; here we shall examine the role of the fast time

$$\tau = \theta^2 t \tag{92}$$

in the stability of the spherical diffusion flame near the extinction point, following Taliaferro, Buckmaster, & Nachman (1981).

The unsteady counterpart of the steady-state structure equation is

$$\partial T/\partial\tau - \partial^2 T/\partial\xi^2 = D(T^2 - \xi^2)\,e^{T+\gamma\xi}, \tag{93}$$

the boundary conditions (70) being unchanged. Here we have dropped all sub- and superscripts and allowed τ to absorb various parameters, including the value of the density at the flame sheet. Infinitesimal disturbances of the steady state T_0 therefore satisfy

$$d^2T_1/d\xi^2 + [V(\xi) - \lambda]T_1 = 0 \quad \text{with } V = D(T_0^2 - \xi^2 + 2T_0)\,e^{T_0+\gamma\xi}, \tag{94}$$

$$T_1 \to 0 \quad \text{as } \xi \to \pm\infty, \tag{95}$$

where a time dependence $e^{\lambda\tau}$ has been assumed. (The use of λ as growth coefficient, rather than the $\tilde{\alpha}$ of Chapter 11, not only conforms with standard eigenvalue notation but also emphasizes that τ is different, being a fast time here.) This singular eigenvalue problem is of a type familiar in quantum mechanics, V playing the role of the potential. Stability at a point on the response curve is determined by the sign of λ (which is necessarily real) for the corresponding potential.

The steady-state problem determines the local response, i.e. M versus D, the latter representing the Damköhler number. Asymptotic analysis as $D \to \infty$ leads to the Burke–Schumann limit and the top of the middle branch of the full response, where there are three possibilities accordingly as $\gamma \gtreqless 0$ (i.e. $T_\infty \lesseqgtr T_a - (1 + X_\infty)$; cf. section 5 remembering the different meaning of γ). The Burke–Schumann limit is found to be stable, while all three possibilities are unstable.

The asymptotics for the Burke–Schumann limit consist in setting

$$\bar{T} = T/\varepsilon, \quad \bar{\xi} = \xi/\varepsilon, \quad \bar{\tau} = \tau/\varepsilon^2, \quad \bar{\lambda} = \varepsilon^2\lambda \quad \text{with } \varepsilon = D^{-\frac{1}{3}} \tag{96}$$

to obtain

$$\partial\bar{T}/\partial\bar{\tau} + \partial^2\bar{T}/\partial\bar{\xi}^2 = \bar{T}^2 - \bar{\xi}^2 \tag{97}$$

to leading order. The steady-state equation should be compared with the limit equation (27); the only property of its solution that we need is $\bar{T}_0 < -|\bar{\xi}|$, to assure that T_0 is everywhere negative in the perturbation equation

$$d^2\bar{T}_1/d\bar{\xi}^2 + (2\bar{T}_0 - \bar{\lambda})\bar{T}_1 = 0. \tag{98}$$

Standard treatments (Titchmarsh 1946, p. 107) then show that the spectrum of λ, which is both continuous and discrete, is negative. Thus, the Burke–Schumann limit is stable to the class of disturbances considered here.

Of the three possibilities corresponding to $\gamma \gtreqless 0$, only the proof of instability for $\gamma = 0$ is simple. While this is a very special case in the present context, the details are worth giving because the problem is also that for the whole middle branch when X_* and Y_* are nonzero (see the end of section 5). The asymptotics consist in setting

$$\bar{T}=T-\ln\varepsilon \quad \text{with } (\ln\varepsilon)^2\varepsilon=D^{-1} \tag{99}$$

to obtain

$$\partial\bar{T}/\partial\tau-\partial^2\bar{T}/\partial\xi^2=e^{\bar{T}} \tag{100}$$

to leading order, an equation that also arises in the induction phase of a thermal explosion (section 12.4).

The steady state is described by

$$\bar{T}_0=-\ln\{2\cosh^2\tfrac{1}{2}(\xi-\xi_m)\} \tag{101}$$

where ξ_m is an integration constant giving the location of the maximum $\bar{T}_0$; the associated perturbation satisfies

$$d^2\bar{T}_1/d\xi^2+\{\tfrac{1}{2}\operatorname{sech}^2[\tfrac{1}{2}(\xi-\xi_m)]-\lambda\}\bar{T}_1=0, \tag{102}$$

an equation also encountered by Matalon & Ludford (1980) in a stability study of chambered diffusion flames. They cite Morse & Feshbach (1953, p. 768), who show that for $\lambda=\frac{1}{4}$ there is an eigenfunction

$$\bar{T}_1=\operatorname{sech}\tfrac{1}{2}(\xi-\xi_m). \tag{103}$$

Thus instability is assured.

Sufficiently far to the right, then, the eigenvalue problem (94), (95) predicts stability for points on the upper branch of the local response curve and instability on the lower. Between these two extremes lies a point of neutral stability; Taliaferro, Buckmaster, & Nachman (1981) show analytically that it lies exactly at the steady-state extinction point, a conclusion that was foreshadowed by the numerical work of Peters (1978).

A similar stability analysis, with the same conclusion, was made by Matalon & Ludford (1980). Their intrinsically more complicated eigenvalue problem ultimately forced them to use numerical methods, so that a mathematical proof of their results is still needed.

7

CYLINDRICAL AND SPHERICAL PREMIXED FLAMES

1 Cylindrical flames

Although it has not been studied to the same extent as the plane premixed flame, the cylindrical premixed flame is (in principle) almost as easy to produce. The reacting mixture is supplied through the surface of a circular cylinder and is induced to flow radially by means of end plates sufficiently close together. The flame then forms a coaxial cylinder that can be observed through the end plates, which should be transparent and good thermal insulators.

Analytically the cylindrical flame lies between the plane and spherical flames. The structure of its reaction zone is the same as that of the plane flame, with temperature constant beyond, so that there is none of the curvature effect of the spherical flame. Like the spherical flame it does not exhibit the cold-boundary difficulty: the mixture must be introduced at a finite radius, which can be so small that a line source is effectively formed. In their attempt to treat curved flames Spalding & Jain (1959) use plane-flame analysis on the spherical flame, where it is never valid, and neglect the cylindrical flame, where it is always valid. The results, however, are qualitatively correct (Ludford 1976).

Sections 2 and 3 show how cylindrical geometry modifies a premixed flame (Ludford 1980). For simplicity we shall consider a single reactant undergoing a first-order decomposition and take $\mathscr{L}=1$. Spherical premixed flames, which have many similarities to spherical diffusion flames, will be treated in sections 4–6.

2 Planar character

We shall consider only the basic features of cylindrical flames, but the discussion can easily be extended to other questions answered for plane flames in Chapters 2 and 3. When making comparisons with the latter, bear in mind that the nondimensionalization is based here on the radius, a, of the supply cylinder. As a consequence, the representative mass flux (at the supply cylinder) is dimensionless and appears explicitly, rather than dimensional and hidden in the units, as it is for the plane flame.

If v is the radial velocity, the equation of continuity shows $r\rho v$ is a constant. When M is the mass flux at the supply cylinder we may, therefore, write

$$r\rho v = M \quad \text{or } \rho v = M/r, \tag{1}$$

the latter giving the mass flux at every other radius r. Once the temperature T has been determined, the density $\rho = 1/T$ and velocity $v = MT/r$ follow immediately, while the small variations in pressure about its constant level can be calculated from the momentum equation of the mixture. There remain then the energy and reactant-species equations

$$\mathfrak{L}(T) = -\mathfrak{L}(Y) = -DY\mathrm{e}^{-\theta/T}, \tag{2}$$

where

$$\mathfrak{L} \equiv \frac{1}{r}\frac{\mathrm{d}}{\mathrm{d}r}\left(r\frac{\mathrm{d}}{\mathrm{d}r}\right) - \frac{M}{r}\frac{\mathrm{d}}{\mathrm{d}r} \tag{3}$$

in the cylindrical geometry.

Part of the relationship of cylindrical flames to plane and spherical ones stems from their convection–diffusion operators, which may all be written in the form

$$\mathfrak{L}_\nu \equiv \frac{1}{r^\nu}\frac{\partial}{\partial r}\left(r^\nu\frac{\partial}{\partial r}\right) - \frac{M}{r^\nu}\frac{\partial}{\partial r}; \tag{4}$$

here $\nu = 0$ (plane, when $r = x$), 1 (cylindrical), or 2 (spherical). The homogeneous equation

$$\mathfrak{L}_\nu(T) = 0, \tag{5}$$

governing the temperature field behind the flame sheet, has the general solution

$$T = A + B\,\mathrm{e}^{Mr}, \quad A + Br^M, \quad A + B\,\mathrm{e}^{-M/r} \tag{6}$$

for $\nu = 0, 1, 2$, respectively. The boundedness of T forces the choice $B = 0$ in the first two cases but not the third, where T_∞ becomes an assignable parameter in addition to any others. Thus a common feature of cylindrical and plane flames is that T_∞ may not be assigned. It should also be noted that the uniformity of the far field implied by the cylindrical operator precludes the possibility of a flux of oxidant from infinity so that, as in the plane case, such a diffusion flame cannot exist.

Since the temperature does not have to remain constant behind a spherical flame sheet, the possibility arises that it may actually decrease; in which case the equilibrium argument of section 2.4 cannot be applied and there is no contradiction in supposing that the reactant is not completely consumed at the flame sheet. Partial-burning flames will be needed in section 6; the corresponding diffusion flames were introduced in section 6.5 without comment.

First note that the Shvab–Zeldovich variable satisfies

$$\mathfrak{L}(H)=0 \quad \text{for } 1<r<\infty. \tag{7}$$

Since the only solutions that remain bounded at infinity are constants, we have

$$H \equiv T+Y=T_s+Y_s \tag{8}$$

where s denotes conditions at the supply $r=1$. The fact that the T, Y-relation is identical to that for plane flames (section 2.2) is responsible for the structures of the reaction zones being the same.

The asymptotic analysis of the equations (2) proceeds as for a plane flame. The temperature beyond the flame sheet must be constant, i.e.

$$T=T_b \equiv T_s+Y_s, \quad Y=0 \quad \text{for } r>r_*, \tag{9}$$

while up to the flame sheet the reaction is frozen, i.e. $\mathfrak{L}(T)=0$, so that

$$T=T_s+L(r^M-1) \quad \text{for } 1<r<r_*. \tag{10}$$

Here $L=M^{-1}T_s'$ is the heat conducted back into the supply per unit mass of mixture, T_s' being the temperature gradient at the supply. The two formulas (9*a*) and (10) give the same values at

$$r_*=[1+(T_b-T_s)/L]^{1/M}; \tag{11}$$

as expected, the stand-off distance for a plane flame is recovered as the radius a of the supply cylinder tends to infinity when due attention is paid to a mass-flux unit proportional to $1/a$. Consistency requires $r_*>1$, i.e. $L>0$, which means the supply must be a conductive heat sink.

As usual, the interior of the flame sheet is investigated by writing

$$\xi=\theta(r-r_*), \quad T_0=T_b. \tag{12}$$

The structure is thereby found to satisfy the equation

$$d^2\phi/d\xi^2=\tilde{D}\phi\, e^{-\phi} \quad \text{with } \tilde{D}=e^{-\theta/T_b}D/\theta^2, \quad \phi=(1/T)_1=-T_1/T_b^2 \tag{13}$$

as well as the boundary conditions

$$\phi=-MJ_s\xi/T_b^2 r_*+o(1) \quad \text{as } \xi\to-\infty, \quad \phi=o(1) \quad \text{as } \xi\to+\infty, \tag{14}$$

which come from matching with the expressions (10) and (9*a*) outside the flame sheet. Here

$$J_s = Y_s - M^{-1} Y_s' \tag{15}$$

is the reactant flux fraction $Y - M^{-1} r Y'$ at the supply. Exactly the same problem is obtained for a plane flame except that the coefficient $MJ_s/T_b^2 r_*$ in the condition (14*a*) is replaced by J_s/T_b^2; thus the cylindrical reaction zone is seen to be locally plane. For spherical flames, the T, Y-relation (8) is changed, so that the reaction zone is quite different from its plane counterpart.

From the solution (2.21) of the corresponding plane-flame problem we deduce that

$$\tilde{D} = (MJ_s/r_*)^2/2T_b^2, \quad \text{i.e. } D = [J_s^2 \theta^2 \, e^{\theta/T_b}/2T_b^4 r_*^2] M^2, \tag{16}$$

which is the required M, D-relation. (The presence of r_* in this formula means that it depends on all supply parameters, unlike the plane flame.) If L, T_s and J_s are fixed then so is $T_b(=J_s + T_s - L)$ and this relation has the general shape of the parabola (2.22) obtained for the plane flame, despite the presence of the extra factor $1/r_*^2$. The parabola is useful for determining the speed (i.e. M) with which a plane flame propagates into a fresh mixture at given pressure (i.e. D), but there is no equivalent use here. On the other hand, for the set-up envisaged in the opening paragraph of section 1 both M and D are prescribed (along with T_s and J_s), and the formula determines the final temperature.

3 Near-surface, surface, and remote flames

For the solution to be valid the parameter values must be such that r_*, as given by the formula (11), lies between 1 and ∞. As for the plane flame, three limiting cases arise, two of which are essentially the same as in section 2.6. The third leads to an interesting new phenomenon.

When M and D become large, with all other parameters (i.e. L, T_s, J_s) held fixed, the flame approaches the surface and DM^{-2} is constant. An intense convective–diffusive zone of thickness $O(M^{-1})$ forms near the surface, bounded by a reaction zone of thickness $O(\theta^{-1}M^{-1})$.

By contrast, a true surface flame can be produced for any M by adjusting D to make $T_b \rightarrow T_s$. Similarly, remote flames can be produced by making $T_b \rightarrow J_s + T_s$ (i.e. $L \rightarrow 0$). At both extremes the preceding analysis becomes invalid: either the boundary intrudes into the reaction zone and there is no frozen region between them or the isothermal limit of (10) is not uniformly valid in the unbounded frozen region. In either case the asymptotics must be reworked.

The analysis of the cylindrical surface flame is identical to that for the plane case provided x is changed to $r-1$. We conclude that $\tilde{D}$ will change from $J_s^2M^2/2T_b^2$, the value (16a) when $r_*=1$, to ∞ as the temperature difference T_b-T_s, measured on the θ^{-1}-scale, decreases from ∞ to 0.

Unlike the plane case, the asymptotic analysis of the remote cylindrical flame can be made uniformly valid in L. When $L\to 0$ as $\theta\to\infty$, the reactant flux MJ_s/r_* at the flame tends to zero like $L^{1/M}$ (as $r_*\to\infty$) and the condition (14a) shows the need of a different scale for the structure. Setting

$$\xi = L^{1/M}\theta(r-r_*) \tag{17}$$

gives the new condition

$$\phi = -MJ_s^{1-1/M}\xi/T_b^2 + o(1) \quad \text{as } \xi\to-\infty; \tag{18}$$

and then the corresponding change

$$\tilde{D} = \theta^{-2}L^{-2/M}\,\mathrm{e}^{-\theta/T_b}D \tag{19}$$

must be made to keep the structure equation balanced. Since we end with the same problem, D continues to be given by the formula (16b), in which r_* is $(J_s/L)^{1/M}$ for L small. Clearly, as L vanishes, $D\to 0$ even on scales small compared to $\theta^2\,\mathrm{e}^{\theta/T_b}$.

The most interesting feature is the spreading of the zone in which there is chemical activity as M is decreased. The transformation (17) implies that its thickness is $O(\theta^{-1}L^{-1/M})$, so that at the value $M=-\ln L/\ln\theta$ it ceases to be a sheet and for smaller values has infinite extent (remember $r_*=O(L^{-1/M})$ is larger still).

4 Spherical flames; D-asymptotics

If general values of ν_1 and ν_2 had been retained in Chapter 6, most of the remaining results in this chapter could now be obtained by setting $\nu_1=1$ and $\nu_2=0$, i.e. $\alpha_1=-1$ and $\alpha_2=0$ (see Normandia & Ludford 1980). The exceptions concern strong burning with $M\to\infty$. The analysis, however, need only be sketched, with omitted details being similar to those in Chapter 6.

We shall suppose that

$$T_s \quad \text{and } L=M^{-1}T_s'>0 \tag{20}$$

are prescribed at the supply and that

$$J_s \equiv Y_s - M^{-1}Y_s'/\mathscr{L} = 1 \tag{21}$$

there, so that it is purely a source of reactant. Prescribing T_∞ and requiring

$$Y_\infty = 0 \tag{22}$$

then places five boundary conditions on the fifth-order system (6.1), (6.2a) for v, T, and $Y_1=Y$.

Consider steady states and take $\mathscr{L}=1$. Then the mixture velocity is given by equation (6.10) and the single Shvab–Zeldovich variable is

$$H \equiv T + Y = T_\infty + (T_a - T_\infty)(1 - e^{-M/r}), \tag{23}$$

where

$$T_a = T_s - L + 1 \tag{24}$$

is the adiabatic flame temperature. (See the discussion of the definition (6.14).) The problem again reduces to one for T alone, namely

$$\mathfrak{L}(T) = -DY\,e^{-\theta/T} \quad \text{for } 1 < r < \infty \quad \text{with } T_s, L, T_\infty \text{ prescribed.} \tag{25}$$

A comprehensive treatment in the limit $\theta \to \infty$ has been given by Kapila, Ludford, & Buckmaster (1975) and Ludford, Yannitell, & Buckmaster (1976*a*, *b*); the most important features were simultaneously found by Liñán (1975).

We first turn to the limits $D \to 0, \infty$, which had previously been considered by Fendell (1969). Frozen combustion is described by the same formulas as for the diffusion flame, provided the new definition

$$\varepsilon^2 = D\,e^{-\theta/T_\infty} \tag{26}$$

is made, and the 2 in (6.19) is replaced by 1; in particular, the burning rate (6.16*b*) holds. The corresponding phenomenon in the plane case is the remote flame of section 2.6 (see also the cylindrical case in section 3). Equilibrium limits ($D \to \infty$) require

$$Y = 0 \tag{27}$$

outside the reaction zone, a state that cannot hold next to the source: the zone must, therefore, lie at the surface of the source. There are two possibilities, accordingly as M is finite or not; these are the counterparts of the Buckmaster and Burke–Schumann limits (section 6.2) for the diffusion flame, respectively, as we shall see. (The same terms will be used.) The corresponding phenomena in the plane case are the surface and near-surface flames of section 2.6 (see also the cylindrical case in section 3).

If general stoichiometric coefficients had been retained, the formulas (6.23), (6.24) would have had $-X_\infty/2\alpha_2$ in place of X_∞. Then $M_e \to \infty$ and $r_* \to 1$ as $\alpha_2 \to 0$: the Burke–Schumann solution tends to a surface flame with an indefinitely large burning rate as the oxidant becomes inactive. The limit behavior cannot be deduced from generalizations of section 6.2, but must be determined *ab initio*.

For that purpose, note that the Shvab–Zeldovich relation (23) gives

$$T = T_a \quad \text{for } r > 1 \tag{28}$$

with an adjustment to T_∞ at distances $O(M)$. The temperature must, therefore, vary to leading order in the reaction zone; that also follows from the boundary condition $M^{-1}T_s'=L$, which, in addition, points to

$$x=M(r-1) \tag{29}$$

as the appropriate coordinate there. We conclude that the reaction zone is governed by

$$(\mathrm{d}^2/\mathrm{d}x^2-\mathrm{d}/\mathrm{d}x)T=-DM^{-2}Y\,\mathrm{e}^{-\theta/T},\quad T+Y=T_a, \tag{30}$$

$$T=T_s,\quad \mathrm{d}T/\mathrm{d}x=L\quad \text{for } x=0,\quad T\to T_a\quad \text{as } x\to\infty, \tag{31}$$

i.e. the plane problem for a vaporizing liquid considered in section 2.3. It follows that the result

$$DM^{-2}=\mathscr{D}(T_s, L, \theta), \tag{32}$$

where $\mathscr{D}$ is the eigenvalue introduced there, determines how M tends to infinity with D.

The determination of $\mathscr{D}$ for finite θ is a numerical question on which much effort has been spent in the past. (Its asymptotic form (2.22) as $\theta\to\infty$ was the goal of section 2.4.) For the diffusion flame the Burke–Schumann solution can be written in simple analytical terms for any θ, but here we end with the corresponding plane problem. The asymptotic behavior of M, the counterpart of the formula (6.23), is a numerical question here. One simple feature does remain, however: the plane solution only exists for

$$L<1 \tag{33}$$

(the counterpart of the condition (6.29)), as was shown in section 2.4.

We turn now to the Buckmaster limit. The burning rate remains finite, the temperature is constant to leading order in the reaction zone, and the Shvab–Zeldovich relation (23) gives

$$M_s=\ln[1+(T_\infty-T_s)/(L-1)] \tag{34}$$

in place of the result (6.30). The expansions (6.32) and (6.33) are replaced by

$$M=M_s[1+D_*^{-\frac{1}{2}}/(1-L)+\ldots],$$

$$T=T_s+D_*^{-\frac{1}{2}}M_s[1-\mathrm{e}^{-\xi}+(L-1)\xi]+\ldots \tag{35}$$

if the new definition

$$D_*=D\,\mathrm{e}^{-\theta/T_*}\quad \text{with } T_*=T_s \tag{36}$$

is made. The changes amount to 2 being replaced by 1 to account for the different (nondimensional) heat release at the flame. Restrictions on the parameters come from M_s, which must be positive if the (small) mass fraction of reactant is to be positive:

$$T_\infty \gtrless T_s\quad \text{accordingly as } L\gtrless 1. \tag{37}$$

The conditions for the existence of the three limits (frozen, Burke–Schumann, and Buckmaster) determine three regions in the parameter plane (Figure 1) in each of which two of the limits hold. In the fourth region there is no solution for any values of D and θ: the Shvab–Zeldovich relation (23) would require Y_s to be negative there.

5 Ignition and extinction

We turn now to activation-energy asymptotics, starting with the nearly adiabatic flame for $\mathscr{L} \neq 1$, i.e. the analog of section 6.3. Again, different values of k in the defining condition (6.37) will be generated by varying T_s, keeping T_∞ and L fixed with

$$L < 1. \tag{38}$$

The frozen and infinite-M limits then exist.

Slow variations are governed by the analog

$$M(T_a - H_*) = -r^2\left(\frac{\partial T}{\partial r} + \mathscr{L}^{-1}\frac{\partial Y}{\partial r}\right)\bigg|_{r_*+0} + \theta^{-1}\int_1^{r^*} r^2 \frac{\partial}{\partial \tau}[\rho(Y - H_*)]\,\mathrm{d}r \tag{39}$$

of the result (6.40). To use this formula the combustion field must be

1 Parameter plane for spherical premixed flame, showing form of response in the various regions. Cf. Figure 6.3.

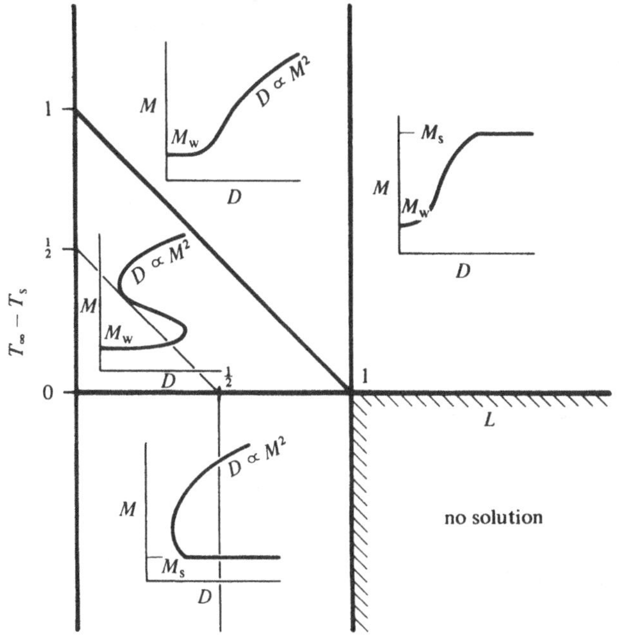

determined to appropriate accuracy. Noting that, as before, it is frozen ahead of the flame and in equilibrium behind we find

$$T=\begin{cases} T_s-L+L\,e^{M(1-1/r)} \\ T_\infty+\theta^{-1}T_{1*}(1-e^{-M/r})/(1-e^{-M/r_*}), \end{cases} \quad Y=\begin{cases} 1-e^{\mathscr{L}M(1/r_*-1/r)} \\ 0 \end{cases}$$

$$\text{for } r \lessgtr r_*, \quad (40)$$

where

$$r_*=M/(M+\ln L), \quad (41)$$

as replacements for the results (6.41), (6.42), (6.43), and (6.44). Substitution in the basic equation (39) yields the same results (6.46), where now

$$b=\int_0^{\ln(1/L)} \frac{L\,e^l-(L\,e^l)^{\mathscr{L}}}{T_s-L+L\,e^l}\,\frac{d}{dl}\left[\frac{l}{(M-l)^4}\right]dl. \quad (42)$$

As before, a second relation between T_{1*} and M comes from the structure, the analysis of which is very similar to that of the diffusion flame in section 6.3. We find

$$T_{1*}=T_\infty^2 \ln(M^2\theta^2\,e^{\theta/T_\infty}/2D\mathscr{L}r_*^4T_\infty^4), \quad (43)$$

so that elimination of T_{1*} yields the basic equation

$$bM\,dM/d\tau=(1-e^{-M/r_*})^{-1}T_\infty^2 \ln(M^2\theta^2\,e^{\theta/T_\infty}/2D\mathscr{L}r_*^4T_\infty^4)-k, \quad (44)$$

where $r_*(M)$ is the function (41). The steady-state response for $k=0$ has a simple algebraic form in which D is proportional to $(M+\ln L)^4/M^2$. This result was essentially found by Fendell (1972); otherwise these results have not been previously published.

Steady-state responses are shown in Figure 2. They differ from those

2 Steady-state responses when the combustion is nearly adiabatic (T_a-T_∞ small), as measured by k in the distinguished limit (6.37); drawn for $L=0.5$.

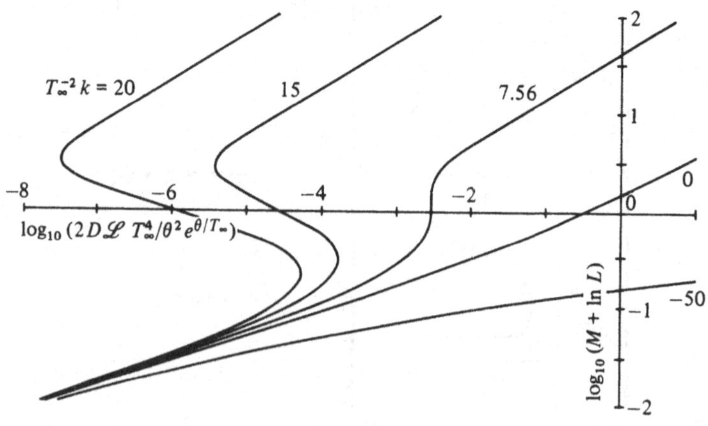

for the diffusion flame (Figure 6.4) because of the factor X_* there, which vanishes at the Burke–Schumann limit; here M tends to infinity (like $D^{\ddagger}$) on all curves. For large values of k the turning points are

$$M \sim \ln(1/L) + 4T_\infty^2/k, \quad D \sim (4/e)^4 T_\infty^4 \theta^2 \, e^{\theta/T_\infty}/2(\ln L)^2 \mathscr{L} k^4 \tag{45}$$

for ignition, and

$$M \sim \ln(k/LT_\infty^2) + \ln\ln(k/LT_\infty^2) - \ln 2,$$
$$D \sim (\theta^2 \, e^{\theta/T_\infty}/2\mathscr{L}T_\infty^4)\, e^{-k/T_\infty^2}[M^2 + (2 + 4\ln L)M + (2 + 4\ln L + 6\ln^2 L)] \tag{46}$$

for extinction. Conditions for instability, predicted by the full equation (44) are the same as for the fuel drop: for $\mathscr{L} < 1$ the middle branch of an S-response is unstable; for $\mathscr{L} > 1$ the other two branches, as well as the whole of a monotonic curve, are unstable.

These results and those in the previous section suggest that an S-response will be obtained in the triangular region

$$T_s < T_\infty < T_a \tag{47}$$

of the parameter plane (Figure 1). The first inequality ensures that a weak-burning limit exists, while the second comes from the existence of S-shaped near-adiabatic responses for k sufficiently large. As for the diffusion flame, the lower branch (including the ignition point) will be obtained for values of M within $O(\theta^{-1})$ of its frozen value. A similar analysis cannot be made of the upper branch, however, because the Burke–Schumann limit is now infinite. The required treatment has been given by Kapila, Ludford, & Buckmaster (1975).

Again take $\mathscr{L} = 1$. At any point on the lower branch the temperature is still given by the formulas (6.58). Moreover, at large distances the equation (6.60) and boundary conditions (6.61) still hold, provided only that the replacement

$$D_w = D\theta^2 \, e^{-\theta/T_\infty} \tag{48}$$

is made. The lower branch is, therefore, again given by Figure 6.6, which may be used to read off the ignition point.

For large values of M the combustion field has the limiting structure (30), (31). In the limit $\theta \to \infty$ it shows a flame sheet located a distance $O(M^{-1})$ from the surface, with equilibrium beyond. (Since the decomposition zone has thickness $O(\theta^{-1})$ on the x-scale, it stands clear of the surface however large M is.) As M is decreased through large values, the first effects of finite M are felt through changes in the flame temperature, with extinction occurring when this has decreased an $O(\theta^{-1})$ amount below

T_a. Behind the flame the temperature varies according to equation (23) with $Y=0$ (forced by the flame-sheet analysis) and r replaced by $1+xM^{-1}$. As a consequence, the flame temperature is given by the formula

$$T_* = T_a + (T_\infty - T_a)\,e^{-M}/L \tag{49}$$

but temperature gradients behind the flame are $o(1)$. We have the plane problem of section 2.4, therefore, and the result (2.22), with $J_s=1$ and T_b replaced by this T_*, gives the response curve

$$\tilde{D} = M^2 \exp(\tilde{\theta}\, e^{-M})$$
$$\text{with } \tilde{D} = [2T_a^4\theta^{-2}\, e^{-\theta/T_a}]D, \quad \tilde{\theta} = [(T_a - T_\infty)/T_a^2 L]\theta \tag{50}$$

to leading order.

The response has a turning point at $M = M_c(\tilde{\theta})$ where

$$e^{M_c} = (\tfrac{1}{2}\tilde{\theta})M_c, \quad \text{i.e. } M_c = \ln(\tfrac{1}{2}\tilde{\theta}) + \ln\ln(\tfrac{1}{2}\tilde{\theta}) + o(1). \tag{51}$$

(Note that this result only makes sense if $\tilde{\theta} \to +\infty$, i.e. for $T_\infty < T_a$, as supposed.) Near the turning point we write

$$M = M_c + m \quad \text{to obtain } \tilde{D} = M_c^2 + 2M_c(m + e^{-m}) + O(1) \tag{52}$$

when $m = O(1)$. The function $m + e^{-m}$ is graphed in Figure 3. The results (51*b*) and (52*b*), with $m=0$, are explicit asymptotic formulas for the location of the extinction point.

6 Other aspects of responses

We shall now briefly describe the middle branch of the S-response, the two types of monotonic response (depending on the nature of the equilibrium limit), and the C-response. A detailed treatment of all these has been given by Ludford, Yannitell, & Buckmaster (1976*a*, *b*), in par-

3 Universal steady-state response near extinction for unit Lewis numbers; M_c has the asymptotic form (51*b*).

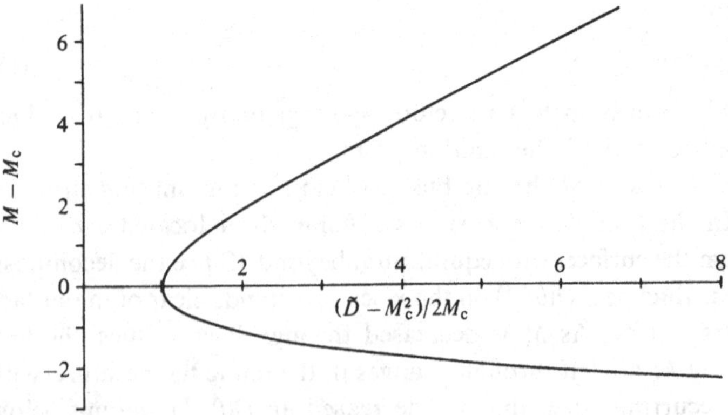

ticular for the C-response where no analogous treatment has been given for spherical diffusion flames (see section 6.6).

The flame temperature on the middle branch, which ranges from T_∞ to T_a, is always the maximum temperature in the combustion field, so that the reaction is frozen on either side of the flame sheet. While Y could still vanish to leading order beyond the flame, we shall first suppose that is not the case. As in section 6.5, we shall consider M to be specified and anticipate the equal conduction of heat to the two sides of the flame. The temperature profiles (6.72) still hold and lead to the results (6.73), and the structure equation (6.74) still governs, provided the new definition

$$D_m = D^0 Y_* \theta^{-1} e^{-\theta/T_*} \tag{53}$$

is used. As for the diffusion flame, details of the determination of D_m are not needed to draw the conclusion $dD^0/dM < 0$.

The bound (6.77), with 2 replaced by 1, shows that the solution now holds for

$$T_\infty < T_a - \tfrac{1}{2}; \tag{54}$$

there are clearly no parameter values for which it covers the whole range of M, since the range is infinite here. Over the rest of the middle branch, i.e. the whole of it when the inequality (54) is not satisfied, Y_* is zero and the complete burning structure of the monotonic response is needed. Examples of computed S-responses are shown in Figure 4.

We turn now to the monotonic responses, for which the inequality (6.80) holds and the temperature increases beyond the flame sheet. The formulas (6.81), (6.82), and (6.83) are still valid, so that dT_*/dM is again negative and D is again an increasing function of M to leading order. Once more the monotonicity of the response is established without discussing the flame-sheet structure. The latter is, of course, needed to construct the curves (Figure 5).

As M increases from the frozen value, r_* decreases from ∞; there are two alternatives for M as $r_* \to 1$. If L is less than 1, so that the limit (34) does not exist, then M increases without bound with T_* approaching T_a as the flame sheet approaches the surface. The frozen limit is joined to the Burke–Schumann limit of infinitely rapid burning. If L is greater than 1, so that the limit (34) does exist, then T_* approaches T_s as the flame sheet settles down on the supply sphere. The frozen limit is joined to the Buckmaster limit; the approach is similar to that for diffusion flames (section 6.6).

Finally, we come to supplies that are hotter than the ambient atmosphere, i.e. inequality (6.86). The frozen limit does not exist but both equilibrium limits do. The response takes the shape of a C whose upper part always corresponds to complete burning, but whose lower part may correspond to incomplete burning. The division (if it occurs) lies below the turning point of the C, which corresponds to both ignition and extinction

4 S-response for $\theta=50$: (*a*) $T_\infty=1$, $T_s=0.75$, $L=0.5$; (*b*) $T_\infty=1$, $T_s=0.98$, $L=0.02$. In (*b*) the region near $M=3.87$ corresponds to a transition from one distinguished limit to another; ---- denotes faring in of the central portion. In each case D is eventually proportional to M^2.

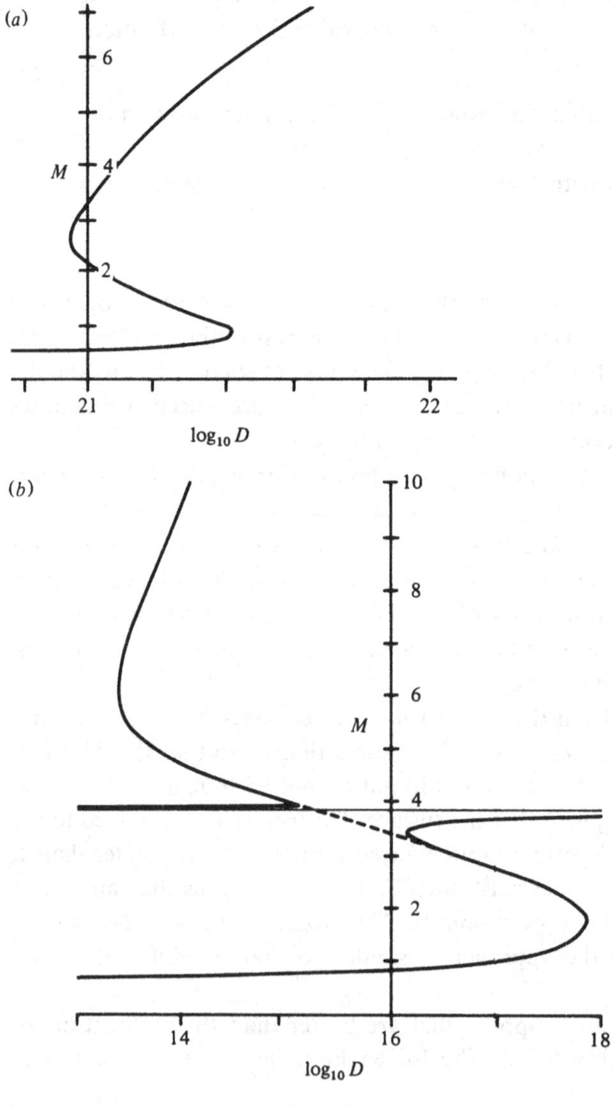

conditions, so that these conditions are determined by the formulas (51*b*), (52*b*). Examples of such a response are shown in Figure 6.

The application of the results in this chapter to the burning of 'monopropellant' drops follows the same lines as section 6.7 (Normandia & Ludford 1980). The Damköhler number is again varied via the radius of the drop and not the pressure. As a consequence, the sudden evaporation found in section 4.7 does not occur: lowering the Damköhler number there eventually brought the pressure down to the level at which the liquid boiled. Even when the Damköhler number is varied via the pressure

5 Monotonic response for $\theta = 50$: (*a*) $T_\infty = 1$, $T_s = 0.5$, $L = 2$, showing plateau $M = 0.41$ corresponding to surface flame; (*b*) $T_\infty = 1$, $T_s = 0.5$, $L = 0.75$, where D is eventually proportional to M^2.

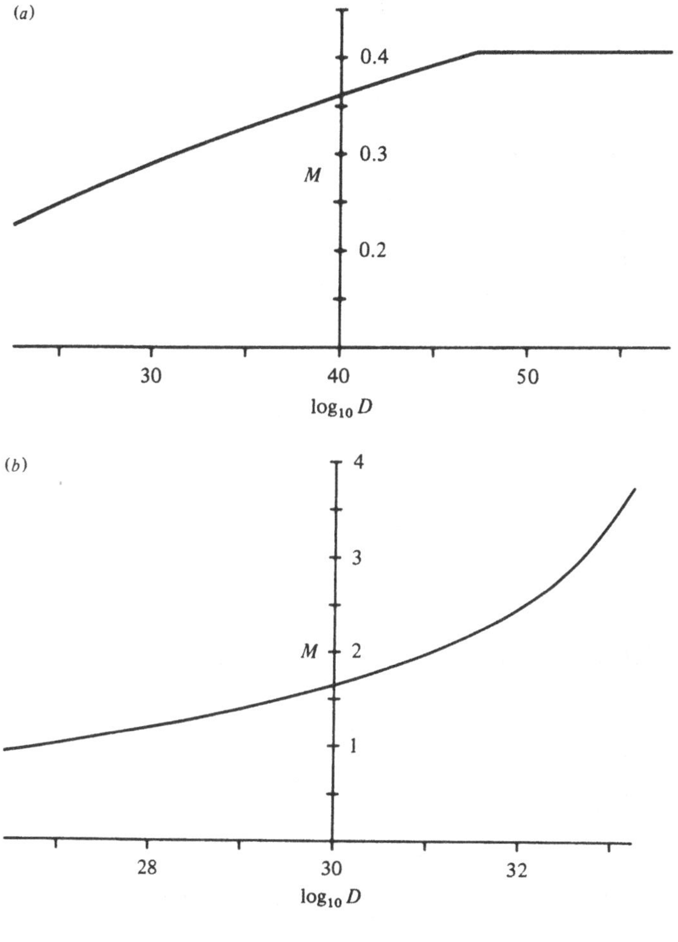

6 C-response for $\theta=50$; (*a*) $T_\infty=1$, $T_s=1.25$, $L=0.75$; (*b*) $T_\infty=1$, $T_s=1.25$, $L=0.25$. Both show plateaus, at $M=0.69$ and 0.29 respectively, and D is eventually proportional to M^2. In (*b*) the region near $M=1.39$ corresponds to a transition from one distinguished limit to another; ---- denotes faring in the central portion.

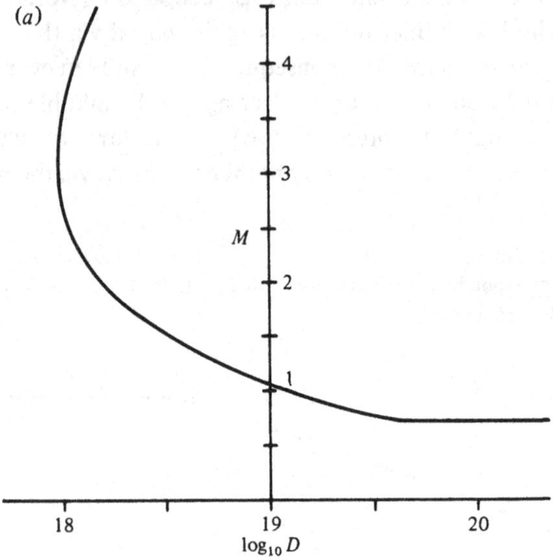

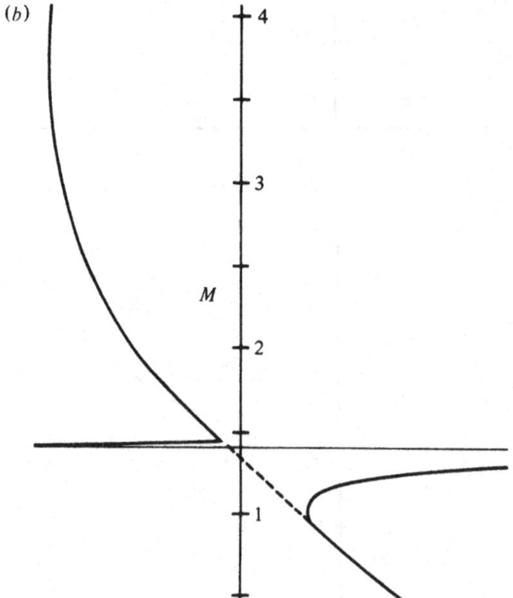

there is no sudden evaporation of the drop (Janssen 1981) because its temperature is not maintained, as is that at the remote end of a linear condensate.

While we have restricted our attention to inert atmospheres, the problem of the ambient atmosphere being an oxidizing one for the product of the decomposition has also been investigated (McConnaughey 1982). There is now a diffusion flame in addition to the premixed one, so that features from Chapter 6 are also involved. The Damköhler-number asymptotics had already been considered by Fendell (1969) and, more completely, by Buckmaster, Kapila, & Ludford (1978), but much remained to be done.

8

MULTIDIMENSIONAL THEORY OF PREMIXED FLAMES

1 The flame as a hydrodynamic discontinuity

The plane premixed flame discussed in Chapters 2 and 3 is an idealization seldom approximated, since in practice the flame is usually curved. A Bunsen burner flame is inherently so; but even under circumstances carefully chosen to nurture a plane state, instabilities can precipitate a multidimensional structure. Such flames have been extensively studied using a hydrodynamical approach (Markstein 1964, p. 7); a brief description of it will provide an introduction to our subject.

On a scale that is large compared to its nominal thickness $\lambda/c_p M_r$, the flame is simply a surface across which there are jumps in temperature and density subject to Charles's law (as appropriate for an isobaric process). Deformation of the surface from a plane is associated with pressure variations in the hydrodynamic fields of the order of the square of the Mach number (see section 1.5). These small pressures jump across the surface to conserve normal momentum flux. Because Euler's equations for small Mach number hold outside the flame (cf. the end of section 3), the temperature and density do not change along particle paths; so that for a flame traveling into a uniform gas the temperature and density ahead are constant and the flow is irrotational. The flow behind is stratified, however, since flame curvature generates both vorticity and nonuniform temperature jumps. Variations in temperature from the adiabatic flame temperature are usually neglected everywhere (a matter we shall treat later); but vorticity generation cannot be ignored so that, even though Euler's equations apply behind the flame, the flow is not potential there.

The flow fields on the two sides of the flame are coupled by conservation of mass and momentum fluxes through the surface. If V is the speed of the discontinuity back along its normal $\boldsymbol{n}$ (Figure 1) we have

$$\rho_f(v_{nf}+V)=\rho_b(v_{nb}+V), \quad \boldsymbol{v}_{\perp f}=\boldsymbol{v}_{\perp b}, \quad p_f+\rho_f(v_{nf}+V)^2=p_b+\rho_b(v_{nb}+V)^2 \quad (1)$$

where $v_n=\boldsymbol{v}\cdot\boldsymbol{n}$ and $\boldsymbol{v}_\perp=\boldsymbol{v}-v_n\boldsymbol{n}$. The flame speed W is defined to be the normal gas speed immediately ahead, as measured in a local frame fixed at the discontinuity, i.e.

$$W\equiv v_{nf}+V. \quad (2)$$

If this is specified the hydrodynamic problem can, in principle, be solved and the locus of the flame determined. Equations (1) become four scalar equations for $\boldsymbol{v}_b$ and p_b once ρ_b is known, and that follows from the flame temperature

$$T_b=T_f+Y_f. \quad (3)$$

Evaluation of the flame speed from a combustion analysis of the flame has often been sidestepped, and replaced by one of several hypotheses. The simplest is that the speed is constant when the thermodynamic state ahead is uniform, the constant being determined (if necessary) by a local plane-flame analysis. In this way Darrieus (1938) and Landau (1944) considered infinitesimal disturbances of a plane flame and concluded that it is always unstable, a result that will be examined in Chapter 11. (By contrast, for a shock wave, the speed is determined by an external condition, such as prescribed deflection of the streamlines.)

The already nonlinear flow field is only complicated by the combustion, even for a constant-speed flame. Accordingly, an additional simplification often adopted is that the flow is not affected by the flame. This can be justified, in general, only when the heat released by the combustion is small compared to the thermal energy of the fresh mixture (section 1.5), so that the temperature jump across the sheet is negligible. Small heat

1 Notation for flame as hydrodynamic discontinuity.

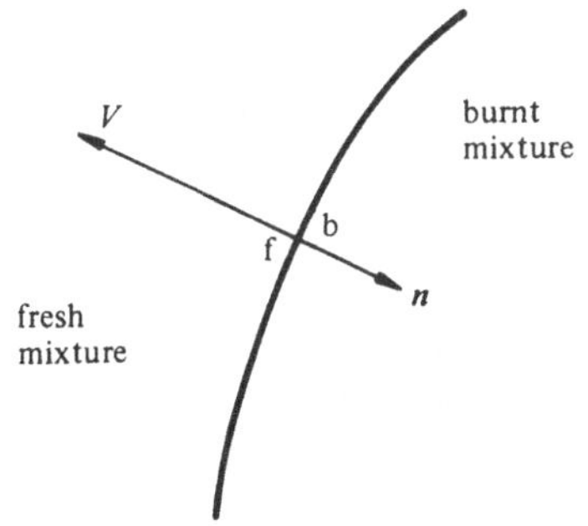

release is not a common characteristic of actual flames, but the approach can provide an insight into their qualitative nature.

Consider, for example, a stationary flame located in the shear flow of Figure 2, representative of that from a Bunsen burner. Since v_n is equal to the flame speed, assumed constant, the angle ψ decreases as v increases and there is a cusp on the centerline. The resulting flame shape has some resemblance to that of the inner cone of a Bunsen flame. In general, the shape of the flame can only be obtained numerically; for a parabolic velocity distribution (a good approximation to the flow from the burner tube when that is unaffected by the combustion) the shape is given by elliptic functions (Lewis & von Elbe 1961, p. 268).

In practice, the tip is usually rounded so that the flame speed at the centerline is equal to that of the flow and must have a smaller value elsewhere. (A detailed study of flame tips is presented in Chapter 9.) Observations of this kind and dissatisfaction with the stability prediction of Darrieus and Landau have prompted more sophisticated hypotheses, in particular that of Markstein (see section 3.6). We shall not pursue these since rational analysis of the burning zone (embedded within the hydrodynamic discontinuity) determines the flame speed from first principles.

Not all flames can be treated as hydrodynamic discontinuities: when the cylindrical or spherical flame of Chapter 7 is viewed on a scale large compared to its thickness, it shrinks to a line or a point. An exception occurs if $L \to 0$ (so that $r_* \to \infty$) and $M \to \infty$ in such a way that the mass

2 Bunsen flame for small heat release when the flame speed is assumed constant.

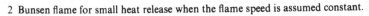

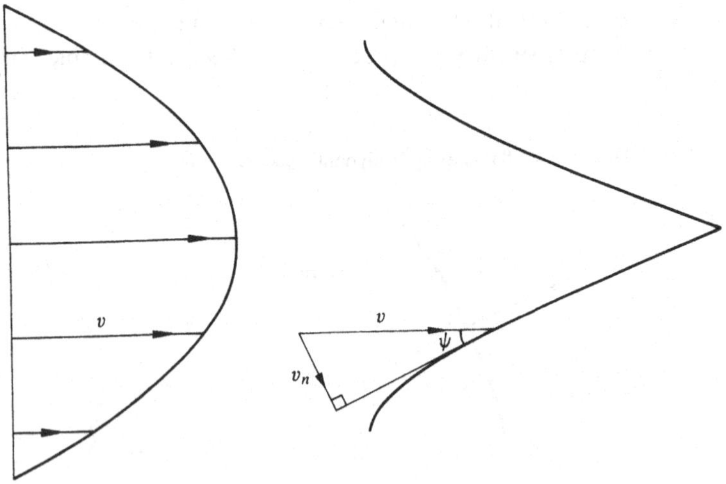

flux at the flame sheet (M/r_*^ν) is fixed and vigorous burning is sustained. The radius a of the source then plays no role, corresponding to a point-source solution. The partial burning encountered in Chapter 7 for spherical flames becomes complete burning in this limit, in agreement with the equilibrium that is assumed behind a hydrodynamic discontinuity.

2 Slow variation and near-equidiffusion

As discussed in section 2.8, the combustion of a two-component mixture that is well removed from stoichiometry can be characterized by a single mass fraction Y related to the component that is consumed at the sheet. The argument was given for plane flames, equal molecular masses, and unit Lewis numbers and stoichiometric coefficients; but all the restrictions are readily seen to be inessential (including the number of species). The single Lewis number is then defined by the deficient component – oxidant in a rich flame, fuel in a lean one.

Appropriate equations for our discussion of three-dimensional unsteady motion are, therefore,

$$\rho T=1, \quad \partial\rho/\partial t+\nabla\cdot(\rho\boldsymbol{v})=0, \quad \rho(\partial Y/\partial t+\boldsymbol{v}\cdot\nabla Y)-\mathscr{L}^{-1}\nabla^2 Y=-\Omega, \tag{4}$$

$$\rho(\partial \boldsymbol{v}/\partial t+\boldsymbol{v}\cdot\nabla\boldsymbol{v})=-\nabla p+\mathscr{P}[\nabla^2\boldsymbol{v}+\nabla(\tfrac{1}{3}\nabla\cdot \boldsymbol{v})],$$

$$\rho(\partial T/\partial t+\boldsymbol{v}\cdot\nabla T)-\nabla^2 T=\Omega, \tag{5}$$

where

$$\Omega=\mathscr{D}Y\,\mathrm{e}^{-\theta/T}. \tag{6}$$

These have been rendered dimensionless as in section 1.6, and the undisturbed pressure has been taken constant (i.e. $p_c=1$), with p the variations from it.

On the length scale λ/c_pM_r now being used, the flame is no longer a surface but, for large activation energy, the reaction zone within it can be considered so; accordingly, we use the term flame sheet. We now take the x-axis instantaneously parallel to the normal of this sheet (but not necessarily of the hydrodynamic discontinuity) at the point of interest and introduce the new variable

$$n=x-F(0,0,t) \tag{7}$$

as was done in the discussion of plane deflagration waves (section 3.3). Here $F(y, z, t)$ denotes the position of the sheet. Everywhere in the equations (4) and (5) $\partial/\partial t$ and $\partial/\partial x$ should now be replaced by

$$\partial/\partial t+V\partial/\partial n \quad \text{and } \partial/\partial n, \tag{8}$$

respectively, where $V=-\dot{F}(0,0,t)$ is the speed of the sheet back along its normal at the instant considered.

In analyzing the resulting equations for the limit $\theta \to \infty$ some thought must be given to the flame temperature, i.e. the temperature immediately behind the sheet. It is clear from the results of Chapters 2 and 3 (see section 3.1) that there must be a general correlation between this temperature and the flame speed such that $O(\theta^{-1})$ changes in the former are associated with $O(1)$ changes in the latter. We may, therefore, expect that if the flame temperature varies by larger amounts, either temporally or spatially, then the flame will behave in an extreme fashion (large acceleration or curvature); such behavior will introduce serious difficulties into the analysis. Only in the discussion of explosions in Chapter 12 will these difficulties be confronted, and then only in relatively simple circumstances. Here, as in Chapter 3, we shall restrict our attention to physical situations where the temperature of the burnt mixture in the neighborhood of the flame is everywhere, and at all times, within $O(\theta^{-1})$ of the adiabatic flame temperature. Necessary restrictions are established by the following argument.

As in section 3.3, we can calculate the flame temperature from the overall change in enthalpy $H = T + Y$ in the flame up to and including the reaction zone. Thus, equations (4c) and (5b), with the modifications (8), are added to eliminate the reaction terms and then integrated from $n = -\infty$ to $0+$ to obtain

$$\int_{-\infty}^{0+} \rho\left[\frac{\partial}{\partial t} + (v_n + V)\frac{\partial}{\partial n} + \boldsymbol{v}_\perp \cdot \boldsymbol{\nabla}_\perp\right] H \, dn = \left.\frac{\partial T}{\partial n}\right|_{0+} + \int_{-\infty}^{0+} \nabla_\perp^2 (T + \mathscr{L}^{-1} Y)\, dn. \quad (9)$$

As before the subscript '$\perp$' denotes the component perpendicular to $\boldsymbol{n}$, i.e. in the tangent plane to the flame sheet, and we have assumed that Y vanishes behind the sheet. The middle term on the left side can be integrated by parts to yield

$$\rho H(v_n + V)\Big|_{-\infty}^{0+} - \int_{-\infty}^{0+} H \frac{\partial}{\partial n}[\rho(v_n + V)]\, dn,$$

in which

$$\frac{\partial}{\partial n}[\rho(v_n + V)] = -\frac{\partial \rho}{\partial t} - \boldsymbol{\nabla}_\perp \cdot (\rho \boldsymbol{v}_\perp),$$

$$\rho(v_n + V)|_{0+} = \rho_f(v_{nf} + V) - \int_{-\infty}^{0+}\left[\frac{\partial \rho}{\partial t} + \boldsymbol{\nabla}_\perp \cdot (\rho \boldsymbol{v}_\perp)\right] dn$$

according to the equation of continuity (4b). Equation (9) may, therefore, be rewritten

$$\rho_f(v_{nf}+V)(T_*-H_f)-\left.\frac{\partial T}{\partial n}\right|_{0+}=\int_{-\infty}^{0+}\left[\nabla_\perp^2(T+\mathscr{L}^{-1}Y)+T_*\left(\frac{\partial \rho}{\partial t}+\nabla_\perp\cdot\rho\boldsymbol{v}_\perp\right)\right.$$
$$\left.-\frac{\partial}{\partial t}(\rho H)-\nabla_\perp\cdot(\rho H\boldsymbol{v}_\perp)\right]\mathrm{d}n \qquad (10)$$

where, as usual, T_* denotes the temperature at the flame sheet.

By assumption the left side of equation (10) is at most $O(\theta^{-1})$ so that the right side must be similarly bounded. The only way to guarantee this within a general framework is to make the terms in $\nabla_\perp^2$, $\nabla_\perp$, and $\partial/\partial t$ separately small, which has been accomplished in two ways. Either we confine our attention to disturbances that vary over times and distances of order θ. Such SVFs are a generalization of those introduced in section 3.5 (where their precise role is described). Or we restrict the Lewis number to values close to 1, i.e. we write

$$\mathscr{L}^{-1}=1-l/\theta \quad \text{with } l=O(1), \qquad (11)$$

and impose the integral

$$H=H_f+O(\theta^{-1}), \qquad (12)$$

which is then a possibility when the state f does not vary along the flame. Such NEFs are a generalization of those introduced in section 3.8.

The full generality of this analysis has not appeared elsewhere although its essence, for shear flows, is to be found in Buckmaster (1979*a*). In short, small changes in enthalpy on and along the sheet can be ensured either by considering perturbations of steady plane flames or by taking a Lewis number close to 1 with appropriate boundary and initial conditions (when changes are small everywhere). The former was first recognized by Sivashinsky (1973, 1974*a*, 1974*b*, 1975); the latter by Matkowsky & Sivashinsky (1979*b*), following Sivashinsky (1977*a*). The rest of the chapter is concerned with the detailed formulation of each of these two kinds of flame.

It should be emphasized that NEFs are a restricted class of solutions: even though $\mathscr{L}$ may be close to 1 in the sense (11), the boundary and initial conditions may be incompatible with the assumption (12). The spherical flame of section 7.4 (where $\mathscr{L}=1$) is an NEF only when conditions are nearly adiabatic, i.e. $T_\infty=T_a+O(\theta^{-1})$, as equation (7.23) demonstrates. The spherical flame is an exception to the general rules of this section: symmetry ensures that the temperature does not vary over its flame sheet, so that it need not be either an SVF or an NEF.

3 The basic equation for SVFs

Slowly varying flames can be characterized by the scaling

$$(y, z, t) = \theta(\eta, \zeta, \tau), \tag{13}$$

but it should be recognized from the outset that the actual lengths involved can be quite moderate, given a preheat zone thickness of 0.5 mm (cf. section 2.4). Of course, for flames that are 1 cm thick (easily contrived in the laboratory) the lengths involved are quite large.

Flame-sheet disturbances on the θ-scale will be associated with field disturbances at distances $n = O(\theta)$, where the governing equations are fundamentally different from those where $n = O(1)$. The limit $\theta \rightarrow \infty$ generates, therefore, a structure with at least three zones: the flame sheet described on the scale $n = O(\theta^{-1})$, to which all the reaction is confined; a diffusion zone described on a scale $n = O(1)$, called a preheat zone in front of the flame; and an ideal-fluid region described on the scale $n = O(\theta)$. Description of the latter constitutes the hydrodynamic problem discussed in section 1. The $O(\theta^{-1})$ stratification behind the flame, which is an essential feature of this description, must eventually be smoothed out by diffusion, so that additional structure is described on a scale much larger than $n = O(\theta)$; but that will not be considered, since it does not influence the hydrodynamic structure (Sivashinsky 1976).

Anticipation of the matching conditions imposed by the ideal-fluid region enables the diffusion zone to be analyzed. These conditions are

$$n \rightarrow -\infty: \quad \rho, T, Y, \mathbf{v}, p \rightarrow \rho_f, T_f, Y_f, \mathbf{v}_f, p_f; \tag{14}$$

$$n \rightarrow +\infty: \quad \text{exponential growth not permitted.} \tag{15}$$

The state f is that immediately ahead of the hydrodynamic discontinuity, and is to be considered a known function of η, ζ, and τ. Our aim is the same as that in section 3.5, namely to calculate various quantities in the overall enthalpy balance (10) and, hence, derive an expression, in terms of the state f, for the perturbation of the flame temperature.

The perturbation in the flame temperature may be calculated from the enthalpy balance (10) as soon as all other quantities have been determined, in terms of V, to sufficient accuracy. The result is quite complicated, involving terms in $\partial/\partial\tau$ and $\mathbf{\nabla}_{\perp}$ of ρ_f, T_f, Y_f and $\mathbf{v}_f$. Considerable simplification occurs when the upstream temperature and composition of the mixture (on the hydrodynamic scale) are uniform, since then ρ_f, T_f, and Y_f are constant, as we shall see at the end of this section. Only the simpler result will be written here. The constancy of ρ_f also allows us to change the density unit from ρ_c to ρ_c/T_f, as in section 3.3.

To leading order the governing equations become

$$\rho T = T_f,\quad \partial[\rho(V+v_n)]/\partial n = 0,\quad \rho(V+v_n)\partial Y/\partial n - \mathscr{L}^{-1}\partial^2 Y/\partial n^2 = -\Omega, \tag{16}$$

$$\rho(V+v_n)\partial v_n/\partial n = -\partial p/\partial n + (4\mathscr{P}/3)\partial^2 v_n/\partial n^2, \tag{17}$$

$$\rho(V+v_n)\partial \boldsymbol{v}_\perp/\partial n = \mathscr{P}\partial^2 \boldsymbol{v}_\perp/\partial n^2,\quad \rho(V+v_n)\partial T/\partial n - \partial^2 T/\partial n^2 = \Omega, \tag{18}$$

where V can now be identified with the speed of the hydrodynamic discontinuity. Apart from notation, equations (16), (17), and (18) are identical to those for a plane flame so that (cf. sections 3.4 and 3.5) we may immediately write

$$\rho(V+v_n) = V + v_{nf} = W, \tag{19}$$

$$T = \begin{cases} T_f + Y_f e^{Wn} \\ T_b - \theta^{-1} T_b^2 \phi_*(\eta, \zeta, \tau) \end{cases},\quad Y = \begin{cases} Y_f(1 - e^{\mathscr{L}Wn}) \\ 0 \end{cases},\quad \rho = \begin{cases} T_f/(T_f + Y_f e^{Wn}) \\ T_f/T_b \end{cases} \quad \text{for } n \lessgtr 0. \tag{20}$$

Note that the perturbation $-T_b^2\phi_*$ to the uniform temperature T_b behind the flame sheet satisfies the leading-order equation (18*b*), as required; and that the normal mass flux (also called the burning rate) is equal to the flame speed in this nondimensionalization. Here ϕ is the perturbation of $1/T$, introduced in section 2.4. Expressions for v_n and p can also be written, but we shall not need them (so that the normal momentum equation plays no role). On the other hand, $\boldsymbol{v}_\perp$ is now needed; since it does not jump across the hydrodynamic discontinuity (which has the same normal as the flame sheet for SVFs) it can be expected to stay constant over the whole range of n, i.e.

$$\boldsymbol{v}_\perp = \boldsymbol{v}_{\perp f}. \tag{21}$$

Formal proof comes from equation (18*a*), which holds even in the flame sheet and possesses only constant vectors as bounded solutions.

Determining the right side of equation (10) to order θ^{-1} requires some care, especially in evaluating $\nabla_\perp^2(T + \mathscr{L}^{-1}Y)$ and $\nabla_\perp \cdot \boldsymbol{v}_\perp$. It is important to recognize that the leading-order structure (19)–(21) is valid everywhere along the flame provided n is given its local meaning. Now, second derivatives along the flame sheet are $O(\theta^{-2})$, but in evaluating $\nabla_\perp^2$ the curvature of the preheat zone, in which there are $O(1)$ normal gradients, produces an $O(\theta^{-1})$ contribution. In the neighborhood of the point of interest at the instant concerned, the shape of the flame sheet is paraboloidal with principal radii of curvature R_1 and R_2 (say), of order θ. Evaluation of the Laplacian $\nabla_\perp^2$ of a function that depends on n and the slow variables yields, therefore,

$$\nabla_\perp^2 = -\theta^{-1}\kappa\partial/\partial n \quad \text{with } \kappa = \theta(1/R_1 + 1/R_2), \tag{22}$$

the θ-multiplied, first (or mean) curvature of the sheet (see (24)). For a flame sheet that is concave when viewed from the burnt side, κ is positive. On extending the scaling to F, so that the location of the flame sheet is now

$$x=\theta F(\eta,\zeta,\tau), \tag{23}$$

we may write

$$\kappa=\nabla_{\perp}^{2}F \tag{24}$$

where the gradient operator relates to η, ζ. This expression is to be used in the result

$$\nabla_{1}^{2}(T+\mathscr{L}^{-1}Y)=-\theta^{-1}\kappa\partial(T+\mathscr{L}^{-1}Y)/\partial n. \tag{25}$$

Curvature also affects the term $\nabla_{\perp}\cdot\boldsymbol{v}_{\perp}$, simply because the normal component of velocity at neighboring points makes a contribution to the velocity in the tangent plane. The divergence of this contribution is

$$-\theta^{-1}\kappa v_n=\theta^{-1}\kappa(V-W/\rho) \tag{26}$$

so that, again taking gradients in η and ζ,

$$\nabla_{\perp}\cdot\boldsymbol{v}_{\perp}=\nabla_{\perp}\cdot\boldsymbol{v}_{\mathrm{p}}+\kappa(V-W/\rho)=\nabla_{\perp}\cdot\boldsymbol{v}_{\perp\mathrm{f}}\,]+\kappa W(\rho-1)(T_{\mathrm{b}}-H)\} \tag{27}$$

where $\boldsymbol{v}_{\mathrm{p}}$ is the tangential component at neighboring points projected onto the tangent plane at the point of interest, and the second equality follows from replacing $\boldsymbol{v}_{\mathrm{p}}$ by $\boldsymbol{v}_{\mathrm{pf}}$ according to the result (21) and using the first equality for the state f.

The remaining terms in equation (10) are time derivatives, which are due not only to variations in W but also to rotation of the normal. The latter makes no contribution at the instant considered, so that on replacing $\partial/\partial t$ by $\theta^{-1}\partial/\partial\tau$ we may treat n as fixed in the formulas (20). To sufficient accuracy, therefore, the integrand (10) is θ^{-1} times

$$\begin{aligned}\{-\kappa\partial(T+\mathscr{L}^{-1}Y)/\partial n+\partial[\rho(T_{\mathrm{b}}-H)]/\partial\tau \\ +\nabla_{\perp}\cdot[\rho(T_{\mathrm{b}}-H)\boldsymbol{v}_{\perp\mathrm{f}}]+\kappa W(\rho-1)(T_{\mathrm{b}}-H)\} \\ =[(\dot{W}+\boldsymbol{v}_{\perp\mathrm{f}}\cdot\nabla_{\perp}W)W^{-1}n\partial/\partial n+\nabla_{\perp}\cdot\boldsymbol{v}_{\perp\mathrm{f}}+\kappa W][\rho(T_{\mathrm{b}}-H)]\end{aligned}$$

where

$$\rho(T_{\mathrm{b}}-H)=T_{\mathrm{f}}Y_{\mathrm{f}}(e^{\mathscr{L}Wn}-e^{Wn})/(T_{\mathrm{f}}+Y_{\mathrm{f}}e^{Wn}),$$

which is replaced by

$$[-W^{-1}(\dot{W}+\boldsymbol{v}_{\perp\mathrm{f}}\cdot\nabla_{\perp}W)+\nabla_{\perp}\cdot\boldsymbol{v}_{\perp\mathrm{f}}+\kappa W][\rho(T_{\mathrm{b}}-H)]$$

on integration by parts. All derivatives are taken with respect to slow variables and the analysis requires those of W to be formed from the normal flux and not the flux in the n-direction. For $\boldsymbol{v}_{\perp\mathrm{f}}=\kappa=0$ the calculation is the same as for a plane flame, so that it is not surprising to find the generalization

$$\phi_{*}=-bW^{-3}[\dot{W}+\boldsymbol{v}_{\perp\mathrm{f}}\cdot\nabla_{\perp}W-W(\nabla_{\perp}\cdot\boldsymbol{v}_{\perp\mathrm{f}}+\kappa W)] \tag{28}$$

of the result (3.43), with the same parameter b, from the integration (10).

Another relation between the wave speed W and the perturbation $-T_b^2\phi_*$ of the flame temperature follows from the structure of the flame sheet, which is described on the scale $n=O(\theta^{-1})$. The analysis is the same as in Chapter 3 since, to leading order, the equations for T, Y, and ρ are those for a plane flame. We conclude that

$$W=e^{-\frac{1}{2}\phi_*} \tag{29}$$

(cf. equation (3.25)) if the mass flux M_r on which units have been based is chosen to be the burning rate of a plane flame moving into the state Y_f, T_f, i.e.

$$M_r=\sqrt{2\mathscr{L}D}\,T_b^2\,e^{-\theta/2T_b}/Y_f\theta \tag{30}$$

(cf. equation (3.14)). The two results (28) and (29) now give us the basic equation of SVFs,

$$b(\dot{W}+\boldsymbol{v}_{\perp f}\cdot\boldsymbol{\nabla}_{\perp}W-W\boldsymbol{\nabla}_{\perp}\cdot\boldsymbol{v}_{\perp f}-\kappa W^2)=W^3\ln W^2, \tag{31}$$

which should be compared with its plane form (3.44). Note that the normal momentum equation (17) has not been used in its derivation, a fact that will be employed in section 11.3.

The equation is due to Sivashinsky (1976) and Buckmaster (1977), whose derivations have been simplified. Restricted forms of it have been given by Sivashinsky (1974*a*, *b*, 1975) and, in a linearized context, by Eckhaus (1961). In the limit $\mathscr{L}\to 1$ (i.e. $b\to 0$) it gives $W=1$, then substantiating the constant-speed hypothesis of section 1. Its interpretation involves the concept of flame stretch, introduced in the next section. We shall finish this section with a brief discussion of the ideal-fluid region where $n=O(\theta)$.

To the slow variables η, ζ, and τ we now add

$$\chi=x/\theta. \tag{32}$$

Then T, Y, and $\rho(=T_f/T)$ are found to be constant on particle lines, to leading order, while the velocity and pressure fields satisfy

$$\boldsymbol{\nabla}\cdot\boldsymbol{v}=0,\quad \rho(\partial\boldsymbol{v}/\partial\tau+\boldsymbol{v}\cdot\boldsymbol{\nabla}\boldsymbol{v})=-\boldsymbol{\nabla}p, \tag{33}$$

where $\boldsymbol{\nabla}$ is now the gradient operator with respect to χ, η, and ζ. These are Euler's equations for an incompressible (but not necessarily constant-density), ideal fluid, i.e. one devoid of viscosity and heat conduction. Particles retain the values T_f, Y_f, and $\rho_f(=1)$ they acquire far upstream until they reach the hydrodynamic discontinuity, where they exchange them for the values

$$T_b=H_f,\quad Y_b=0,\quad \rho_b=T_f/H_f, \tag{34}$$

that they then carry downstream. The ideal-fluid region is coupled to the flame through the velocity and pressure, in a manner that will now be seen in detail for uniform temperature and composition upstream.

The state (T, Y, and ρ) is constant along either side of the hydrodynamic discontinuity. In particular, there is no $O(1)$ thermal stratification on the downstream side, in accordance with a common assumption mentioned in section 1. The problem is to find $\mathbf{v}$ and p subject to the jump conditions (1), (3), and the evolution equation (31), which may be thought of as five equations for $\mathbf{v}_b$, p_b, and F in terms of the state f. Its complete analysis would clearly be formidable.

4 Flame stretch

Considering the difficulties of giving a complete analysis of a multidimensional flame (even when it is slowly varying), it is natural to try to identify special characteristics that play particularly important roles in the understanding of flame behavior. Flame speed and temperature are elementary examples of such characteristics that have already been identified. A more subtle characteristic, attributed to Karlovitz, *et al.* (1953), is flame stretch.

To define this and for that matter the flame speed in an unambiguous fashion it is first necessary to define a flame surface, i.e. a surface that characterizes the location of the reaction. For large activation energy the reaction zone is such a surface when viewed on the scale of the preheat zone. For finite θ, Lewis & von Elbe (1961, p. 225) use the locus of inflection points of the temperature field near the burning zone, a definition that coincides with the preceding one in the limit $\theta \to \infty$. If the flame can be viewed as a hydrodynamic discontinuity, as in section 1, then the discontinuity itself is a flame surface. In all of these examples and in any other for which our discussion is meaningful, a flow velocity is defined on each side of the surface such that $\mathbf{v}_\perp$ is continuous across the surface.

Consider a point on the flame surface that moves in it with the velocity $\mathbf{v}_\perp$. A set of such points forming an infinitesimal surface element of area S will, in general, deform during the motion, so that S will vary with time. If S increases the flame is said to be stretched, whereas if S decreases the flame experiences negative stretch and is said to be compressed. A precise measure of the stretch is the proportionate rate of change introduced by Karlovitz, which in terms of the slow time is

$$K \equiv S^{-1}\, dS/d\tau. \qquad (35)$$

Certain configurations for which stretch is an important characteristic will be discussed in detail in Chapter 10. Here we shall give two elementary examples.

Consider a stationary flame surface inclined at an angle ψ to the horizontal (Figure 3). If the flame is located in a horizontal shear flow $U(\eta)$, it undergoes extensional stretch

$$K=\sin\psi\, \mathrm{d}[U(\eta)\cos\psi]/\mathrm{d}\eta=\sin\psi\cos\psi\, U'(\eta) \tag{36}$$

the last equality holding wherever the curvature $\sin\psi\, \mathrm{d}\psi/\mathrm{d}\eta$ is negligible. (Note that if $\psi=0$ or $\pi/2$ there is then no stretch because the tangential velocity does not change along the surface.) The second example is an expanding cylindrical flame. If the radius is $R(\tau)$, then the resulting dilatational stretch is

$$K=\dot{R}/R. \tag{37}$$

Stretch of this type is discussed more generally in connection with the quasi-plane flames of section 3.8.

In these two examples the stretch could be determined by inspection, but we now need a general expression for it. That comes from recognizing it to be a combination of extensional and dilatational stretches, of which these examples are prototypes. Consider the trace of the flame surface on a plane through the normal at the point of interest. An element of arc at the point undergoes extensional stretch, due to the difference in tangential gas velocity between its ends, and dilatational stretch, due to the motion of the element along its normals. If neither the flow nor flame varies (locally) at right angles to the plane, the resultant stretch is the sum of terms like (36) and (37); otherwise elements in two orthogonal planes must be considered. Then the extensional stretch of a surface element is the sum of two terms like (36) and the dilatational stretch is the sum of two terms like (37), so that in total the stretch may be written

$$K=\nabla_{\perp}\cdot \mathbf{v}_{\mathrm{pf}}+\kappa V=\nabla_{\perp}\cdot \mathbf{v}_{\perp\mathrm{f}}+\kappa W \tag{38}$$

(cf. equation (27*b*), which holds for slow variations).

In the first example $\nabla_{\perp}\cdot \mathbf{v}_{\mathrm{pf}}=\partial(U\cos\psi)/\partial s$, where s is the slow distance along the flame, from which the result (36) follows. In the second example

3 Flame in a slowly varying shear flow.

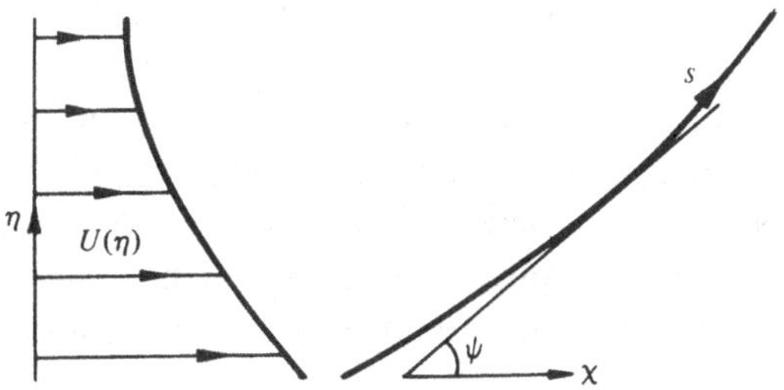

$v_{pf}=0$, $\kappa=1/R$, and $V=\dot{R}$, giving the result (37). For a spherical, rather than a cylindrical, flame the curvature is $2/R$ so that

$$K=2\dot{R}/R, \tag{39}$$

as is obvious.

Stretch is a kinematical concept. For an SVF with uniform temperature and composition upstream it may be expressed in terms of the flame speed. Thus, the basic equation (31) reads

$$K=W^{-1}\,\mathrm{d}W/\mathrm{d}\tau-b^{-1}W^2\ln W^2, \tag{40}$$

where now the time derivative is evaluated following an element of the flame surface. Note that stretch is the same for the hydrodynamic discontinuity as for the reaction zone, but flame speed relates to the former and not the latter. The result (3.53) is a special case of equation (40).

Flame thickness is another concept that can usefully characterize a flame. Although a plane deflagration wave is not confined to a zone of finite thickness, to say that it has a thickness λ/c_pM obviously has meaning. The implication is that, while an infinite distance is required for the temperature to rise from the cold value to the hot, most of the change takes place over a small multiple of λ/c_pM.

For a general flame, it is natural to define the local thickness as λ/c_pM, where M is the normal mass flux. Indeed, for the SVF the dimensionless normal mass flux W is the decay coefficient for T in the preheat zone, as formula (20*a*) shows, so that W^{-1} is properly called the nondimensional thickness. If now the elemental volume

$$\Delta=S/W \tag{41}$$

generated by S is introduced, so that $\Delta^{-1}\,\mathrm{d}\Delta/\mathrm{d}\tau=K-W^{-1}\,\mathrm{d}W/\mathrm{d}\tau$, the basic equation becomes

$$\Delta^{-1}\,\mathrm{d}\Delta/\mathrm{d}\tau=-b^{-1}W^2\ln W^2, \tag{42}$$

expressing what may be called voluminal stretch in terms of flame speed or thickness. The results (40) and (42) were originally derived by Buckmaster (1979*a*).

Note that the sign of b depends on the Lewis number $\mathscr{L}$, being positive for $\mathscr{L}>1$ and negative for $\mathscr{L}<1$. When $\mathscr{L}=1+O(\theta^{-1})$ so that $b=O(\theta^{-1})$, equation (42) suggests that changes occur on the scale of t rather than τ, although it then fails to provide a valid description. In section 5 we shall develop the theory of near-equidiffusional flames (NEFs) to fill the void.

We shall end the section with a few remarks about the contemporary significance of flame stretch. Although the concept leads to an elegant description of SVFs (especially when extended to voluminal stretch), the importance attached to it derives from its far-ranging use as an intuitive tool in the prediction and explanation of flame behavior (Lewis & von Elbe 1961). This is not the appropriate place to discuss the arguments used, which are, in fact, no more than speculation. We just mention here the fundamental claim that stretching a flame causes it to decelerate. The mathematical evidence for such a simple picture is mixed. Certainly the results for SVFs are not favorable: equation (40) shows that rapid deceleration is associated with negative stretch. But there is supportive evidence from NEFs in certain special circumstances. These will be discussed in Chapter 10.

5 The basic equations for NEFs

The nonuniformity of SVFs as $\mathscr{L}\to 1$ (revealed by equation (31)) makes it clear that a different formulation is needed when heat and reactant diffuse at comparable rates. The essential characteristics of such a formulation have been identified in section 2, in particular equations (11) and (12); a detailed mathematical treatment follows. The starting point is the system of equations (4)–(6) with

$$\mathscr{D}=(Y_f^2\theta^2/2T_b^2)\,e^{\theta/T_b}, \tag{43}$$

corresponding to the choice (30), with $\mathscr{L}=1$, for M_r. Recall that the state f does not vary along the flame.

We seek an asymptotic development in which spatial and temporal variations in the flame-sheet location are $O(1)$. As discussed in section 2, limitation of the allowable variation in flame temperature then imposes the restriction (11) and leads to the integral (12). Equations (5b) and (4c) may then be written

$$\rho(\partial/\partial t+\boldsymbol{v}\cdot\nabla)T=\nabla^2 T,\quad \rho(\partial/\partial t+\boldsymbol{v}\cdot\nabla)h=\nabla^2 h+l\nabla^2 T \tag{44}$$

outside of the reaction zone, where ρ, T, and $\boldsymbol{v}$ now stand for the leading terms in their own expansions for θ large and H_1 has been replaced by h. The same convention (including p) in the remaining equations (4a, b) and (5a) completes the system. The relevant solution of equation (44a) behind the reaction zone is

$$T=T_b=H_f. \tag{45}$$

Note that the equations are essentially multidimensional. Nonplanar effects are no longer removed to distances $O(\theta)$, where diffusion is negligible; and the problem is correspondingly more difficult. Note also that the actual distances over which changes take place can be very small, being measured on the scale of the preheat zone.

To complete the formulation we need jump conditions across the reaction zone. While some of these could be deduced from section 3.8, the generalization in notation here makes it more convenient to obtain all of them *ab initio*. Introduce the coordinate (7) at the point of interest; set

$$\xi = \theta n \quad \text{and} \; T_0 = T_b, \quad \rho_0 = T_f/T_b, \quad Y_0 = 0, \quad \mathbf{v}_0 = \mathbf{v}_*(y, z, t) \tag{46}$$

in the flame zone, implying that the temperature, density, mass fraction, and velocity are continuous to leading order across the flame sheet. Their perturbations and the leading-order pressure are taken to be functions of ξ (as well as y, z, and t), implying that they may jump across the flame sheet. The goal of the reaction-zone analysis is to find explicit expressions for this jump in pressure and those in the normal derivatives of the other variables.

The variables T_1 and Y_1 satisfy

$$\partial^2 T_1/\partial\xi^2 = -\partial^2 Y_1/\partial\xi^2 = -(Y_f^2/2T_b^4)Y_1 \, e^{T_1/T_b^2} \tag{47}$$

so that, in particular, $T_1 + Y_1$ is a linear function of ξ, which matching shows to be independent of ξ (to leading order the enthalpy gradient vanishes outside). Hence, h is continuous across the flame sheet and we may write

$$T_1 + Y_1 = -T_b^2 \phi_*(y, z, t), \tag{48}$$

where ϕ has the same meaning as in section 3. In the usual way, integration of the equation for T_1 now yields an expression for the temperature gradient immediately ahead of the reaction zone (coming from $\partial T_1/\partial\xi$ as $\xi \to -\infty$), which may be written

$$\delta(\partial T/\partial n) = -Y_f \, e^{-\frac{1}{2}\phi_*}. \tag{49}$$

Here

$$\delta(\;) = (\;)_{0+} - (\;)_{0-} \tag{50}$$

denotes the jump across the flame sheet at $n = 0$.

We turn now to the continuity and momentum equations, and immediately note that, to leading order, the former confirms the continuity of v_n across the flame sheet, while the latter confirms that of $\mathbf{v}_\perp$ (see equation (21)). To next order, the continuity equation yields

$$(V + v_{n*})\partial T_1/\partial\xi = H_f(\partial v_{n1}/\partial\xi + \nabla_\perp \cdot \mathbf{v}_{\perp *})$$

so that

$$H_f \delta(\partial v_n/\partial n) = (V + v_{n*})\delta(\partial T/\partial n). \tag{51}$$

Similarly, the normal component of the momentum equation yields

$$3\partial p_0/\partial\xi = 4\mathscr{P}\partial^2 v_{n1}/\partial\xi^2 \quad \text{so that (by integration)} \; 3\delta(p) = 4\mathscr{P}\delta(\partial v_n/\partial n). \tag{52}$$

Finally, the other components of the momentum equation yield

$$\partial^2 \boldsymbol{v}_{\perp 1}/\partial\xi^2 = 0 \quad \text{so that } \delta(\partial \boldsymbol{v}_\perp/\partial n) = 0. \tag{53}$$

The jump conditions are completed by noting that

$$\rho(\partial/\partial t + \boldsymbol{v}\cdot\nabla)H = (1-\theta^{-1}l)\nabla^2 H + \theta^{-1} l \nabla^2 T$$

is an exact equation everywhere, so that

$$\delta(\delta h/\partial n) = -l\delta(\partial T/\partial n). \tag{54}$$

In particular, we see that when the Lewis number is exactly one ($l=0$) the function h vanishes identically, in the absence of nonhomogeneous initial or boundary data.

For SVFs it does not matter, as far as stretch is concerned, whether the flame surface is taken to be the hydrodynamic discontinuity or the reaction zone, since $\boldsymbol{v}_\perp$ does not change across the preheat zone. The flame speed, however, differs for these two possibilities since v_n does change across the preheat zone; convention adopts the hydrodynamic discontinuity. For NEFs, for which a hydrodynamic far field might not exist, it is natural to select the flame sheet. The determinations of speed would be consistent if calculated relative to the burnt gas, but that would flout long-established convention.

6 Reduction to Stefan problems

The problem described in section 5, for the first time in full generality, presents formidable mathematical difficulties. In general, a numerical treatment would be needed, involving the simultaneous solution of six first- or second-order, partial differential equations (some nonlinear) on each side of a (possibly) moving discontinuity that has to be found. Even the description of perturbations of a plane flame (section 3.8) demands the solution of equations with complicated variable coefficients.

These difficulties are significantly reduced when the fluid mechanics is eliminated, as was done for Figure 2 (Buckmaster 1979*a*, *b*, *c*). Then only equations (44) remain with $\rho=1$ and $\boldsymbol{v}$ prescribed as a solution of the continuity and (perhaps) momentum equations. This is the constant-density approximation of section 1.5, leading to the system

$$(\partial/\partial t + \boldsymbol{v}\cdot\nabla)(T, h) = \nabla^2(T, h + lT). \tag{55}$$

The accompanying jump conditions are (49) and (54), h and T being continuous across the flame sheet.

We pause here to make some remarks about the constant-density approximation. The system (49), (54), (55), as it stands, can be regarded as an irrational approximation, somewhat in the spirit of an Oseen linearization. It can be used to predict the behavior of flames for which there is significant heat release, the normal scheme of things. Alternatively, with some slight modification, it can be regarded as a rational approximation valid in the limit of small heat release: either the reactant is scarce ($Y_f \to 0$) or the heat of combustion is small ($T_f \to \infty$). In either case use of lY_f/T_f^2 and the variables $(T - T_f)/Y_f$, h/T_f^2 instead of l and T, h leads to a proper limit. The relevant parameter is the coefficient of thermal expansion

$$\sigma = \rho_f/\rho_b = T_b/T_f = 1 + Y_f/T_f \tag{56}$$

(with $\rho_f = 1$ in our nondimensionalization), which tends to 1. Equations (49), (54), (55), as modified, can be obtained by expansion in $\sigma - 1$, i.e. they result from taking $\lim_{\sigma \to 1} \lim_{\theta \to \infty}$; that formalizes the approach. Matkowsky & Sivashinsky (1979*b*) derive the same equations by an analysis in which the limits $\theta \to \infty$, $\sigma \to 1$ are taken simultaneously, rather than consecutively. They are led to the requirement $T_f^2 \theta^{-1} \ll Y_f \ll T_f^{\frac{4}{3}} \theta^{-\frac{1}{3}}$ with $\theta Y_f T_f^{-2}(1 - \mathscr{L})$ an $O(1)$ quantity.

An attractive feature of the equations (55) is their linearity; but we are still faced with a spatially elliptic, free-boundary problem, even for the case of a uniform flow

$$\boldsymbol{v} = (U, 0, 0) \quad \text{with } U \text{ constant.} \tag{57}$$

For simplicity, consider steady plane flow. If U is large, i.e. the gas speed is much greater than the flame speed, then a flame sheet would have to be almost horizontal (with slope of order U^{-1}); under the right circumstances, changes in the x-direction would then be much smaller than those in the y-direction. An example, discussed in detail in section 9.4, is that of a flame next to an adiabatic wall parallel to the x-axis. Under such circumstances the field equations become parabolic; for, by writing

$$\chi = U^{-1}x \tag{58}$$

in equations (55), we obtain

$$\mathfrak{L}(T, h) = -l\partial(0, T)/\partial\chi \quad \text{with } \mathfrak{L} \equiv \partial^2/\partial y^2 - \partial/\partial\chi \tag{59}$$

in the limit $U \to \infty$. Moreover, the normal $\boldsymbol{n}$ is almost vertical so that the n-derivatives in the jump conditions (49) and (54) may be replaced by y-derivatives. The result is a generalized Stefan problem, readily amenable to numerical analysis (Buckmaster 1979*a*). Clearly the argument is not affected in substance if U is a function of y; the χ-derivatives in equations (59) acquire coefficients depending on y, the same for each. This will form the basis of the discussion of shear flows presented in Chapter 10.

A similar reduction is possible for steady straining flows (Buckmaster 1979*b*); the simplest representative is plane stagnation-point flow. The velocity field is

$$\boldsymbol{v} = \varepsilon(x, -y), \tag{60}$$

where the rate of strain ε is proportional to U/R, U being the free-stream velocity and R the radius of curvature of the cylindrical body at the stagnation point. In the limit $U \to \infty$ with ε fixed (so that $R \to \infty$ also) changes along the body (i.e. in the x-direction) are $O(U^{-1})$ and the governing equations on the axis $x = 0$ become,

$$\mathfrak{L}(T, h) = l\varepsilon y \partial(0, T)/\partial y \quad \text{with } \mathfrak{L} \equiv \partial^2/\partial y^2 + \varepsilon y \partial/\partial y. \tag{61}$$

This system of ordinary differential equations and its generalization to other straining flows are discussed in sections 10.3 and 10.4, respectively.

9

BURNERS

1 Features of a tube flame

Both inside and outside the laboratory it is commonly observed that a premixed flame can be stabilized at the mouth of a tube through which the mixture passes. Such a flame, usually conical in shape though not necessarily so, can be conveniently divided into three parts: the tip, the base (near the rim of the tube), and the bulk of the flame in between.

Elementary considerations of the flame speed and the nature of the flow explain the conical shape (see Figure 8.2 and the accompanying discussion). Simple hydrodynamic arguments provide salient features of the associated flow field, as we shall see in section 2; additional details are outlined in section 6.

An understanding of the nature of the combustion field in the vicinity of the rim is crucial in questions of existence and stability of the flame. Gas speeds near the tube wall are small because of viscous effects, so that if the flame could penetrate there it would be able to propagate against the flow, traveling down the tube in a phenomenon known as flashback. In actuality, the flame is quenched at some distance from the wall through heat loss by conduction to the tube (for a stationary flame); this prevents it from reaching the low-speed region. Such quenching enables unburnt gas to escape between the flame and the wall through the so-called dead space, a phenomenon that is described mathematically in section 5.

The heat loss to the tube also plays an important role in the global stability of the combustion field. If the flame is disturbed so that its speed increases it will move closer to the rim, threatening flashback. But heat loss will then increase, causing a decrease in the temperature of the flame and hence its speed, so that it will be swept away again. Similarly, if there is a decrease in flame speed so that the flame is swept away from the rim, threatening what is known as blow-off, heat conduction to the wall decreases, the speed increases, and the flame moves back again. Thus, heat loss acts as a restoring force.

The third part of a burner flame, the tip, is similar in one respect to the base in that here also the flame approaches a boundary (the centerline), but it is now an adiabatic one. Experiment shows that the tip may be closed, the flame cutting the centerline at right angles, or open, the flame being quenched (though not by heat loss) at a finite distance from the centerline leaving a dead space or hole through which the mixture escapes unburnt (Figure 1). Which of these two possibilities occurs depends on the

1 Flame tips: (*a*) methane/air mixture, circular tube (courtesy R. Günther); (*b*) natural gas/air mixture, rectangular tube (courtesy B. Lewis); (*c*) hydrogen/air mixture, rectangular tube (courtesy B. Lewis).

(*a*)

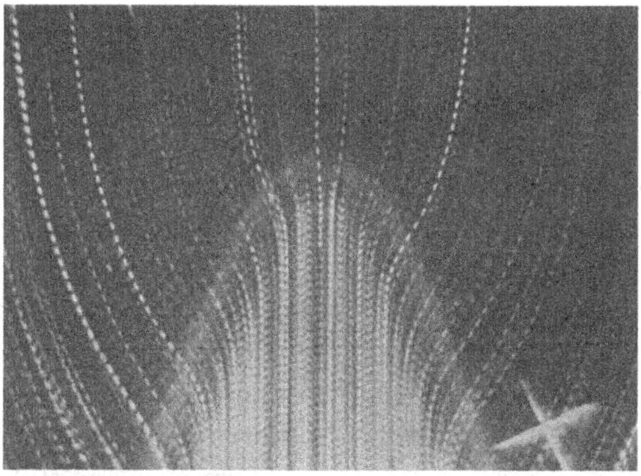

(*b*)

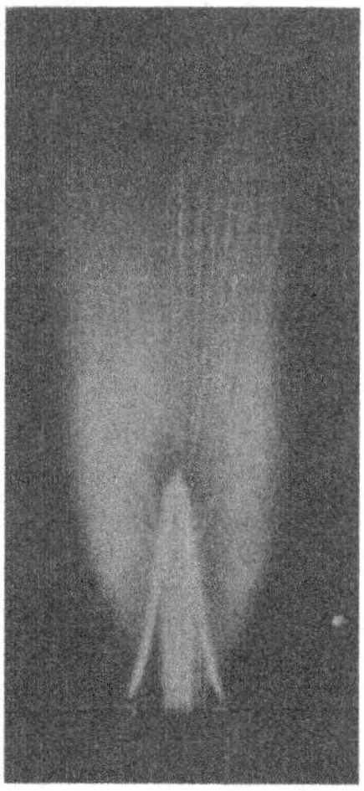

(*c*)

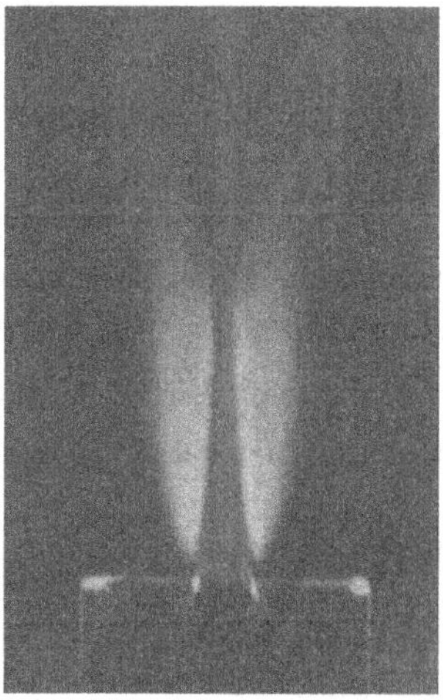

composition of the mixture and flow conditions. Flame tips are treated in sections 3 and 4.

2 Hydrodynamic considerations

Consider a weak plane flame, of speed εU (with $0<\varepsilon \ll 1$), lying stationary in an infinite uniform flow

$$\boldsymbol{v}=(U, 0). \tag{1}$$

The hydrodynamic jump conditions (8.1), (8.3) are satisfied by the undisturbed flow (1) ahead of the flame and

$$\boldsymbol{v}=(U, \pm\varepsilon U(\sigma-1))+O(\varepsilon^2, \varepsilon^2) \tag{2}$$

behind, where the sign is opposite to that of the slope of the flame; here σ is the expansion coefficient (8.56) of the mixture, representing the total heat released at the flame sheet. There is simply a refraction of the streamlines, and we may expect that to be the primary characteristic of the flow field in the neighborhood of any point on the conical part of the burner flame. A more detailed description requires a hydrodynamic analysis, which we shall pursue at length assuming that the flame speed is constant. Such an assumption is correct if the burner is sufficiently large; in the context of SVFs the voluminal stretch in equation (8.42) is then small, so that $W=1$.

Following Buckmaster & Crowley (1982) we shall first seek a plane flame tip in which the uniform flows (1) and (2) hold far upstream; elsewhere it is necessary to solve Euler's equations on each side of the constant-speed flame. A Poiseuille profile might be more realistic for the upstream flow but the analysis is much easier for a uniform flow and still uncovers the essential characteristics. (The efflux from a nozzle burner is reasonably uniform.)

Although we are not necessarily dealing with an SVF, the notation χ, η of section 8.3 will be used to emphasize that the distance unit is large compared to the flame thickness. The χ-axis is now taken along the symmetry line (Figure 2); velocity components are u and v. Since the flame generally slopes gently, we may write

$$\eta=F(\mu)=F_0(\mu)+O(\varepsilon) \quad \text{with } \mu=\varepsilon\chi \tag{3}$$

for its upper surface. In front u, p, and $\partial/\partial\eta$ are $O(1)$, while v and $\partial/\partial\chi$ are $O(\varepsilon)$.

It is natural to seek a solution for the interior problem in the form

$$u=U+\varepsilon u_1(\mu, \eta)+\ldots, \quad v=\varepsilon^2 v_1(\mu, \eta)+\ldots, \tag{4}$$

where the relative magnitude of the perturbations is fixed by the continuity equation. The momentum equation then shows that the pressure perturbation is a function of μ alone; from Bernoulli's equation it follows that the same is true of u_1. Continuity now allows us to write

$$u_1 = f(\mu), \quad v_1 = -\eta f'(\mu), \tag{5}$$

where $f(\mu)$ is unknown. Moreover, since v is $O(\varepsilon^2)$ it makes no contribution to the $O(\varepsilon)$ velocity component normal to the flame (i.e. the flame speed) so that

$$-\varepsilon F_0' U = \varepsilon U \quad \text{or } F_0 = -\mu, \tag{6}$$

corresponding to a wedge.

We now consider the exterior problem, where the refraction conditions (8.1), (8.3) impose the boundary conditions

$$u = U + \varepsilon f(\mu) + \ldots, \quad v = \varepsilon U(\sigma - 1) + \ldots \quad \text{for } \eta = -\mu \quad \text{and } \mu < 0. \tag{7}$$

Symmetry about the centerline requires that

$$v = 0 \quad \text{for } \eta = 0 \quad \text{and } \mu > 0. \tag{8}$$

We write, therefore,

$$u = U + \varepsilon u_1(\mu, \nu) + \ldots, \quad v = \varepsilon v_1(\mu, \nu) + \ldots \quad \text{with } \nu = \varepsilon\eta, \tag{9}$$

the perturbations satisfying linear versions of Euler's equations. (Note that μ and ν are on the same scale.)

A potential solution of this boundary-value problem is

$$u_{1p} = (U/\pi)(1 - \sigma) \ln r, \quad v_{1p} = (U/\pi)(\sigma - 1)\phi, \tag{10}$$

where r and ϕ are polar coordinates in the μ, ν-plane. Its adoption implies

$$f(\mu) = (U/\pi)(1 - \sigma) \ln(-\mu) \tag{11}$$

for the interior flow, which is perfectly acceptable. The weak singularity at infinity merely implies that the states (1) and (2) do not hold precisely in the far field, i.e. account must be taken of holding the flame somewhere. A similar difficulty occurs in considering the potential flow in the neighborhood of the wedge-shaped trailing edge of a slender body (cf. Van Dyke (1975, p. 68)). The singularity at the origin is also of no great concern.

Nevertheless, it is not necessary for the exterior flow to be potential. Noting that the leading-order vorticity must be convected unchanged along lines of constant ν shows that a (rotational) shear flow

$$u_{1r} = g(\nu), \quad v_{1r} = 0 \quad \text{with } g \text{ arbitrary} \tag{12}$$

may be added to (10), so that now

$$f(\mu) = g(-\varepsilon\mu) + (U/\pi)(1 - \sigma) \ln(-\mu). \tag{13}$$

An arbitrary f cannot be so obtained since the corresponding g (being a function of ν/ε) has, in general, an $O(\varepsilon^{-1})$ derivative inconsistent with the exterior linearization. But a multiple of $\ln(-\mu)$ cannot be ruled out on this account, and a particularly interesting choice is

$$g(v) = (U/\pi)(\sigma - 1)\ln(v/\varepsilon), \tag{14}$$

for which $f \equiv 0$ and

$$u_1 = u_{1p} + u_{1r} = (U/\pi)(\sigma - 1)\ln(\sin\phi/\varepsilon). \tag{15}$$

The advantage is boundedness at infinity, but now there is a wake-like singularity on $\phi = 0$.

Which of these choices, if either, is appropriate is not evident; fortunately the ambiguity only affects the perturbation of u, so that the flow field may be sketched (Figure 2).

A similar treatment is possible for axisymmetric flames; only the changes in the previous formulas need be mentioned. In the interior

$$u = U + \varepsilon f(\mu) + \dots, \quad v = -\tfrac{1}{2}\varepsilon^2 \eta f'(\mu) + \dots, \tag{16}$$

so that the flame function is as before (now giving a conical shape), and in the exterior we may take

$$u = U + o(\varepsilon), \quad v = \tfrac{1}{2}\varepsilon^2 U(\sigma - 1)\tan(\phi/2) + \dots, \tag{17}$$

corresponding to $f \equiv 0$. The addition of a shear flow, as in the plane case, generates nonuniformities at infinity (particularly through f) and on the centerline, so that it has little attraction. Note that the result (17*b*) is $O(\varepsilon)$, in an angular neighborhood of the flame, but is otherwise $O(\varepsilon^2)$, a fact reflected in Figure 3 where the streamlines straighten out much faster than in Figure 2 after undergoing the same deflection at the flame, namely $\varepsilon U(\sigma - 1)$. The difference can be seen in the photographs (Figure 1).

The streamlines of actual flames (Figure 1) differ in two respects. The undisturbed flow ahead of the flame may not be uniform; and finite values of ε give rise to a pressure gradient associated with the divergence of streamlines there. The first effect can be taken into account (at least insofar as it changes the flame shape) and, for a Poiseuille flow

$$(u, v) = U(1 - \eta^2/a^2, 0), \tag{18}$$

2 Plane tip with small, constant flame speed; drawn for $\varepsilon(\sigma - 1)/\pi = 0.1$, $\varepsilon = \tan 10°$.

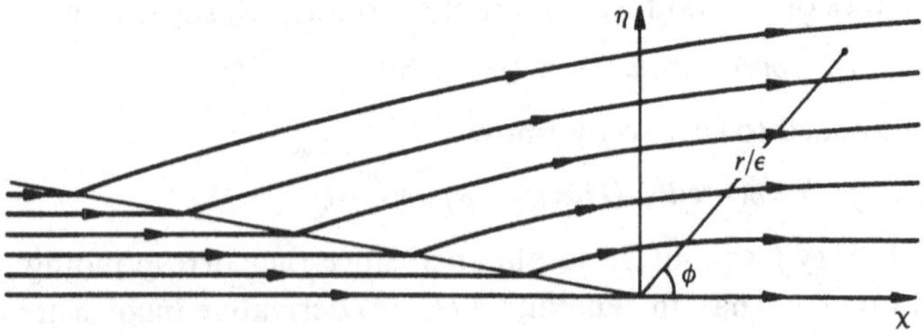

a simple analytical result is obtained. Thus, the interior analysis, which always establishes that u is unchanged to $O(1)$ and v is $O(\varepsilon^2)$, leads to the flame-sheet condition

$$F'_0=(F_0^2/a^2-1)^{-1}, \quad \text{i.e. } F_0-F_0^3/3a^2=-\mu \tag{19}$$

irrespective of whether plane or axisymmetric flames are considered. For slender flames the approximation introduced in section 8.1 is justified, therefore, while the computational difficulties mentioned there do not appear.

3 SVF tips

A hydrodynamic analysis is appropriate for the bulk of the flame (provided it is large enough); close to the tip combustion must be included. The question that we examine here is whether the equation (8.31) governing SVFs admits solutions corresponding to the tips (Sivashinsky 1974*b*, 1975, 1976). The difficulty of the hydrodynamic problem, even when the burning rate is prescribed (section 2), leads us to sever the coupling with Euler's equations by setting

$$\mathbf{v}_f=(U,0,0), \tag{20}$$

in accordance with the constant-density approximation adopted for NEFs (section 8.6).

If ψ is the inclination of the flame (Figure 4(*b*)), then

$$v_{\perp f}=U\cos\psi, \quad W=U\sin\psi, \quad \text{with } 0\leqslant\psi\leqslant\pi, \tag{21}$$

so that, in the plane case,

$$\mathbf{v}_{\perp f}\cdot\nabla_\perp W=U^2\cos^2\psi\,\mathrm{d}\psi/\mathrm{d}s, \quad \nabla_\perp\cdot\mathbf{v}_{\perp f}=0, \quad \kappa=-\mathrm{d}\psi/\mathrm{d}s \tag{22}$$

and the governing equation becomes

$$\mathrm{d}\psi/\mathrm{d}s=(k\sin^3\psi)\ln(\sin\psi/\sin\alpha) \quad \text{with } k=2U/b, \quad \alpha=\operatorname{cosec}^{-1}U. \tag{23}$$

3 Axisymmetric tip with small, constant flame speed and same parameter values as for Figure 2.

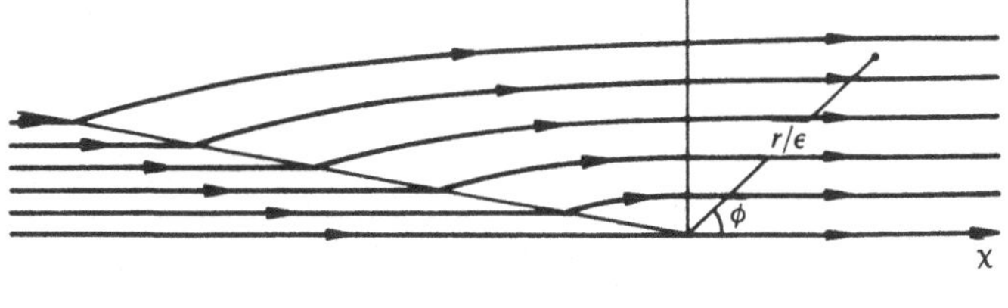

4 Plane SVF tips, drawn for $\alpha=\pi/6$ and $k=\pm 1$: (*a*) integral curves of equation (24); (*b*) possible shapes for $\mathscr{L}>1 (k>0)$; (*c*) only possible shape for $\mathscr{L}<1 (k<0)$.

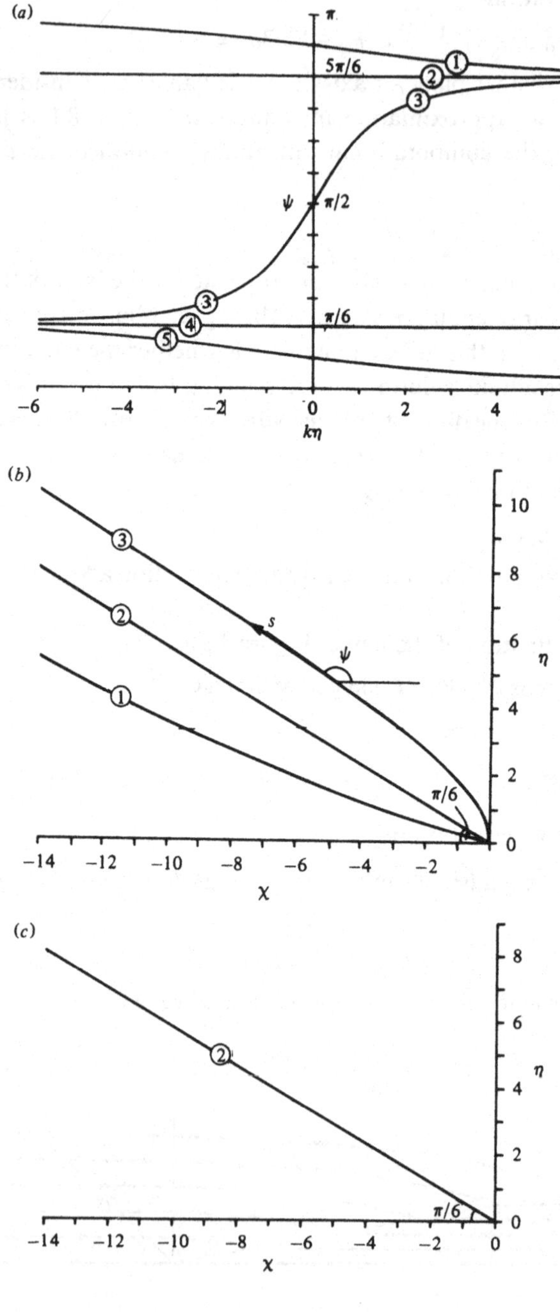

The implicit requirement $U>1$ simply reflects the fact that a stationary flame can only exist in a flow that moves faster than the adiabatic flame speed. For comparison with the axially symmetric case that follows, it is more convenient to use η as the independent variable, in terms of which the equation becomes

$$d\psi/d\eta=(k\sin^2\psi)\ln(\sin\psi/\sin\alpha). \tag{24}$$

Apart from $\psi=\alpha$ and $\pi-\alpha$, only three integral curves are shown in Figure 4(*a*). The remainder can be determined by horizontal translation. Those that lead to shapes with $\psi\rightarrow\alpha$ or $\pi-\alpha$ as $\chi\rightarrow-\infty$, as required, give the possibilities in Figure 4(*b*, *c*). In Figure 4(*b*) the curve ① and its reflection ⑤ in the χ-axis (not shown) represent a family filling in the wedge between the straight line ② and its reflection ④. In Figure 4(*c*), only these lines are obtained.

For $\mathscr{L}>1$ (i.e. $b, k>0$) there is a smooth tip, namely ③, and presumably that should be chosen rather than one of the family of cusped tips. For $\mathscr{L}<1$ (i.e. $b, k<0$), there is no smooth tip; only the wedge-shaped tip formed by ② and its reflection ④ results from the theory.

Turning to the axisymmetric case, we find that equations (21) and (22*a*, *b*) still hold, but that (22*c*) is replaced by

$$\kappa=-d\psi/ds+\cos\psi/\eta \tag{25}$$

because of the azimuthal curvature of the flame sheet. Hence, the governing equation now becomes

$$d\psi/d\eta-\sin\psi\cos\psi/\eta=(k\sin^2\psi)\ln(\sin\psi/\sin\alpha), \tag{26}$$

with k and α defined as before, where η ranges over positive values only.

Because of the new term the integral curves, shown in Figure 5(*a*), no longer have the translation property, but they have the same shape as in the plane case as $\eta\rightarrow\infty$. Only those asymptoting to $\pi-\alpha$ as $\eta\rightarrow\infty$ are acceptable, which leads to the open (including re-entrant) shapes of Figure 5(*b*), as well as one closed, and to the single open shape of Figure 5(*c*). Representative curves are shown, as marked; all open curves approach the centerline as $\chi\rightarrow\pm\infty$.

For $\mathscr{L}>1$ (i.e. $b, k>0$) a closed tip is possible, namely ②, and presumably that occurs rather than either a re-entrant or open one. (See also later remarks concerning flame temperature.) For $\mathscr{L}<1$ (i.e. $b, k<0$) only one tip is possible and that is open. In short, the theory of slowly varying flames predicts closed tips for $\mathscr{L}>1$ and open tips for $\mathscr{L}<1$. It is noteworthy that open tips are found experimentally (Figure 1) for both lean hydrogen and rich, heavy-hydrocarbon mixtures (Lewis & von Elbe

5 Axisymmetric SVF tips, with same parameter values as for Figure 3: (*a*) integral curves of equation (26); (*b*) possible shapes for $\mathscr{L} > 1 (k > 0)$; (*c*) only possible shape for $\mathscr{L} < 1 (k < 0)$.

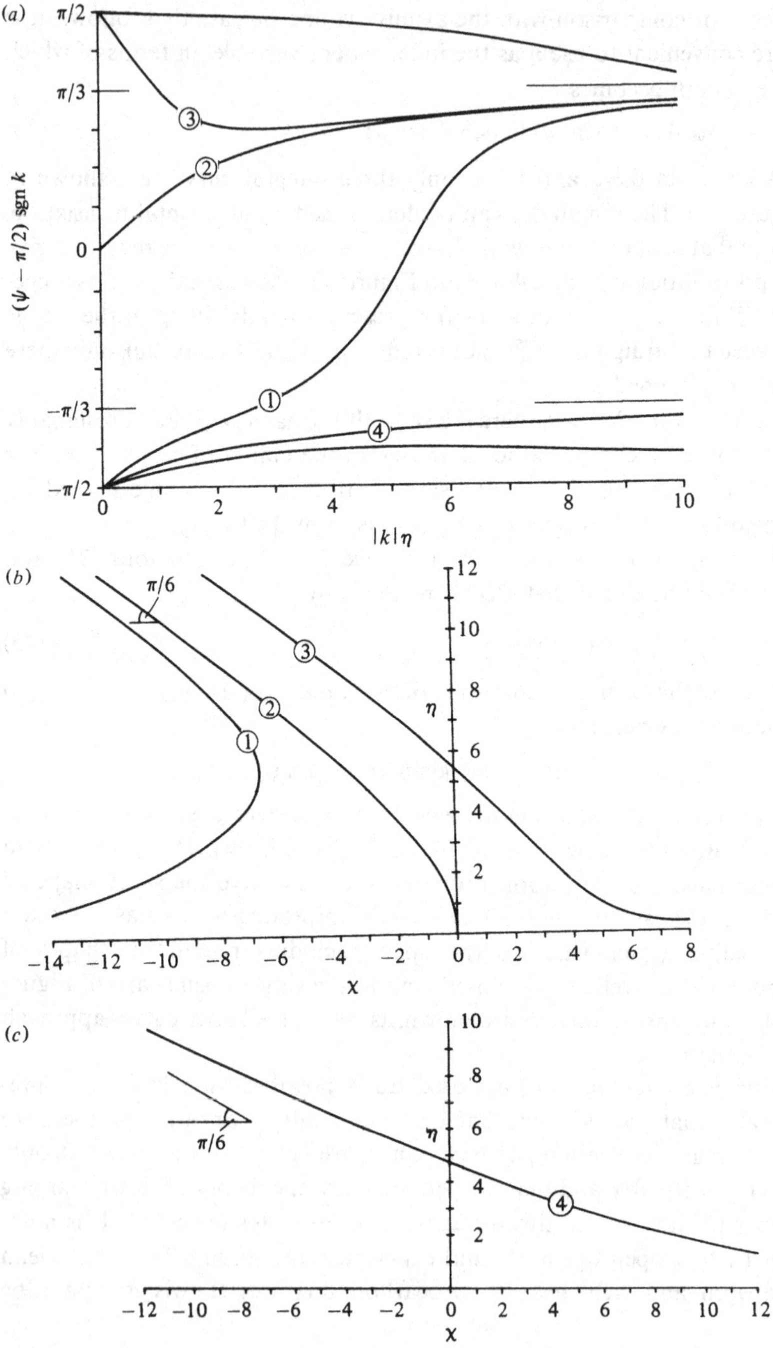

1961, p. 309). In each case the deficient component diffuses much more readily than the component in excess; a calculation based on values given by Kanury (1975, p. 387) shows that $\mathscr{L}=0.3$ for air containing small amounts of hydrogen at 1 atmosphere and 298 K. In this case the binary diffusion coefficient is that of H_2/air and all other thermodynamic data are those of air.

The flame temperature is readily deduced from the formula (8.29) as

$$T_b + 2\theta^{-1}T_b^2 \ln W = T_b + 2\theta^{-1}T_b^2 \ln(\sin\psi/\sin\alpha). \tag{27}$$

In all cases the adiabatic flame temperature T_b is approached, of course, as $\psi \to \pi - \alpha$ at large distances. For $\mathscr{L} > 1$, the temperature tends to a maximum $T_b + 2\theta^{-1}T_b^2 \ln \operatorname{cosec}\alpha$ at the very tip, irrespective of whether the flame is plane or axisymmetric. On the other hand, for $\mathscr{L} < 1$ the temperature does not change along a plane tip but decreases like

$$T_b + 2\theta^{-1}T_b^2 \ln(\pi - \psi) \tag{28}$$

for an axisymmetric tip, giving an unbounded perturbation. (The same result holds for the open tips in Figure 5(*b*), so that the closed tip selected is the only one without this singularity.) Such unboundedness means that the temperature differs by more than $O(\theta^{-1})$ from the adiabatic value, so the asymptotic structure is not uniformly valid and a different kind of analysis is required. While such an analysis has not yet been attempted, the drop in temperature suggests that the flame is extinguished, so that the profile of Figure 5(*c*) should be truncated at some large value of χ, giving a flame more in accord with experimentally observed open tips. For the plane tip, nonuniformity manifests itself geometrically rather than through a singularity in the flame temperature: the pointed tip violates the assumption of slow variation on which the analysis is based.

The dramatic change in the nature of the solution as $\mathscr{L}$ passes through 1, together with the breakdown of the analysis for $\mathscr{L} - 1 = O(\theta^{-1})$, suggests that flame tips should be investigated when the diffusion of heat is almost the same as that of the reactant, i.e. under the NEF assumption. That has been done by Buckmaster (1979*c*), whose work is considered in the next section.

4 NEF tips

The general formulation of NEFs as elliptic free-boundary problems is given in section 8.5. If the constant-density approximation is used to uncouple the fluid mechanics and the flame is immersed in a uniform flow whose speed is much greater than the flame speed, then these problems reduce to ones of Stefan type (as is shown in section 8.6).

To describe flame tips, take the χ-axis along the centerline. Far upstream the flame is either wedge shaped or conical with half-angle $\pi/4$, corresponding to the adiabatic value 1 for the flame speed. Only $y>0$ is considered, because of symmetry in the plane case and by definition in the axially symmetric case. Take

$$y=F(\chi) \tag{29}$$

as the locus of the flame, choosing the origin of χ sufficiently far upstream for $F(0)$ to be large. Near $\chi=0$ the combustion field is that of a plane flame, so that equations (3.84) hold with n measured vertically up, i.e. on $\chi=0$ we have

$$T=\begin{cases} T_f+Y_f\,e^{y-F(0)} \\ T_b=H_f \end{cases}, \quad h=\begin{cases} lY_f[F(0)-y]\,e^{y-F(0)} \\ 0 \end{cases} \quad \text{for } y \lessgtr F(0), \tag{30}$$

which will be taken as initial conditions. These conditions are valid only in the limit $F(0)\to\infty$, but any desired accuracy can be obtained by taking $F(0)$ sufficiently large. In the plane case the error is exponential (boundary conditions at $y=0$ being involved); for the axisymmetric problem it is algebraic (the structure neglecting terms such as $y^{-1}\partial T/\partial y$).

The equations to be solved are

$$\mathfrak{L}_\nu(T,h)=-lf(y)\partial(0,T)/\partial\chi \quad \text{for } 0<y<F(\chi), \tag{31}$$

$$T=H_f, \quad \mathfrak{L}_\nu(h)=0 \quad \text{for } y>F(\chi) \tag{32}$$

(cf. section 8.6), where

$$\mathfrak{L}_\nu \equiv \frac{1}{y^\nu}\frac{\partial}{\partial y}y^\nu\frac{\partial}{\partial y}-f(y)\frac{\partial}{\partial\chi} \quad \text{with } f(y)=1 \tag{33}$$

and $\nu=0$ or 1 accordingly as the flame is plane or axisymmetric. The accompanying jump conditions are

$$l\delta(\partial T/\partial y)=-\delta(\partial h/\partial y)=-lY_f\,e^{-\frac{1}{2}\phi_*} \quad \text{at } y=F(\chi), \tag{34}$$

T and h being continuous there. To these we must add the symmetry conditions

$$\partial T/\partial y=\partial h/\partial y=0 \quad \text{at } y=0 \tag{35}$$

and, finally,

$$h\to 0 \quad \text{as } y\to\infty, \tag{36}$$

consistent with the result (30*b*). Use of new variables $(T-T_f)/Y_f$ and h/T_b^2 shows that the problem depends on the single parameter lY_f/T_b^2, and not on l, Y_f, and T_f separately. This motivates the introduction of a parameter

$$\bar{l} \equiv l/l_s \quad \text{with } l_s = 2T_b^2/Y_f. \tag{37}$$

(The subscript s reflects the importance of l_s in flame stability.)

At first glance it seems questionable that this formulation, based as it is on the assumption that the gas speed is much greater than the flame speed, could provide an adequate description of flame tips. At a smooth, closed tip the two speeds are equal on the centerline. But there is no such difficulty for open tips; if the flame speed is small in the far field, we may expect it to remain small everywhere so that uniformly valid results are obtained. Our analysis could, therefore, lead to open-tip solutions only, whatever the parameter values. In fact, it also generates closed-tip solutions, albeit ones that are not valid near the centerline. Since the reduction to a parabolic problem requires only that the slope $U^{-1}F'(\chi)$ of the flame be small, the region of nonuniformity is small.

The analysis is completed by numerical integration, which for such parabolic systems involves marching downstream, i.e. in the direction of the time-like variable χ. Both plane and axisymmetric flames have been computed by Buckmaster (1979*c*), and these solutions can be expressed in terms of the single parameter $\bar{l}$.

Results for plane flames are shown in Figure 6. Note that there is no appreciable curvature until F has decreased to about 3, at least for $\bar{l} < 3.47$. Then the cold fringe of the preheat zone reaches the centerline and interaction with the conditions (35) begins, the effect of the interaction depending on l. For non-negative values the speed of the flame increases

6 Plane NEF tips.

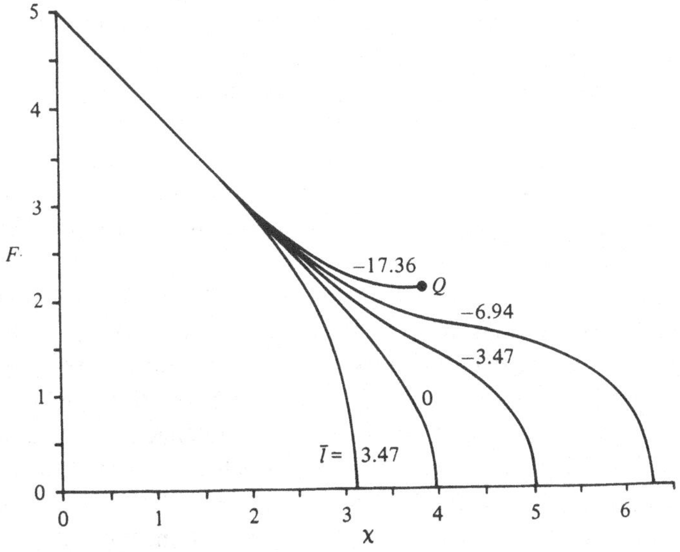

monotonically, bending it around towards the centerline, which it intersects at right angles. Apparently the tip is smooth with F proportional to $(\chi_0-\chi)^{\frac{1}{2}}$ near the intersection point χ_0, a conclusion that Buckmaster (1979*a*) supports by a similarity solution there for $l=0$. (Then the region of nonuniformity is characterized by $\chi_0-\chi=O(U^{-2})$.) These values of l give $\mathscr{L}>1$, for which smooth closed tips were also found for SVFs. Moreover, as $\bar{l}$ increases above 3.47 the flame begins to curve at progressively larger values of F, as if it were trying to get onto the scale of the SVF.

For small negative values of $\bar{l}$ the flame still closes smoothly, although its speed initially decreases, resulting in a somewhat extruded tip. For larger negative values the flame speed decreases to zero at the point labeled Q where the curve has a horizontal tangent. The curve for $\bar{l}=-17.36$ typifies this behavior. Again, as $|\bar{l}|$ increases, the flame curves at larger values of F, although Buckmaster (1979*c*) has proposed a limit solution as $\bar{l}\to-\infty$.

The perturbation temperature tends to follow the changes in flame speed. It has a finite negative value at Q, showing once more (cf. Chapter 7) that curved flames cannot necessarily be treated as locally plane ones (when ϕ_* would tend to ∞). The numerical integration can be continued downstream from Q without difficulty, the profile first curving away from the centerline and then back to it to form a closed tip. Buckmaster, however, has proposed to terminate the solution at χ_Q, so that Q is identified with quenching. There are several reasons why this hypothesis is attractive: there is no experimental evidence for the bulbous tip that would otherwise be obtained; the shape then resembles experimentally observed, open tips; the same hypothesis in a different context explains flame quenching due to the proximity of a cooled surface (section 5); termination of the reaction at the point where the flame speed vanishes is consistent with the views of Lewis & von Elbe (1961, p. 213) founded on experimental observation; and the disturbing reverse flow (from hot to cold) through the flame sheet immediately downstream of Q is avoided. All these reasons provide some hope that a mathematical justification may be forthcoming, based perhaps on questions of stability or existence. No sound evidence, however, is available at this time, and it must be noted that there are no suggestive mathematical clues either, unlike the situation for SVF tips. Others have adopted an equivalent hypothesis in related numerical problems for finite activation energy, also without clear mathematical justification (e.g. Westbrook, Adamczyk, & Lavoie 1981). A good understanding of the mechanism by which an evolving flame is quenched remains elusive.

If the hypothesis is correct, then downstream of χ_Q there is no chemical reaction; the unburnt gas below Q simply mixes with the burnt gas above.

That is, the values of T and h on $\chi=\chi_Q$ are used as initial conditions for a smooth solution of the parabolic equations (31), which now hold for all values of y. Such a reactionless solution is consistent in that T immediately drops below H_f, the temperature at which reaction occurs: while the initial temperature for $y>F(\chi_Q)$ is equal to H_f, that for $0<y<F(\chi_Q)$ is less. Agreement with results for SVFs for $\mathscr{L}<1$ can be obtained by supposing that the wedge shape found there is actually open at the vertex.

The axisymmetric results (Figure 7) are qualitatively similar, but there is one sharp quantitative difference. Azimuthal curvature affects the flame speed (through the terms in $y^{-1}\partial/\partial y$) long before the preheat zone reaches the centerline. As a consequence, the flame starts to curve much further from the centerline (at about $F=40$) than for the plane flame (moreover, Buckmaster's description of a limit solution as $\bar{l}\to-\infty$ is certainly incorrect in this case). Likewise, the dead space between Q and the centerline is more than five times larger. In short, open plane tips are much narrower than their axisymmetric counterparts, a result also found for SVFs with $\mathscr{L}<1$ if the wedge is supposed to be open. A similar comparison holds for $\mathscr{L}>1$, so that the NEF analysis provides a smooth transition from the general characteristics of SVFs for $\mathscr{L}<1$ to those for $\mathscr{L}>1$.

A word of caution is needed. The transition from a closed to an open tip occurs for large negative values of $\bar{l}$ between -6.94 and -17.36 for a plane flame and -17.36 and -34.72 for an axisymmetric one. Since $\bar{l}/\theta$ is assumed to be small, the result will probably be quantitatively meaningful only for large values of the activation energy θ: values so large that they might never be realized in practice.

7 Axisymmetric NEF tips.

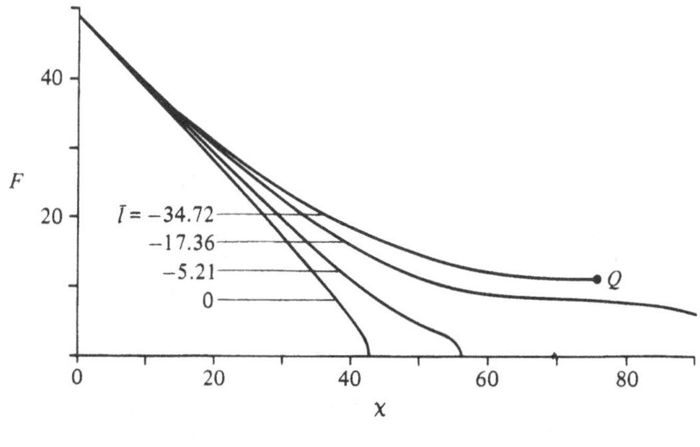

5 Quenching by a cold surface

Quenching near its base plays a crucial role in the existence and stability of a burner flame, as we have already remarked in section 1. The situation near the rim of a burner is complicated, however, involving not only heat transfer but also a multidimensional flow field with mixing between the reactants and the cold ambient atmosphere. In this section quenching by a cold wall is examined using a simple model (Buckmaster 1979*a*), which admittedly discards much that is relevant to the combustion field near a burner rim. Since heat transfer by conduction plays the central role in this phenomenon, NEFs provide an appropriate framework.

Consider a parallel flow of combustible gas moving above the plane $y=0$ with a speed Uy. A linear shear flow is the simplest representative of realistic flows near the surface, but any other could be handled just as easily. (Buckmaster also considers a uniform flow; the problem is then identical to that of a plane deflagration wave moving towards a parallel cold wall, with χ playing the role of time.) Located in the flow is a stationary premixed flame (Figure 8), which is influenced by the presence of the wall as a diffusion inhibitor and heat sink.

The flame locus is again described by equation (29) with the origin chosen to make $F(0)$ large. Far upstream, where the flame is not influenced by the wall, it still propagates with unit velocity but now into a flow with speed $UF(\chi)$. Consequently, it follows the parabola

$$F=[F^2(0)-2\chi]^{\frac{1}{2}}, \tag{38}$$

although this fact does not change the initial conditions (30) but is solely a check on the subsequent computations. (The reason that the flame speed is unaffected by the shear in the far field is that relative changes in gas velocity are small there – the stretch is weak.)

In focusing on the effect of heat loss we shall, for simplicity, take $l=0$.

8 Notation for NEF approaching cold wall.

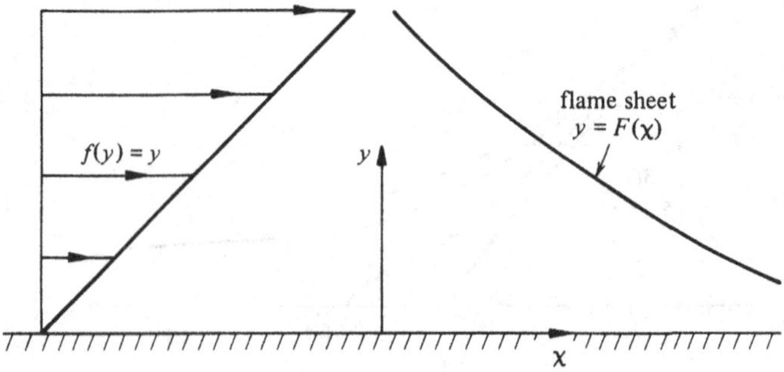

In the limit $U \to \infty$ the equations with which we have to deal are (31), (32) with $v=0$ and

$$f(y)=y \tag{39}$$

to take account of the shear flow. The accompanying jump conditions are

$$\delta(\partial T/\partial y)=-Y_f e^{-\frac{1}{2}\phi_*}, \quad \delta(\partial h/\partial y)=0 \quad \text{at } y=F(\chi) \tag{40}$$

while T and h are continuous there. The condition (36) still holds, of course.

Finally we come to conditions at the wall. The heat flux there is assumed small to avoid excessive variations in the flame temperature (see section 8.2); specifically

$$\partial T/\partial y=k(T-T_f)/\theta \quad \text{at } y=0 \tag{41}$$

to all orders, where k is a prescribed positive constant and T_f is, as before, the temperature of the fresh mixture. Such a condition is appropriate for a thin, poorly conducting wall whose inner surface is maintained at the temperature T_f. Otherwise, it may be considered an assumption that avoids coupling the temperature field in the wall with that in the flow, without compromising the essential physics. We shall also take the wall to be chemically inactive, i.e.

$$\partial Y/\partial y=0 \quad \text{at } y=0 \tag{42}$$

to all orders. The boundary conditions are, therefore,

$$\partial T/\partial y=0, \quad \partial h/\partial y=k(T-T_f) \quad \text{at } y=0. \tag{43}$$

The main simplification for $l=0$ is that h can be removed from the problem. Thus, $\mathfrak{L}_0(h)=0$ everywhere and h has no discontinuities at the flame sheet, so that it can be written explicitly in terms of its normal derivative at the wall and, hence, according to the condition (43b) the surface temperature. At the flame sheet it yields

$$\phi_*=k\frac{3^{\frac{1}{6}}\Gamma(\frac{1}{3})}{2\pi T_b^2}\int_{-\infty}^{\chi}\frac{[T(\bar{\chi},0)-T_f]}{(\chi-\bar{\chi})^{\frac{2}{3}}}\exp\left[-\frac{1}{9}\frac{F^3(\bar{\chi})}{\chi-\bar{\chi}}\right]\mathrm{d}\bar{\chi}, \tag{44}$$

which then determines the temperature gradient (40) on the cold side of the flame in terms of the surface temperature up to that station. We are left with a (numerical) problem for T and F alone. Use of the variable $(T-T_f)/Y_f$ shows that the solution depends on the single parameter $\hat{k} \equiv kY_f/T_b^2$.

Results are shown in Figure 9 for various values of the heat-loss parameter $\hat{k}$. The flame sheet intersects the wall at right angles when $\hat{k}$ is small, F being proportional to $(\chi_0-\chi)^{\frac{1}{3}}$ near the intersection point χ_0 (as Buckmaster shows by a similarity solution there). Exactly what happens very close to the wall cannot be revealed by our analysis, since the assumption of a large flow speed (which reduces the elliptic problem to a parabolic one) is not valid when y is small. A reasonable conclusion from the failure of the heat loss to generate a dead space on the scale of the flame thickness is that flashback occurs.

When $\hat{k}$ is large enough ($\geqslant 9.5$ is sufficient) the flame sheet becomes horizontal at some point Q, which Buckmaster identified with quenching, as in section 4. (Again the numerical solution can be continued beyond Q until ultimately the flame sheet intersects the wall, at least for values of $\hat{k}$ that are not too large. The behavior for large values is not known at the present time.) Failure to adopt such a hypothesis in this context would lead to the unexpected conclusion that $O(\theta^{-1})$ heat loss is incapable of quenching a flame. Its adoption leads to a picture in agreement with Lewis & von Elbe's (1961, p. 213) observation that unburnt mixture can pass downstream, the generation of such a gap being the mechanism by which heat loss prevents flashback. The conclusion is in sharp contrast with that for a plane flame (section 3.4), for which extinction occurs when the loss is sufficient to reduce the flame speed to $e^{-\frac{1}{2}}$.

6 Further hydrodynamic considerations

Uncoupling the fluid mechanics from the combustion process, either completely or nearly so, is a necessary prerequisite to the first attempt at treating multidimensional flames presented here. The physical effects thereby eliminated await future analysis. That they are likely to be important is clear, for example, from the belief that the convection induced

9 Flame-sheet profiles for cold-wall problem.

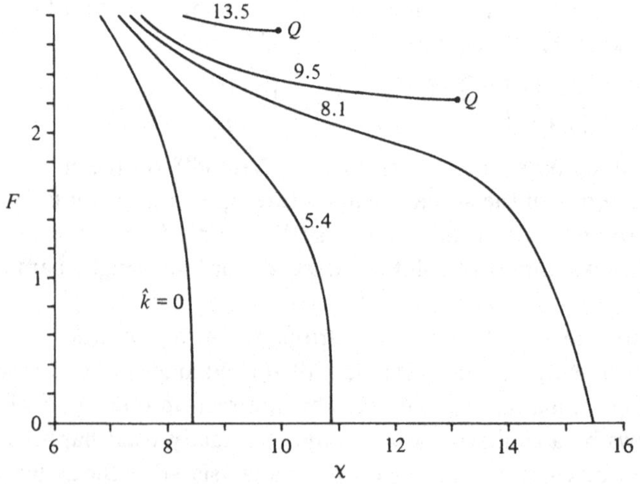

by a flame can be responsible for observed flammability limits in some circumstances, a topic ignored in Chapters 2 and 3.

The purpose of this section is to show, in the context of flame tips, that one simple class of combustion processes admits a full hydrodynamic treatment of no greater difficulty than the approximate one. In doing so, we shall uncover further details of the gas flow for a burner.

An important feature of the flame is the refraction at the hydrodynamic discontinuity (the limiting form (1), (2) was used in section 2). The corresponding structure is plane and so, for NEFs, may be deduced from the equations of section 8.5. It provides a smooth link between the states f and b, which are described by the generalization

$$(u_{\rm f}, v_{\rm f}, p_{\rm f}) = (U, 0, 0),$$

$$(u_{\rm b}, v_{\rm b}, p_{\rm b}) = (U + (\sigma - 1)/U, \quad (\sigma - 1)\sqrt{1 - U^{-2}}, \quad (1 - \sigma)) \tag{45}$$

of equations (1) and (2) when the flame speed is 1. The flame is now inclined to the incident stream at an angle $\sin^{-1}(U^{-1})$.

The actual structure is

$$T = T_{\rm f}[1 + (\sigma - 1)\,{\rm e}^n], \quad \rho = [1 + (\sigma - 1)\,{\rm e}^n]^{-1}, \quad h = -lY_{\rm f}n\,{\rm e}^n, \tag{46}$$

$$u = U + (\sigma - 1)\,{\rm e}^n/U, \quad v = (\sigma - 1)\sqrt{1 - U^{-2}}\,{\rm e}^n, \quad p = (\sigma - 1)(\tfrac{4}{3}\mathscr{P} - 1)\,{\rm e}^n \tag{47}$$

for $n < 0$, where n denotes distance normal to the flame sheet. There is a jump in pressure across the sheet from the value (47*c*) at $n = 0$, to its value in (45*f*), but all the other variables (46), (47) are continuous. Behind the sheet uniformity prevails, implying, in particular, that deflection of the streamlines is completed ahead of the reaction zone, a result in accordance with experimental observations (see Lewis & von Elbe 1961, p. 270)).

Equations (46), (47) give nothing more than the plane structure outlined in section 3.2, except for the addition of a constant velocity parallel to the flame sheet. The normal mass flux is also constant, so that the corresponding velocity increases from 1 ahead of the preheat zone to σ at the flame sheet. This clearly shows the distinction between using the hydrodynamic discontinuity or the flame sheet to define the flame speed when the constant-density approximation is not made. Often the scale is not large enough to give identity to the hydrodynamic discontinuity so that, to preserve the concept of flame speed in a manner consistent with an unbounded plane flame, the normal component of velocity at the flame sheet must be multiplied by $\rho_{\rm b} = 1/\sigma$.

For slender flames, i.e. as $U \to \infty$, the description (46), (47) simplifies and provides appropriate initial conditions for a flame-tip analysis, as in section 4 (see equations (30)). The corresponding full parabolic system is obtained through the scaling

$$\chi = U^{-1}x, \quad \bar{u} = U^{-1}u, \quad \bar{p} = U^{-2}p, \tag{48}$$

which leads in the limit to

$$\frac{\partial}{\partial \chi}(\rho \bar{u}) + \frac{1}{y^\nu}\frac{\partial}{\partial y}(y^\nu \rho v) = 0, \quad \left(\bar{u}\frac{\partial \bar{u}}{\partial \chi} + v\frac{\partial \bar{u}}{\partial y}\right) = -\frac{\partial \bar{p}}{\partial \chi} + \mathscr{P}\mathfrak{D}_\nu(\bar{u}), \tag{49}$$

$$\frac{\partial \bar{p}}{\partial y} = 0, \quad \rho\left(\bar{u}\frac{\partial T}{\partial \chi} + v\frac{\partial T}{\partial y}\right) = \mathfrak{D}_\nu(T), \tag{50}$$

$$\rho\left(\bar{u}\frac{\partial h}{\partial \chi} + v\frac{\partial h}{\partial y}\right) = \mathfrak{D}_\nu(h + lT), \quad \rho T = T_{\mathrm{f}}, \tag{51}$$

where

$$\mathfrak{D}_\nu \equiv \frac{1}{y^\nu}\frac{\partial}{\partial y} y^\nu k(T)\frac{\partial}{\partial y}. \tag{52}$$

The equations are a generalization of those that would be derived from Chapter 1, introduced for this section only. The generalization, represented by $k(T)$, stems from recognizing that in practice the transport coefficients are strongly temperature dependent: the viscosity of the burnt gas can be 3 or 4 times that of the fresh mixture. While the dependence is unlikely to have marked qualitative effects (one reason for its neglect elsewhere in our treatment), quantitative effects may be significant. It is introduced here by supposing that the transport coefficients are proportional, i.e. setting

$$\kappa/\kappa_{\mathrm{b}} = \lambda/\lambda_{\mathrm{b}} = \mu/\mu_{\mathrm{b}} = k(T), \tag{53}$$

where the subscript denotes a reference state, taken to be that for the $O(1)$ flame temperature. The reference values are then used for the nondimensionalization in section 1.6.

Proportionality of the transport coefficients can be justified by kinetic theory (Williams 1965, p. 414), as can the choice

$$k = (T/T_{\mathrm{b}})^\beta, \tag{54}$$

although the value $\beta = \frac{1}{2}$ obtained thereby will not be used here. Instead we shall exploit $\beta = 1$, which has been used extensively in reactionless gas dynamics (Stewartson 1964, p. 5). Of course, any choice for k leaves the adiabatic flame speed unchanged.

The jump conditions at the flame sheet also simplify, but we shall omit the results. Note, however, that $\bar{p}$ is continuous.

An appropriate solution of the two momentum equations is

$$\bar{u} = 1, \quad \bar{p} = 0. \tag{55}$$

Then the continuity and temperature equations simplify to

$$\partial\rho/\partial\chi + y^{-\nu}\partial(y^{\nu}\rho v)/\partial y = 0, \quad \rho(\partial T/\partial\chi + v\partial T/\partial y) = \mathfrak{D}_{\nu}(T), \tag{56}$$

with a corresponding equation for h. The jump conditions (34) still hold.

This simplified system is identical to that describing one-dimensional unsteady propagation of a flame, either plane or cylindrical. Moreover, for the plane case it can be reduced to exactly the same system that arises from the constant-density approximation. Thus, we define a stream function by

$$\rho/\rho_b = \partial\psi/\partial y, \quad \rho v/\rho_b = -\partial\psi/\partial\chi \tag{57}$$

and use $\bar{\chi} = \chi/\rho_b$ and ψ as new independent variables. Equation (56*b*) then becomes

$$\partial T/\partial\bar{\chi} = \partial^2 T/\partial\psi^2 \tag{58}$$

by virtue of Charles's law and the choice (54) with $\beta = 1$; a similar equation holds for h. Since the jump conditions are unchanged (except for ψ replacing y), Figure 6 is still valid provided the ordinate is relabeled ψ_* and the abscissa $\bar{\chi}$. The flame speed is zero at Q, as before, because the streamlines are lines of constant ψ.

Details of the flow in the χ, y-plane have not been calculated, but some general conclusions may be drawn in the axisymmetric context, as well as the plane. If equation (56*b*) is written in divergence form with the help of the continuity equation, and then Charles's law is used, we find

$$\frac{\partial}{\partial y}(y^{\nu}\rho v T) = \frac{\partial}{\partial y}\left(y^{\nu}k\frac{\partial T}{\partial y}\right); \tag{59}$$

therefore

$$v = (k/T_f)\partial T/\partial y + y^{-\nu}f(\chi) \tag{60}$$

for some function f. On the cold side of the flame symmetry conditions imply $f = 0$ so that, in particular,

$$v_* = (\sigma - 1)\,\mathrm{e}^{-\frac{1}{2}\phi_*}, \tag{61}$$

a remarkable relation between the local gas speed and the flame-temperature perturbation $-T_b^2\phi_*$. Note that v_* is positive: heating of the gas as it approaches the flame lowers its density and, hence, the axial mass flux, the deficit appearing as a radial flux.

For a plane flame the relation has implications concerning its shape. Mass conservation in the region bounded by the axes and the flame sheet implies

$$\int_0^{F(0)} \rho_i\,\mathrm{d}y = \int_0^{F(0)} \rho_*\,\mathrm{d}y + \int_0^a \rho_* v_*\,\mathrm{d}\chi, \tag{62}$$

where ρ_i is the density at the initial station $\chi=0$ and a is the length of the flame, i.e. $F(a)=0$. Charles's law relates ρ_i to T_i and, if $F(0)$ is large, T_i is a function of y alone, determined by

$$T_i'=[k(T_i)T_i']'. \tag{63}$$

This equation can be integrated explicitly (it has the solution (30a) for $k\equiv1$) but we only need the first integral in the form $dy=k\,dT_i/(T_i-T_f)$. Rewriting the ρ_i-integral as $F(0)+\int_0^{F(0)}(T_f/T_i-1)\,dy$ now shows that equation (62) is

$$F(0)-a=[\sigma/(\sigma-1)]\int_{T_f}^{T_b} k\,dT/T+\int_0^a (e^{-\phi_*/2}-1)\,d\chi. \tag{64}$$

The left side is the distance between the tip and the downstream intersection of the upstream asymptotes of the flame, i.e. a measure of the bluntness of the flame. The right side can be evaluated for $\mathscr{L}=1$ since then $\phi_*=0$ and the second integral vanishes. In particular,

$$F(0)-a=\begin{cases}\sigma(1-\sigma^{-\beta})/\beta(\sigma-1) & \text{for } \beta\neq0,\\ \sigma\ln\sigma/(\sigma-1) & \text{for } \beta=0\end{cases} \tag{65}$$

under the choice (54). The bluntness decreases as β increases, with $F(0)-a$ passing through 1 for $\beta=1$. Lewis-number effects cannot be explicitly calculated, since, in general, ϕ_* is not known. However, if ϕ_* is negative (positive) the bluntness is increased (decreased) and this is associated with $l>0(<0)$. The result (65b) gives a value 1 in the limit $\sigma\to1$, in agreement with Figure 6 when $\bar{l}=0$.

For axisymmetric tips the far-field perturbations generated by the curvature are not integrable, so that a similar treatment is not useful.

10

EFFECTS OF SHEAR AND STRAIN

1 Nonuniformities

An understanding of the response of a premixed flame to nonuniformities in the gas flow is important in many technological situations. To sustain a flame in a high-velocity stream the turbine engineer must provide anchors, and these generate strong shear. The designer of an internal combustion engine is concerned with the burning rate in the swirling flow of the mixture above the piston. Turbulence is ubiquitous; then the flame is subject to highly unsteady shear and strain. These situations are extremely complicated and it is unlikely that mathematical analysis will ever provide detailed descriptions; those must be left to empirical studies augmented by extensive numerical computations. Nevertheless, analysis of the response of a flame to a simple shear, for example, can provide useful insight into the interaction mechanism in more complex situations.

Moreover, there are simple circumstances in which such an analysis has direct significance. A burner flame is subject to shear in the neighborhood of the rim, and its quenching depends on the local character of that shear. A flame immersed in a laminar boundary layer experiences both shear (due to velocity variations across the layer) and strain (due to streamwise variations) and its quenching will depend on their local values.

In a constant-density flow the relative motion near a point is the superposition of a simple shear and two simple strains (cf. Batchelor 1967, p. 79). These two elementary flows (simple shear and simple strain) are, therefore, basic in any discussion of the effect of nonuniformities on flames; this chapter is mainly devoted to them. Only premixed flames are discussed; except at the end (sections 5 and 6), the formulation adopted is for NEFs, as simplified in section 8.6. The response of a diffusion flame to simple strain (counterflow) has effectively been dealt with in Chapter 6.

A fundamental question is whether or not a nonuniform flow by itself can extinguish a flame. We shall consider whether such flows can reduce the flame speed to zero; in accordance with Buckmaster's hypothesis (section 9.4), this phenomenon will be called quenching (noting again that the idea is unproven). Section 9.5 shows that the proximity of a cold wall can quench an equidiffusional flame, the mechanism involved being enthalpy loss from the combustion field rather than the geometrical constraint imposed on it by the wall. If nonuniformities in the flow can induce a similar enthalpy flux, quenching will occur with the cold unburnt gas replacing the wall as an enthalpy sink. But even if they cannot, we may still expect unequal diffusion of heat and reactant (i.e. $l \neq 0$) to provide other circumstances leading to quenching, since that is the case for the flame tips discussed in section 9.4. When l is sufficiently negative, the presence of a boundary quenches the plane tips and flame curvature quenches axisymmetric ones, without the help of heat loss.

2 Response of NEFs to simple shear

Consider a plane parallel flow that is uniform for $y>0$ but linearly sheared for $y<0$, i.e.

$$\mathbf{v}=(Uf, 0) \tag{1}$$

where, with $\omega>0$,

$$f(y)=\begin{cases}1 \\ 1-\omega y\end{cases} \quad \text{for } y \gtrless 0 \tag{2}$$

(see Figure 1). In the limit $U \to \infty$ the equations with which we have to deal are (9.31) for $-\infty<y<F(\chi)$ and (9.32) for $F(\chi)<y<\infty$, with $v=0$ and f having the definition (2). These equations are augmented by the jump conditions (9.34), together with the boundary conditions (9.36) and

$$T \to T_{\mathrm{f}}, \quad h \to 0 \quad \text{as } y \to -\infty. \tag{3}$$

The uniform part of the flow can support a steady plane deflagration wave inclined at an angle such that the normal component of the gas speed equals the adiabatic flame speed, and this is assumed to be the form of the combustion field for y large and positive (Figure 1). The initial conditions (9.30) still hold, and the first of these for $-\infty<y<F(0)$. As the flame approaches the χ-axis it is influenced by the shear as soon as its preheat zone penetrates the lower half plane. The increased gas speed there may be expected to deflect the flame, so that its slope decreases. Note that from the point of view of the flame there is no such thing as a

simple shear (with no strain): there are inevitably variations in the tangential component of gas speed at the flame that stretch it (see section 8.4), and the response of an SVF can, in principle, be completely understood in this context. For the NEF, shear in the preheat zone combines with this stretch to produce effects quite different from stretch alone, as we shall see.

This formulation is due to Buckmaster (1979*a*), who reaches the following conclusions. For some values of the shear gradient ω and Lewis-number parameter l, but not all, the slope actually decreases to zero. At that point the flame speed is zero and, according to the hypothesis of section 9.4, quenching occurs. Invariably the quenching point lies in the uniform region even though the preheat zone has penetrated the shear region. A rationalization for the absence of quenching in the shear region is that the nonuniformity, as measured by f'/f, has a maximum at $y=0$ so that, if the flame manages to cross the boundary, it can survive the ever-weaker nonuniformity beyond. Certainly there is a solution of the governing equations as $y \to -\infty$ corresponding to a locally plane flame propagating with unit speed (cf. section 9.5).

Clear evidence of the dichotomy is shown for $\omega=5$ by the numerical results plotted in Figure 2. With ω assigned the solution depends only on the single parameter $\bar{l}$ (see the definition (9.37)). For $\bar{l}=0$ or 1.74 the flame penetrates the shear and shows no sign of quenching; indeed no solution shows the flame speed approaching zero in $y<0$. But for $\bar{l}=3.47$

1 Notation for NEF approaching simple shear.

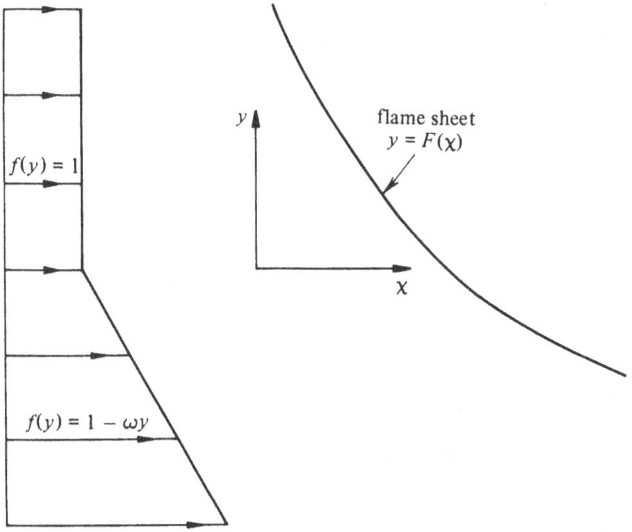

the flame speed falls to zero at the point Q outside the shear region. The figure also shows that the shear has more effect on the flame as $\bar{l}$ increases, in contrast to flame tips (Figure 9.6) where the boundary (i.e. symmetry line) has less. Moreover, this trend is quite different from that for an SVF (section 5).

The effect of increasing ω can be seen from the limit $\omega \to \infty$. There is then a fully developed combustion field, i.e. an asymptotic solution as $\chi \to \infty$, in which the flame lies at a definite level in $y > 0$ with T and h independent of χ. In fact, in the shear region away from the boundary, convection dominates for all χ so that

$$\partial T/\partial \chi = \partial h/\partial \chi = 0 \quad \text{for } y < 0, \tag{4}$$

i.e. there is no dependence on χ anywhere. The problem is, therefore, reduced to one in $y > 0$ only, under the boundary conditions

$$T = T_f, \quad h = 0 \quad \text{on } y = 0. \tag{5}$$

(Such a cold surface does not inevitably quench the flame because of the opposing effect from the reactant supplied there.) In the limit $\chi \to \infty$ the initial conditions may be ignored and the system then has the solution

$$T = \begin{cases} T_f + Y_f y/F(\infty) \\ H_f \end{cases}, \quad h = \begin{cases} -lY_f y/F(\infty) \\ -lY_f \end{cases} \quad \text{for } y \lessgtr F(\infty), \tag{6}$$

where

$$F(\infty) = e^{\bar{l}} \tag{7}$$

is the asymptotic height of the flame. The limit solution (6) violates the boundary condition (9.36) for h, showing that the limits $y \to \infty$ and $\chi \to \infty$ do not commute. There is an adjustment at large values of y that depends on the history of the flame; but that is of no importance for our purposes.

2 Flame-sheet profiles for simple-shear problem when $\omega = 5$.

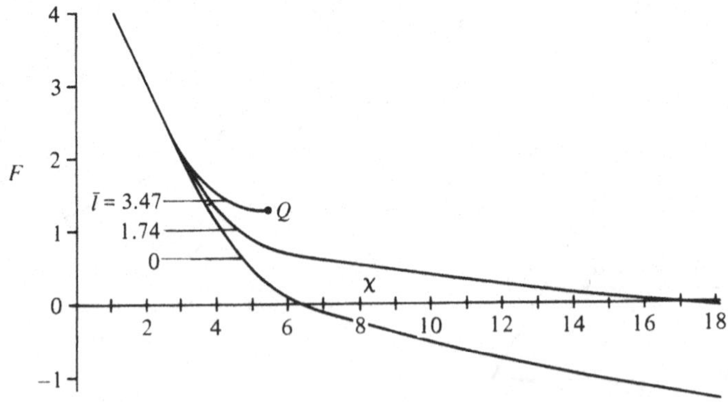

Even for moderate values of $\bar{l}$, $F(\infty)$ is large. On the other hand, inspection of Figure 2 shows that flames are first affected by the shear when F is about 2. The implication when $\omega = \infty$ is that, for sufficiently large $l > 0$, there is a minimum in F, i.e. the flame speed must have fallen to zero at some finite value of χ. Numerical solutions of the limit problem (Figure 3) indicate that $l = 0$ is the critical value: for $l < 0$ the function F decreases monotonically to its asymptotic value (7), whereas for $l > 0$ it has a minimum, corresponding to quenching. We conclude that a sufficiently strong shear flow will prevent a flame from reaching the boundary $y = 0$ and actually quenches it for $\mathscr{L} > 1$.

Eventually the limit equations break down and a new coordinate χ/ω is needed, corresponding to significant diffusion in the shear region. Buckmaster (1979*a*) argues that F increases again when l is sufficiently negative, which implies quenching for $\mathscr{L}$ sufficiently smaller than 1. No numerical results exhibiting such quenching, however, have yet been obtained.

3 Response of NEFs to simple strain

In contrast to simple shear, a simple straining flow involves variations of the velocity in its own direction. The distinguished limit of section 8.6, in which the nondimensional gas velocity and the length scale on which it varies become unbounded but their ratio remains fixed, leads to the governing equations (8.61). The unit of the velocity gradient repre-

3 Flame-sheet profiles for simple-shear problem when $\omega \to \infty$. The asymptotes are at $F = 0.5$, 1, 1.42.

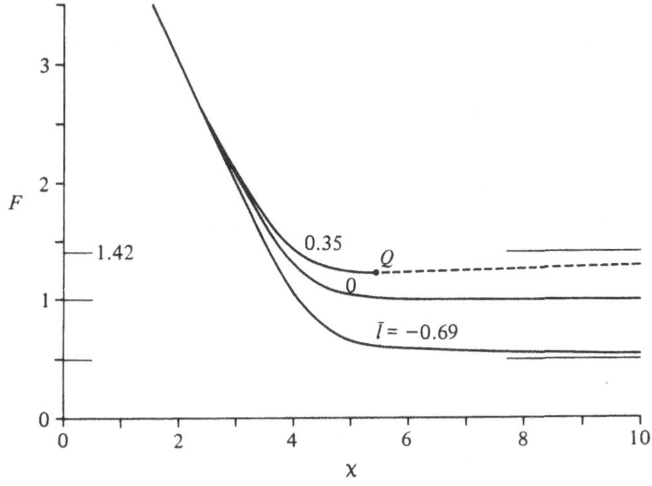

sented by ε is that of the flame speed divided by the thickness under adiabatic conditions. As we shall see, a velocity gradient that is small on this scale has little effect on the flame, whereas a sufficiently large gradient extinguishes it.

Equations (8.61) hold ahead of the flame, i.e. for $y > F(x)$. Behind the flame we have

$$T = H_{\mathrm{f}}, \quad h = -T_{\mathrm{b}}^2 \phi_*, \quad \text{for } 0 < y < F(x) \tag{8}$$

if either (*a*) the wall is insulating or (*b*) the flame is located in equal opposing jets of fresh mixture (Figure 4) so that

$$\partial h/\partial y = 0 \quad \text{on } y = 0. \tag{9}$$

(The condition ensures that wall quenching does not intrude.) For case (*a*) the flame may, in reality, be imbedded in a boundary layer (unless the Prandtl number is very small), so that the model is more directly relevant to case (*b*). We must add the boundary conditions

$$T \to T_{\mathrm{f}}, \quad h \to 0 \quad \text{as } y \to \infty \tag{10}$$

and the jump conditions (9.34).

Since x only appears in F there is a solution with the flame lying parallel to $y = 0$, the constant value y_* being given implicitly by

$$\delta = (\pi^{\frac{1}{2}}/2)\,\mathrm{erfc}(d)\exp\{d^2 + [d\,\mathrm{e}^{-d^2}/\pi^{\frac{1}{2}}\,\mathrm{erfc}(d) - \tfrac{1}{2} - d^2]\bar{l}\}$$
$$\text{with } \delta = (\varepsilon/2)^{\frac{1}{2}}, \quad d = \delta y_*; \tag{11}$$

here $\bar{l}$ has the definition (9.37). The solution ahead of the flame, i.e. for $y > y_*$, is

$$T = T_{\mathrm{f}} + Y_{\mathrm{f}} \frac{\mathrm{erfc}(\delta y)}{\mathrm{erfc}(d)}, \quad h = -T_{\mathrm{b}}^2 \phi_* \frac{\mathrm{erfc}(\delta y)}{\mathrm{erfc}(d)}$$
$$+ \frac{l Y_{\mathrm{f}}}{\pi^{\frac{1}{2}}\,\mathrm{erfc}^2(d)} (\delta y\,\mathrm{e}^{-\delta^2 y^2}\,\mathrm{erfc}(d) - d\,\mathrm{e}^{-d^2}\,\mathrm{erfc}(\delta y)], \tag{12}$$

4 Notation for NEF in simple-strain flow at stagnation point: (*a*) insulating wall; (*b*) equal opposing jets.

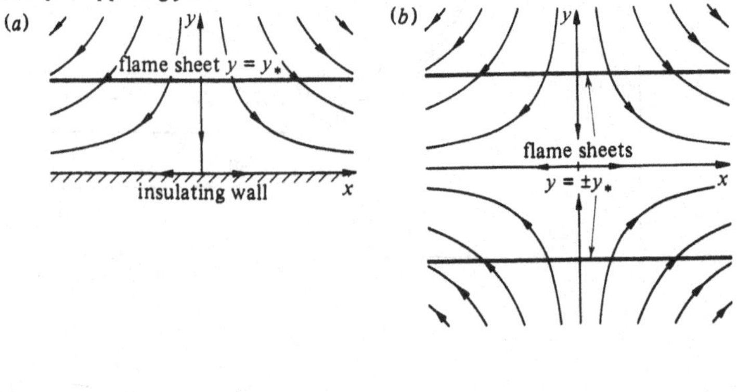

where

$$\phi_* = 2\ \ln[\pi^{\frac{1}{2}}\ e^{d^2}\ \mathrm{erfc}(d)/2\delta]. \tag{13}$$

This determination (giving $\phi_* = 0$ for $l = 0$) completes the description (8).

Of greater interest is the variation in stand-off distance y_* with the straining rate ε, which depends on the single parameter $\bar{l}$ (Figure 5). There are two kinds of response: for $\bar{l} < 2$, it is single valued, with y_* decreasing monotonically to zero at the critical value

$$\varepsilon_c = (\pi/2)\,e^{-\bar{l}}, \tag{14}$$

and for $\varepsilon > \varepsilon_c$, there is no solution; for $\bar{l} > 2$, the response is double valued for some values of ε and there is no solution beyond an extinction value ε_e where $y_* = y_e$ is nonzero. Corresponding values of ε_e and y_e for various values of $\bar{l}$ are plotted in Figure 6.

The picture is now clear. Suppose the strain is increased from a small value sufficiently slowly for the combustion field to be quasi-steady. The increase causes the flame to move closer to the stagnation point. If $\bar{l}$ is less than 2, it reaches the stagnation point for the critical value ε_c, and is extinguished. Otherwise, extinction occurs for ε_e, when it is still a distance y_e away. In the latter case the lower branch of the response (where y_* increases with ε) is unstable, as has been shown by Buckmaster (1979*b*). (His treatment does not rule out other parts of the response being unstable too, so that this extinction picture may have to be modified.)

The flame speed W (Figure 7), i.e. the gas speed normal to the flame

5 Variation of stand-off distance y_* with straining rate ε in stagnation-point flow.

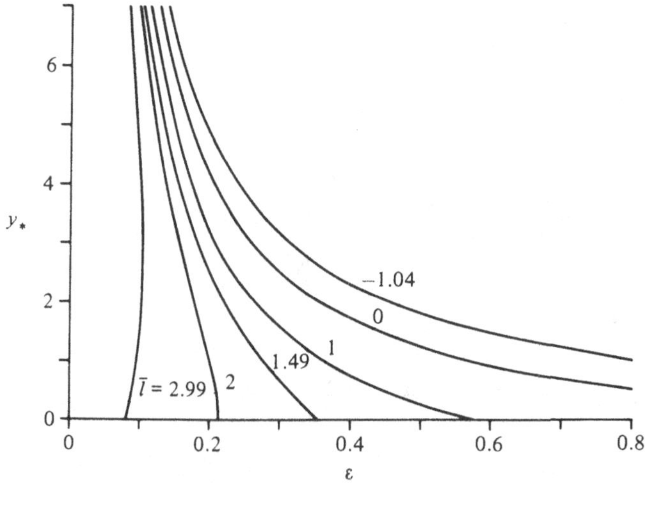

sheet, has the value εy_*. As $\varepsilon \to 0$, so that $y_* \to \infty$, it tends to the adiabatic value 1; when $\varepsilon = \varepsilon_c$ it vanishes, since y_* is zero. Like y_* it is single valued for $\bar{l} < 2$ and partly double valued for $\bar{l} > 2$. There is an initial increase in W for $\bar{l} < -1$; otherwise it decreases monotonically until $\varepsilon = \varepsilon_c$ or ε_e. It is, therefore, possible for the flame to have a speed greater than its

6 Variation of extinction straining rate ε_e and corresponding stand-off distance y_e with Lewis number l in stagnation-point flow.

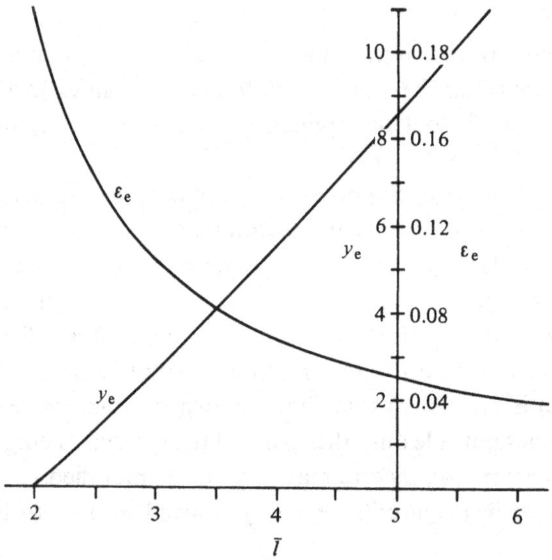

7 Variation of flame speed W with straining rate ε in stagnation-point flow.

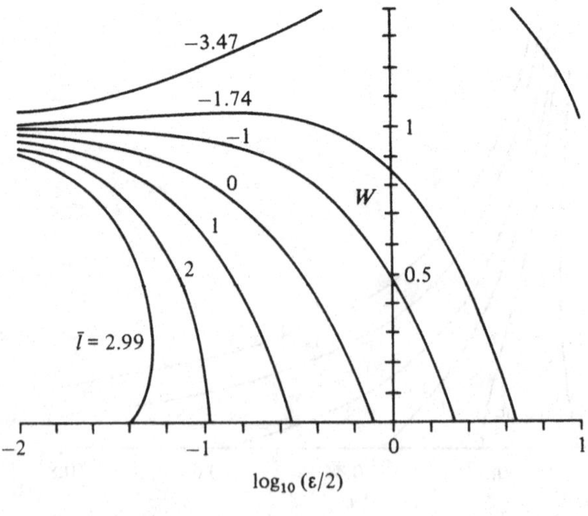

adiabatic value, even though it is being stretched at a positive rate. Such behavior is not solely a feature of stagnation-point flow. For small ε we have

$$W=1-\varepsilon(1+\bar{l})+O(\varepsilon^2), \tag{15}$$

and, when ε is identified with the stretch, the result is universal (Buckmaster 1981). It describes the response of a flame (steady or unsteady) to any nonuniform flow that changes on a scale large compared to the flame thickness (weak stretch) and has important ramifications in flame stability (section 11.4). Equation (3.88) is the equivalent one-dimensional result when the constant-density approximation is not made.

In short, there is a limit (depending on the Lewis number) to the amount of strain that a premixed flame can tolerate. The conclusion is consistent with the experimental fact that a flame held at the front of a bluff body can be blown off by increasing the gas speed sufficiently; attachment is then usually transferred to the rear. Moreover, in an experiment with lean methane/air flames (for which $\mathscr{L}<1$) extinction should occur at the wall, whereas with rich methane/air flames it should occur some distance away (assuming $\bar{l}>2$); this is precisely what has been observed by Tsuji & Yamaoka (1981).

The situation is a little different if the counterflow is burnt gas close to the adiabatic flame temperature, rather than fresh mixture. Then the condition (9) is replaced by

$$h\to h_r \quad \text{as } y\to-\infty, \tag{16}$$

which specifies the temperature of the remote gas. Only the first of the formulas (8) survives, valid now in $y<F(x)$, but all other equations and boundary conditions are unchanged. The situation is relevant perhaps to turbulent premixed flames in which contiguous pockets of burnt gas and fresh mixture are simultaneously stretched, and has been analyzed with that perspective in the special case $l=h_r=0$ (but without the constant density approximation) by Libby & Williams (1981). Within the present framework the location of the flame sheet is given in terms of the three parameters ε, $\bar{l}$ and h_r/T_b^2 by the implicit formula

$$\delta=(\tfrac{1}{2}\pi^{\frac{1}{2}})\,\mathrm{erfc}(d)\exp\{d^2+\tfrac{1}{4}(h_r/T_b^2)\,\mathrm{erfc}(d) \\ +\mathrm{erfc}(-d)[d\,\mathrm{e}^{-d^2}/\pi^{\frac{1}{2}}\,\mathrm{erfc}(d)-\tfrac{1}{2}-d^2](\bar{l}/2)\}. \tag{17}$$

As an example of the kinds of response obtained thereby, Figure 8(*a*) shows variations of y_* with the straining rate ε for different values of h_r/T_b^2 when $l=0$. For large enough straining rates y_* is negative, corresponding to negative flame speeds. Just as for the evolution problems treated in section 2 and Chapter 9 there is no apparent mathematical objection to this situation, fresh mixture reaching the reaction zone by means of a diffusion velocity which is greater in magnitude than the negative mass-average velocity. The steady-state analysis predicts that the flame can tolerate arbitrarily large straining rates, a conclusion that is true for all l (Buckmaster & Mikolaitis 1982). For some values of the parameters (e.g. $l = h_r/T_b^2 = 0$) the flame responds smoothly to a slow increase in ε, but for others ($l = 0$, $h_r/T_b^2 = -8$, say) for which the response

8 Variation of (*a*) stand-off distance y_* and (*b*) flame speed W with straining rate ε when the counterflow is burnt mixture and the Lewis number is exactly $1 (l = 0)$.

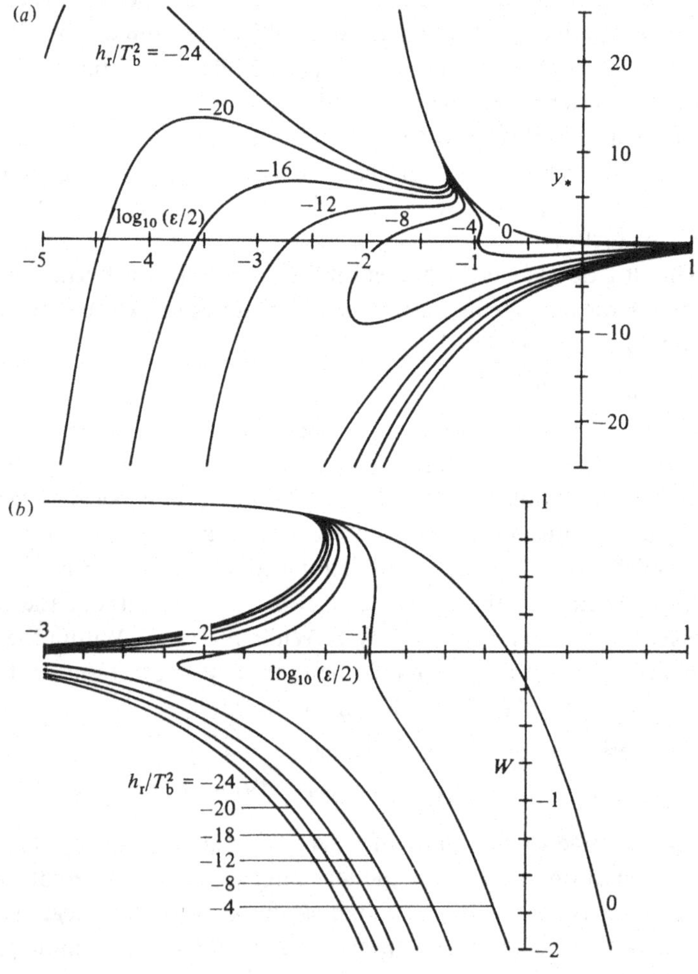

is multivalued, there is a jump from a position ahead of the stagnation point to one behind it as a critical value of ε is passed. Variations of the flame speed W are shown in Figure 8(*b*).

4 More general nonuniformity

Reduction to a parabolic problem occurs for general nonuniform flows in which the velocity is everywhere large but changes proportionately over correspondingly large distances (Buckmaster 1979*b*). An example is the potential motion of a fast stream past a large cylinder, whose front stagnation point was the subject of the last section. Such flows, when plane, are characterized by velocity fields

$$\boldsymbol{v} = U\boldsymbol{q}(x/U, y/U); \tag{18}$$

letting $U \to \infty$ gives a distinguished limit in which $\nabla\boldsymbol{v}$ remains finite. We shall see that the effective part of the limiting motion is, on the x, y-scale, made up of a fast uniform flow and a simple strain, both of which vary slowly. Shear is ineffective, so that the term general straining motion is appropriate.

Take curvilinear coordinates with x measuring distance along one particular streamline and y distance from it. Away from a stagnation point these will be slowly varying Cartesian coordinates, since the curvature of streamlines is vanishingly small. The angle between the flame sheet and an intersecting streamline is also small, so that the variables (8.58) and y are appropriate for T and h. Likewise, the solenoidal velocity field may be approximated by

$$\boldsymbol{v} = (Uq_0, -q_0' y), \tag{19}$$

where $q_0(\chi)$ is the speed on $y=0$. Only leading terms have been retained, so that a simple shear due to variation of the x-component in the y-direction is ignored. Uniformity breaks down at large distances from the flame sheet, but it does not matter there. On the x, y-scale the velocity relative to that at any point is a simple strain with $\varepsilon = q_0'$ (section 8.6). Together with the fast uniform flow $(Uq_0, 0)$ at the point, both of which vary with χ, it makes up the effective part of the limiting motion.

In the limit, only the approximation (19) enters into the governing equations (8.55), which become

$$(q_0\partial/\partial\chi - q_0' y\partial/\partial y)(T, h) = \partial^2(T, h + lT)/\partial y^2. \tag{20}$$

The response to an arbitrary streamwise velocity $q_0(\chi)$ may be determined from the computer as easily as that to simple shear (section 2). Note that equations (8.59), which were used in section 9.4, are the special case $q_0 \equiv 1$

when the strain vanishes everywhere. The simple shear flows discussed in section 9.5 and section 2 are not of the type considered here since changes of the velocity field in the y-direction are large. As already noted, the analysis breaks down at a stagnation point, so that equations (8.61) cannot be recovered. Nevertheless, the stagnation-point solution in section 3

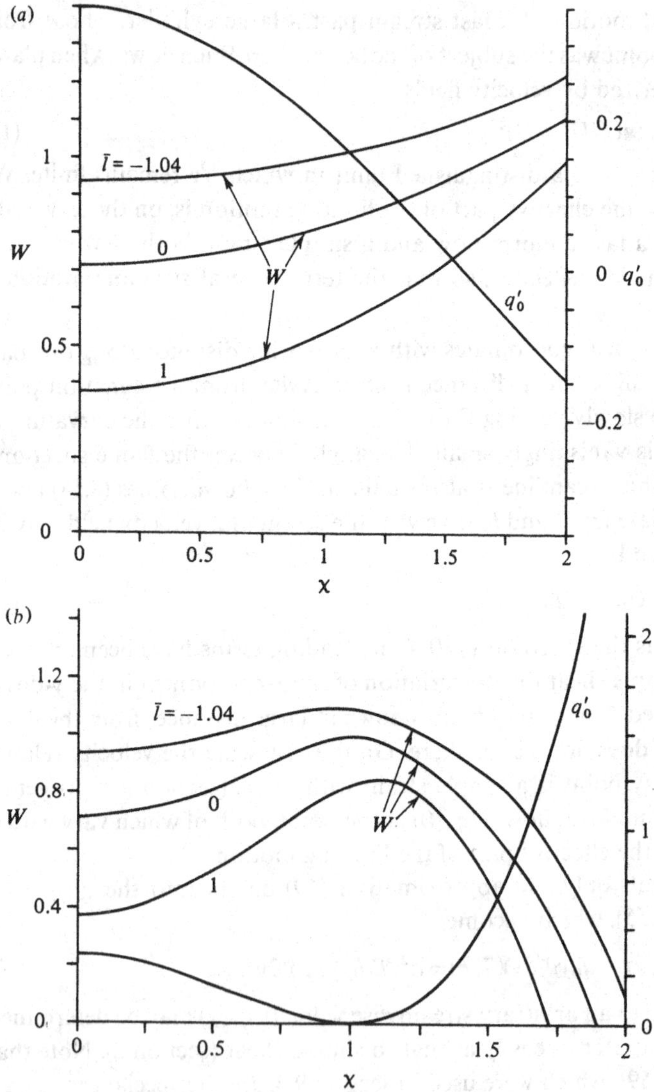

9 Variation of flame speed W and stretch q_0' with χ for the more general non-uniform velocity field (19); drawn for $\varepsilon = 0.35$ and (*a*) $q_0 = \varepsilon \sin \chi$, (*b*) $q_0 = \varepsilon(\chi - \frac{2}{3}\chi^3 + \frac{1}{5}\chi^5)$.

has the required form for large x, with $q_0 = \varepsilon\chi$, and, hence, provides the initial conditions for any $q_0 \sim \varepsilon\chi$ as $\chi \to 0$ when the boundary conditions on T and h are compatible.

Two examples will be considered, either

$$q_0(\chi) = \varepsilon \sin \chi \quad \text{for } 0 < \chi < \pi, \quad \text{or } \varepsilon(\chi - \tfrac{2}{3}\chi^3 + \tfrac{1}{5}\chi^5) \quad \text{for } \chi > 0. \qquad (21)$$

The first corresponds to potential flow around the cylinder and both asymptote $\varepsilon\chi$. If, therefore, the same boundary conditions as in section 3 are imposed, namely

$$\partial h/\partial y = 0 \quad \text{on } y = 0, \quad T \to T_f, \quad h \to 0 \quad \text{as } y \to \infty, \qquad (22)$$

then the results obtained there provide appropriate initial conditions here. Of course, behind the flame the appropriate solution of equations (20) has $T = H_f$.

The flows (21) are quite different in nature. For the first the stretch, i.e. the derivative q_0' (see section 8.4), decreases monotonically from ε to $-\varepsilon$ (Figure 9(a)); but for the second it decreases from ε to a minimum of zero at $\chi = 1$ and then increases indefinitely (Figure 9(b)). These characteristics are particularly relevant to an examination of the flame speed

$$W = (q_0 F)', \qquad (23)$$

which can be deduced from the computations by numerical differentiation.

Figure 9 shows that W increases monotonically for the flow (21a); but that for the flow (21b) it increases initially, reaches a maximum around $\chi = 1$, and then decreases. It is hard to resist the conclusion that there is an inverse correlation between the changes in flame speed and stretch, at least for moderate values of l. The correlation is not precise (as we have already seen in section 3): in Figure 9(b) the flame speed achieves its maximum exactly at $\chi = 1$ for only one value of $\bar{l}$ (which appears to be zero). But the general trend is undeniable.

The computations for the flow (21b) show that the decrease in W from its maximum continues until zero is reached where (according to section 9.4) quenching occurs. The results, therefore, suggest the general conjecture that a flame subjected to sufficiently strong and increasing strain will be extinguished. Hitherto, mathematical evidence for such a conjecture has been lacking.

5 Effect of slowly varying shear

The chapter ends with an examination of SVFs under the influence of simple shear and simple strain.

For simple shear, the χ, η-coordinates of section 8.3 are now taken as shown in Figure 10. Then the velocity field (1) is to be inserted in equation (8.31) where now, with $\omega > 0$,

$$f(\eta)=\begin{cases}1\\ 1-\omega\eta\end{cases} \quad \text{for } \eta \gtrless 0. \tag{24}$$

If

$$\eta = F(\chi) \tag{25}$$

is taken to be the locus of the flame, then for $\eta < 0$ we have

$$W = U(1-\omega F)|F'|/(1+F'^2)^{\frac{1}{2}}, \quad v_{\perp f} = W/F', \quad \kappa = -\operatorname{sgn}(F')F''/(1+F'^2)^{\frac{3}{2}} \tag{26}$$

$$\boldsymbol{v}_{\perp f}\cdot\boldsymbol{\nabla}_{\perp}W = -U\omega WF'/(1+F'^2) - \kappa W^2/F'^2,$$
$$\boldsymbol{\nabla}_{\perp}\cdot\boldsymbol{v}_{\perp f} = -U\omega F'/(1+F'^2). \tag{27}$$

The basic Equation (8.31) becomes, therefore,

$$F'' = k(1-\omega F)F'^3 \ln W, \tag{28}$$

where k is the constant (9.23b).

The corresponding formulas for $\eta > 0$ are obtained on setting $\omega = 0$, and then equation (9.23a), governing the shape of a flame tip in a uniform flow, is recovered (with $\tan\psi = F'$). The appropriate solution is $F' = -\tan\alpha$ corresponding to a plane flame, although a curved flame (corresponding to the tip) is also available when $\mathscr{L}$ is greater than 1. We take, therefore,

$$F(0) = 0, \quad F'(0) = -\tan\alpha \tag{29}$$

as initial conditions for the governing equation (28) in $\eta < 0$.

10 Notation for slowly varying simple shear.

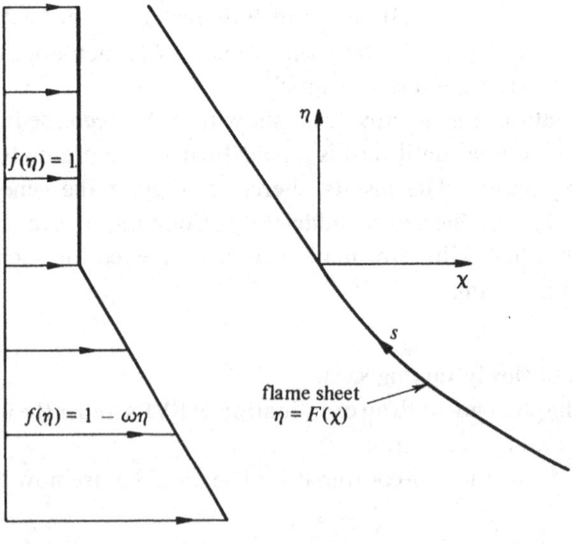

Figure 11 shows the effect of the shear. For $\mathscr{L}<1(k<0)$ the flame is bent toward the flow direction at a rate that increases with ω, but is never actually quenched. On the other hand, the flame bends away from the flow direction for $\mathscr{L}>1(k>0)$, again at a rate that increases with ω. These results are quite different from those for an NEF (section 2), where the turning ability of the shear increases with l, but there is no contradiction. Velocity gradients for an SVF are measured on the θ^{-1}-scale so that agreement between the responses of the two types of flame can only be expected as $|l|\to\infty$ with velocity gradients vanishing, for the one, and $\mathscr{L}\to 1$ with velocity gradients tending to infinity, for the other. A key physical difference is that SVFs only respond to the stretch generated by the shear, while a distinctive role is played by the shearing of the preheat zone in NEFs.

6 Effect of slowly varying strain

We shall consider only simple strain, i.e. stagnation-point flow. With so-called flat flames any amount of strain can be tolerated for $\mathscr{L}<1$, but for $\mathscr{L}>1$ there is a limit.

With the axes shown in Figure 12 we have to deal with the analog of the velocity field (8.60), namely

$$\boldsymbol{v}=\varepsilon(\chi,\ -\eta). \tag{30}$$

For the locus (25) we find

$$v_{\perp f}=-\varepsilon(FF'-\chi)/(1+F'^2)^{1/2},\quad W=\varepsilon(\chi F'+F)/(1+F'^2)^{1/2},$$
$$\kappa=-F''/(1+F'^2)^{3/2}, \tag{31}$$

11 Flame profiles for simple-shear problem, drawn for $U=\sqrt{2}$, i.e. $\alpha=\pi/4$.

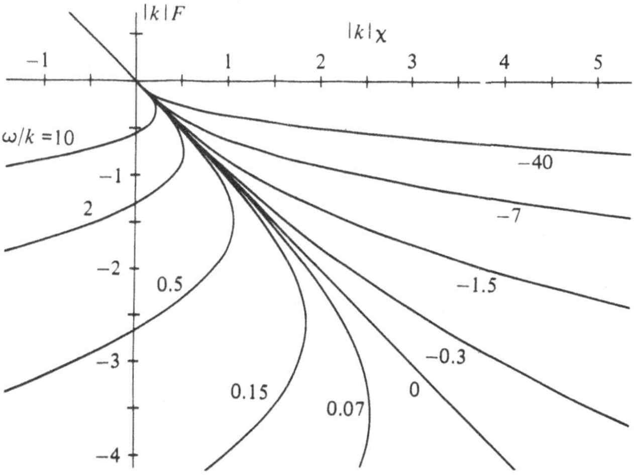

$$\boldsymbol{v}_\perp \cdot \nabla_\perp W = 2\varepsilon v_{\perp f} F'/(1+F'^2) - \kappa v_{\perp f}^2, \quad \nabla \cdot \boldsymbol{v}_{\perp f} = \varepsilon(1-F'^2)/(1+F'^2), \quad (32)$$

so that the basic equation (8.31) becomes

$$(F^2+\chi^2)F'' + (1+F'^2)(\chi F' - F) = k(\chi F' + F)^3 \ln W^2 \quad \text{with } k = \varepsilon/b. \quad (33)$$

Solution of this equation requires two boundary conditions but only one is obvious, namely the symmetry condition

$$F'(0) = 0. \quad (34)$$

The difficulty reflects the elliptic nature of the problem and, for determinacy, some reference must be made to the far field. If, for example, we are dealing with a fast flow past a correspondingly large cylinder (i.e. with ε fixed) then it is appropriate to look for a constant solution, corresponding to a flame that varies on the scale of the cylinder. Such a solution is given by

$$W^2 \ln W^2 = -\varepsilon b \quad \text{with } W = \varepsilon F. \quad (35)$$

Sivashinsky (1976), who considers the axisymmetric counterpart of this problem, proposes that such a choice is correct in all circumstances but there is no evidence for his view. Completely acceptable solutions are obtained by giving $F(0)$ a value that does not satisfy equation (35), see Figure 13. A constant solution is just a separatrix or an asymptote in the figure.

Figure 14 shows that there is always a unique solution of equation (35) for $b<0$ but that the condition

$$\varepsilon \leqslant 1/eb \quad (36)$$

must be met for $b>0$; then there are two solutions. We conclude that there is always a flat flame for $\mathscr{L}<1$ but that there is a limit to the straining rate

12 Notation for slowly varying simple strain.

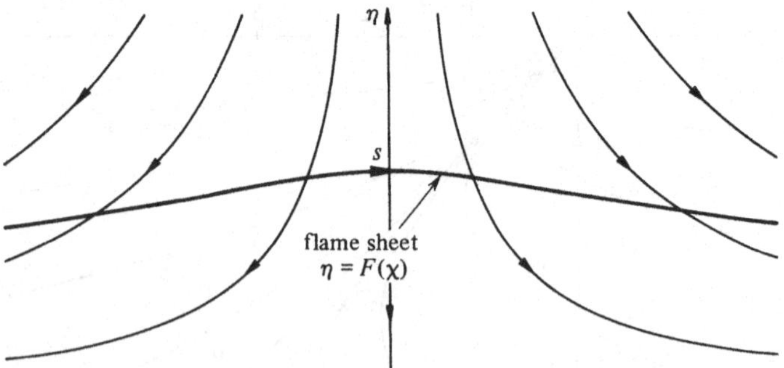

13 Flame profiles for simple-strain problem; drawn for (a) $k/\varepsilon^2 = -4$, (b) $k/\varepsilon^2 = 4$, (c) $k/\varepsilon^2 = 2$.

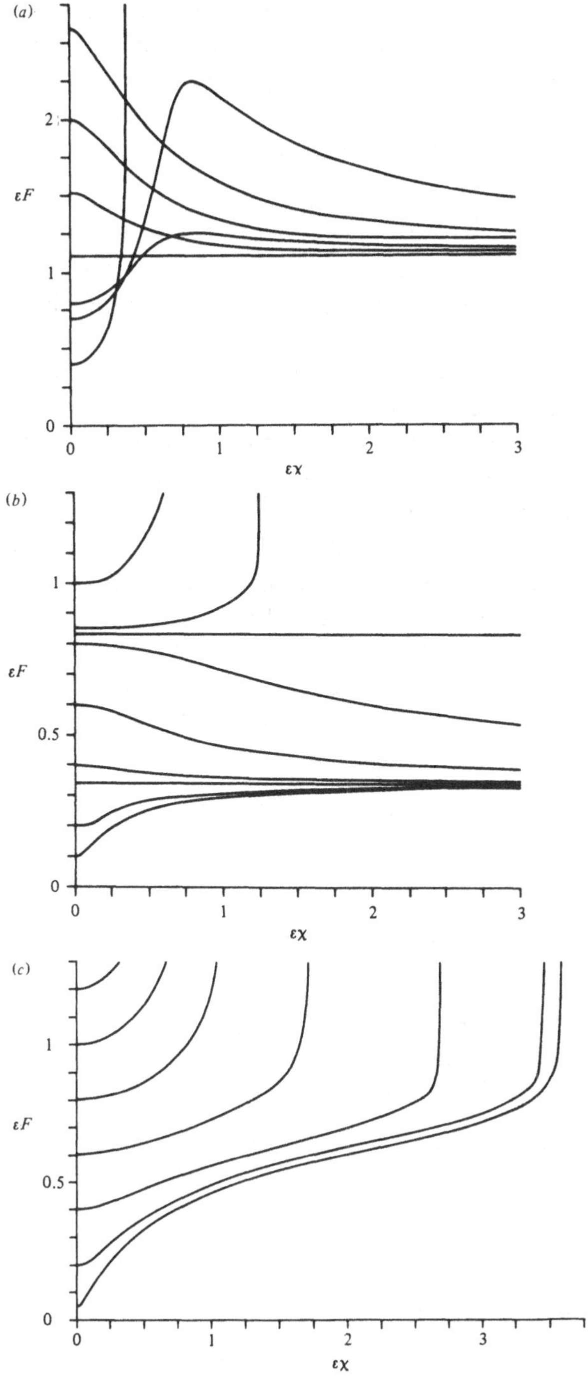

for $\mathscr{L} > 1$. These results are compatible with those for NEFs in the sense mentioned previously. As $l \to -\infty$ no $O(1)$ strain quenches the flame, so that we do not expect a limit as $\mathscr{L} \to 1-0$. On the other hand, as $l \to +\infty$ a vanishing $O(1)$ strain causes extinction (as suggested by Figure 6), consistent with an unbounded strain on the θ^{-1}-scale as $\mathscr{L} \to 1+0$.

Figure 14 could have been introduced after the general relation (8.42) to suggest that extinction will occur whenever

$$\Delta^{-1}\, \mathrm{d}\Delta/\mathrm{d}t \lesseqgtr 1/eb \quad \text{for } b \lesseqgtr 0. \tag{37}$$

Without a guarantee, however, that such voluminal stretch is possible, the speculation would have been futile. Stagnation-point flow provides the first example, and so far the only one. If we restrict attention to flat flames, then the voluminal stretch is constant and equal to ε. The condition (37) for $b<0$ can never be realized but that for $b>0$ can: if, when $\mathscr{L}$ is greater than one, ε is increased sufficiently slowly for the combustion field to be quasi-steady, then extinction occurs when it reaches $1/eb$.

These remarks apply only to the constant solutions in Figure 13. The nonconstant solutions are available whatever the value of ε or the sign of b. No evidence is found of quenching in the computations (cf. section 2), i.e. a curve never becomes tangent to a streamline.

14 Graph determining existence of flat flame in stagnation-point flow according to equation (35). Levels (*a*), (*b*), (*c*) correspond to part of Figure 13.

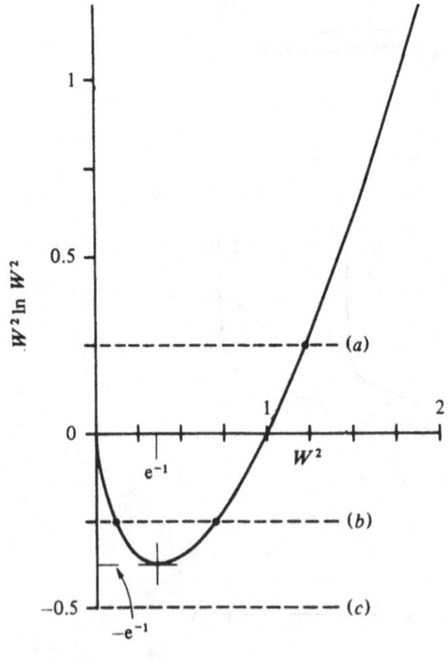

11

STABILITY

1 Scope

Flame stability is a very broad subject; both the mechanisms and manifestations of instability assume many forms. A whole monograph could be devoted to it if a less analytical approach than ours were adopted. It has been studied since the 1777 observations of Higgins (1802) on singing flames, a phenomenon involving interaction between an acoustic field and an oscillating flame of the kind described in Chapter 5. Such interactions were considered to be merely laboratory curiosities for many years, but lately they have assumed technological importance in the development of rocket motors and large furnaces. This type of instability is well understood qualitatively (Chu 1956). Other combustion phenomena thought to be due to instability are, in general, poorly understood; in some cases instability is merely suspected of playing a role but no certain evidence is yet available. The following examples are representative of the abundance of instability phenomena in combustion.

The stability of burner flames depends on an appropriate interaction between the flame and its surroundings, in particular, a flux of heat from the flame to the burner rim. The role of this flux in anchoring the flame and preventing blow-off has already been mentioned in Chapter 9.

Propagation limits are important questions in the study of premixed flames, with steady, sustained combustion being possible only for a certain range of the fuel to oxidant ratio. If a mixture is too rich or too lean it will not burn, and instability may well be involved. A possibly related phenomenon is quenching by heat loss or nonuniformities in the flow (Chapter 10), but whether this is a question of existence or stability (or some combination of the two), is not yet known.

Flow is turbulent in the right circumstances and turbulent combustion is of paramount importance in many technological applications. How the combustion influences the transition from laminar to turbulent flow caused by instabilities is not well understood at present. Instability may also play a role in another transition, that from deflagration to detonation (DDT). A flame propagating through a mixture will, under an appropriate stimulus, accelerate and change from an isobaric deflagration wave ultimately into a detonation wave, which has a shock structure.

Finally, a striking manifestation of instability is provided by cellular flames, which are discussed in detail by Markstein (1964, p. 75). Under certain conditions a nominally plane wave displays an unsteady, quasi-periodic, two-dimensional, transverse pattern of cells (Figure 1). On average the cells have a characteristic dimension, but large cells grow and divide while small cells shrink and may vanish. The corresponding phenomenon in highly curved burner flames leads to beautiful polyhedral tips (Figure 2), which sometimes are quite steady and sometimes rotate, but never exhibit the unsteadiness of plane cellular flames. Cells have also been observed on a spherical flame spreading out from an ignition source (Istratov & Librovich 1969).

All of these examples apply to premixed flames and some to diffusion flames, which introduce their own distinctive questions associated with the separation of the reactants. We shall be concerned, however, only with the stability of the freely propagating, premixed plane flame; cellular flames will play a central role in the development. (A brief discussion of diffusion-flame stability is found in section 6.8.) Such a narrow focus is dictated by the state of the art, analytically speaking: it is the only area

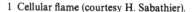

1 Cellular flame (courtesy H. Sabathier).

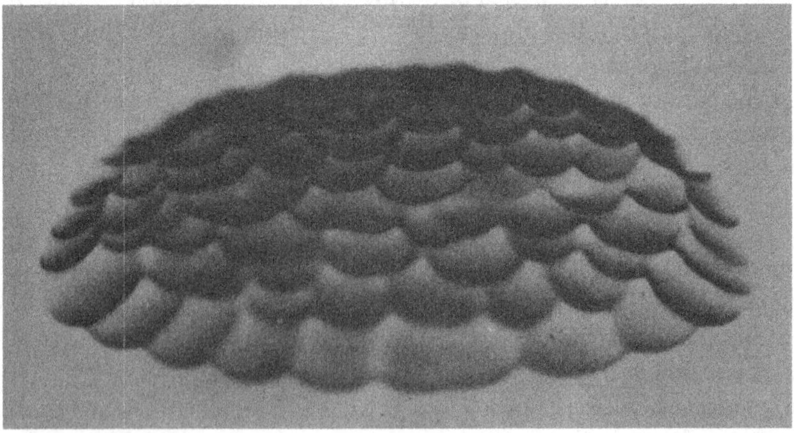

2 Polyhedral tip (courtesy L. Boyer).

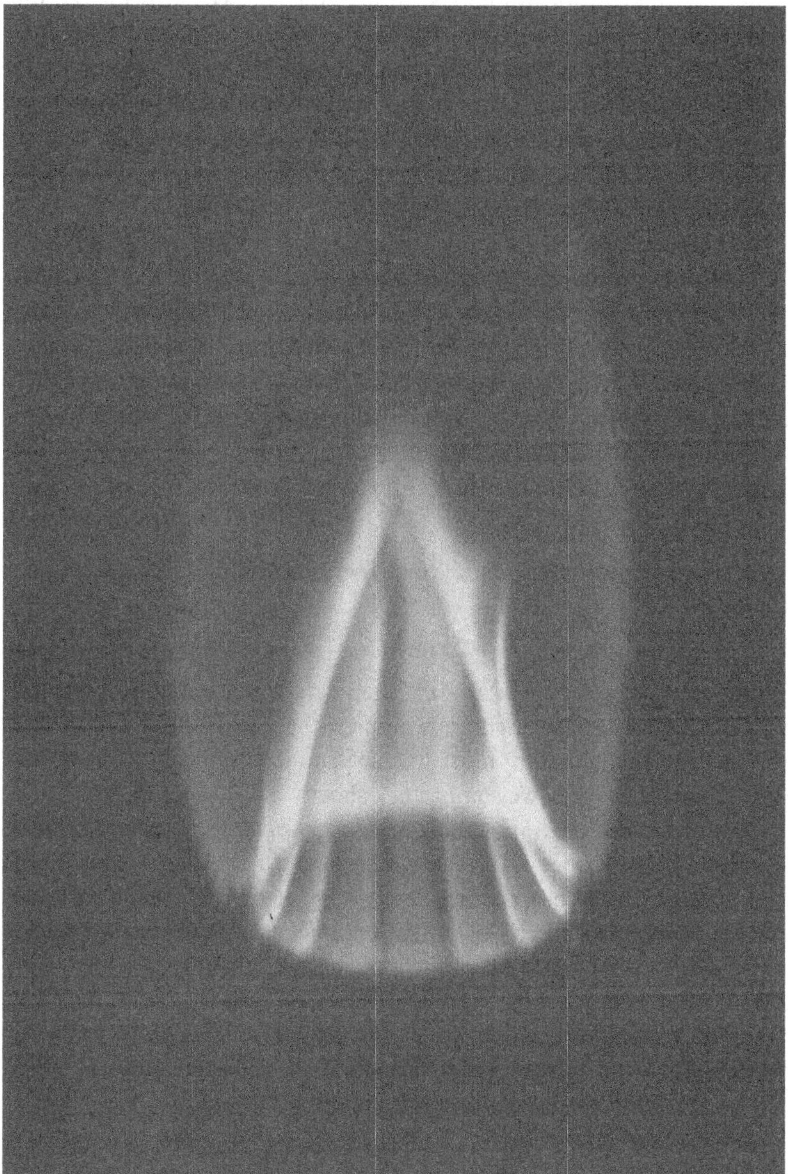

that has been explored with activation-energy asymptotics, probably because no other problem offers such a simple, closed form for the unperturbed steady state. Even then, formidable difficulties remain; that they have been overcome is due to the tenacity with which Sivashinsky, the principal architect of the theory presented, has pursued the problem.

In our opinion, the material in this chapter, answering as it does questions that have long frustrated the combustion scientist, is one of the triumphs of activation-energy asymptotics.

2 Slowly varying perturbations of the plane wave

The motion of a slowly varying flame is governed by the basic equation (8.31), where $\mathbf{v}_f$ is to be determined from a coupled hydrodynamic analysis. There the flame appears as a discontinuity across which the jump conditions (8.1 and 8.3) hold and on each side of which Euler's equations (8.33) are valid.

If the gas is quiescent far ahead of the flame there is a solution with

$$\mathbf{v}_f=0, \quad W=1, \quad \kappa=0, \tag{1}$$

namely the steady plane flame. Small perturbations of this stationary state satisfy

$$\dot{W}_1-\nabla_\perp\cdot\mathbf{v}_{1\perp f}-\nabla_1^2F_1-2b^{-1}W_1=0 \quad \text{with } W_1=v_{1nf}-F_{1\tau}. \tag{2}$$

Without a loss of accuracy, the χ, η, ζ-frame may be fixed at an arbitrary point instantaneously located anywhere on the undisturbed plane wave The velocity perturbation must be determined by a hydrodynamic analysis, a step to which we shall return shortly.

The velocity perturbations can be neglected only for a one-dimensional disturbance of the flame surface, as will become clear. Nevertheless, doing so, in general, leads to a correct description of one of the disturbance modes as the wave number $k\rightarrow0$ or the parameter $b\rightarrow0$. (The latter corresponds to a small heat of combustion, when the unit of temperature becomes small and hence T_b large; or to a Lewis number close to 1.) What remains may be called the diffusional–thermal effect (Sivashinsky 1977*a*), since these are the processes left. With $\mathbf{v}_{1f}=0$ and, hence, $W_1=-F_{1\tau}$, equation (2*a*) admits spatially periodic solutions

$$F_1=A\,e^{\tilde{\alpha}\tau+i\tilde{k}\eta} \tag{3}$$

provided that

$$(\tilde{k}^2-\tilde{\alpha}^2+2b^{-1}\tilde{\alpha})A=0, \quad \text{i.e. } \tilde{\alpha}=b^{-1}\pm\sqrt{b^{-2}+\tilde{k}^2}, \tag{4}$$

where no loss of generality results from considering disturbances independent of ζ. In this chapter we shall set the iω of Chapter 5 (where the emphasis is on forced oscillations) equal to α, since stability considerations are concerned with the sign of Re($i\omega$). Tildes correspond, as in Chapter 5, to slow variables.

For one-dimensional disturbances $\tilde{k}=0$, the two roots are

$$\tilde{\alpha}=0, \quad 2b^{-1}; \tag{5}$$

the flame is unstable if, and only if, b is positive, i.e. the Lewis number is greater than 1. This result was uncovered in our earlier treatment of plane flames (section 3.5). For $\tilde{k}\neq 0$, however, there is a positive root whatever the value of $\mathscr{L}$, so that instability always occurs. We turn now to the hydrodynamic problem.

The jump conditions (8.1 and 8.3) show that the state behind the plane flame is

$$p_b=p_f+1-\sigma, \quad u_b=\sigma-1, \quad v_b=0, \tag{6}$$

while the perturbations on the two sides satisfy

$$p_{1b}=p_{1f}-2(\sigma-1)(u_{1f}-F_{1\tau}), \quad u_{1b}=\sigma u_{1f}-(\sigma-1)F_{1\tau},$$
$$v_{1b}=v_{1f}+(1-\sigma)F_{1\eta}. \tag{7}$$

Here u and v are the velocity components in the χ- and η-directions and σ is the expansion coefficient (8.56). If we suppose that the undisturbed flame is located at $\chi=0$ when $\tau=0$, Euler's equations (8.33) are found to have the solutions

$$[p_1, u_1, v_1]=\begin{cases}[-(\tilde{\alpha}+|\tilde{k}|), |\tilde{k}|, \mathrm{i}\tilde{k}]P_f\,\mathrm{e}^{(\tilde{\alpha}+|\tilde{k}|)\tau+|\tilde{k}|\chi+\mathrm{i}\tilde{k}\eta} \\ [\sigma^{-1}\tilde{\alpha}-|\tilde{k}|, |\tilde{k}|, -\mathrm{i}\tilde{k}]P_b\,\mathrm{e}^{(\tilde{\alpha}-|\tilde{k}|)\tau-|\tilde{k}|\chi+\mathrm{i}\tilde{k}\eta} \end{cases} \quad \text{for } (\chi+\tau)\lessgtr 0.$$
$$+[0, \sigma|\tilde{k}|, \quad -\mathrm{i}\tilde{\alpha}\,\mathrm{sgn}(\tilde{k})]S\,\mathrm{e}^{[(1-\sigma^{-1})\tilde{\alpha}\tau-\sigma^{-1}\tilde{\alpha}\chi+\mathrm{i}\tilde{k}\eta]} \tag{8}$$

The P-terms form potential fields for the velocities while the S-terms, which are only solenoidal, take account of (convected) vorticity behind the flame. These solutions are appropriate for a flame surface (3), and substituting into the jump conditions (7) yields

$$\sigma[\tilde{\alpha}+(2\sigma-1)|\tilde{k}|]P_f+(\tilde{\alpha}-\sigma|\tilde{k}|)P_b-2\sigma(\sigma-1)\tilde{\alpha}A=0, \tag{9}$$
$$\sigma|\tilde{k}|P_f-|\tilde{k}|P_b-\sigma|\tilde{k}|S-(\sigma-1)\tilde{\alpha}A=0,$$
$$|\tilde{k}|P_f+|\tilde{k}|P_b+\tilde{\alpha}S-(\sigma-1)|\tilde{k}|A=0. \tag{10}$$

If the flame displacement A were specified, the equations would determine the hydrodynamic fields on the two sides of the surface. In fact, it is related to P_f for slowly varying flames, as the basic equation (2) shows.

Darrieus (1938) and Landau (1944) independently supposed, in the absence of a theory, that the flame speed is unaffected by the perturbations; which, as we shall see, is true in the theory of SVFs for $\tilde{k}$ or $b \to 0$. The response of the flame is thereby eliminated and it is appropriate to speak of the hydrodynamic effect. Setting the flame-speed perturbation $u_{1\mathrm{f}} - F_{1\tau}$ equal to 0 gives

$$|\tilde{k}|P_{\mathrm{f}} - \tilde{\alpha}A = 0; \tag{11}$$

then the homogeneous system for P_{f}, P_{b}, S, and A has a nontrivial solution if and only if

$$(\sigma+1)\tilde{\alpha}^2 + 2\sigma|\tilde{k}|\tilde{\alpha} - \sigma(\sigma-1)\tilde{k}^2 = 0,$$

$$\text{i.e. } \tilde{\alpha} = \sigma(-1 \pm \sqrt{1+\sigma-\sigma^{-1}})|\tilde{k}|/(\sigma+1). \tag{12}$$

There is always a positive root (in view of $\sigma > 1$) so we conclude that the flame is unstable. Note that the result is quite independent of the theory of SVFs (even though χ, η, and ζ were used): the hydrodynamic effect is always destabilizing even though it may only be present when the variations are slow.

We now come to the stability of the plane wave to slowly varying disturbances, including both the diffusional–thermal and hydrodynamic effects. The basic equation (2) provides

$$(\tilde{k}^2 + \tilde{\alpha}|\tilde{k}| - 2b^{-1}|\tilde{k}|)P_{\mathrm{f}} + (\tilde{k}^2 - \tilde{\alpha}^2 + 2b^{-1}\tilde{\alpha})A = 0 \tag{13}$$

as the fourth equation of the homogeneous system, in place of (11), so that the condition for a nontrivial solution becomes

$$\tilde{\alpha}^3 + B\tilde{\alpha}^2 + C\tilde{\alpha} + D = 0, \tag{14}$$

where

$$B = |\tilde{k}| - 2b^{-1}, \quad C = -\sigma|\tilde{k}|[(3\sigma-1)|\tilde{k}| + 4b^{-1}]/(\sigma+1),$$

$$D = \sigma\tilde{k}^2[-(3\sigma-1)|\tilde{k}| + 2(\sigma-1)b^{-1}]/(\sigma+1). \tag{15}$$

One of the three modes may be called diffusional–thermal and the other two hydrodynamic, based on their behavior as $\tilde{k}$ or $b \to 0$. In either case the cubic becomes

$$(\tilde{\alpha} - 2b^{-1})[(\sigma+1)\tilde{\alpha}^2 + 2\sigma|\tilde{k}|\tilde{\alpha} - \sigma(\sigma-1)\tilde{k}^2] = 0, \tag{16}$$

so that the larger root (4) (in absolute value) and the two roots (12) are obtained. The smaller root (4), for which $\tilde{\alpha} = -\frac{1}{2}b\tilde{k}^2$ for $\tilde{k}$ or b small, appears to be irrelevant; it has been superseded in the complete theory by the hydrodynamic modes, one of which behaves like $[\frac{1}{2}(\sigma-1) + \cdots]|\tilde{k}| - \frac{1}{2}b\tilde{k}^2 + \cdots$ for $\tilde{k}$ and $(\sigma-1)$ small. Nevertheless, the diffusional–thermal effect for $\mathscr{L} < 1$ is responsible for cellular instabilities (Figure 1), as we shall see in section 4.

The cubic shows that steady corrugations ($\tilde{\alpha} = 0$) with wavenumber

$$|\tilde{k}| = 2(\sigma-1)/(3\sigma-1)b \tag{17}$$

are possible when b is positive ($\mathscr{L} > 1$). The result was first obtained by Sivashinsky (1973); it is of historical interest only, since for that value of $\tilde{k}$ (and indeed all values) we shall see that there is at least one root of the cubic with a positive real part, implying instability.

A result of the type (14), for a more complicated model, was first obtained by Eckhaus (1961) from a thin-flame analysis that did not use formal activation-energy asymptotics. Instead, he adopted a formula of Mallard & LeChatelier (see Emmons 1958, p. 585) that leads to the proportionality of the perturbations in flame temperature and speed, a relation consistent with the asymptotic result (8.29) after linearization. The first complete derivation was given heuristically by Sivashinsky (1976) and later, rigorously, by Buckmaster (1977).

Necessary conditions for every root of the cubic (14) to have a negative real part are $B, D > 0$. Stability would, therefore, require $(3\sigma - 1)|\tilde{k}|/(\sigma - 1) < 2b^{-1} < |\tilde{k}|$ which, since σ is greater than 1, is a contradiction. We are forced to conclude that plane flames are unstable.

The instability occurs whatever the wavenumber $\tilde{k}$, but it would be a mistake to infer that all steady SVFs are unstable. Such an extrapolation implies taking the short-wave limit $\tilde{k} \to \infty$ (so as to screen out boundedness and curvature) and, hence, violating the assumption of slow variations on which the conclusion of instability for unbounded plane flames is based. In any event, the consequences of linear instability are not necessarily catastrophic; the result may be a nonlinear, unsteady structure superimposed on the steady state, without obliterating its essential character.

Even the instability of unbounded plane flames to long-wave disturbances cannot be considered definitive, since there may be stabilizing influences that have been ignored so far. Two such influences, gravity and curvature, we now examine.

3 Buoyancy and curvature

The influence that gravity has on flames through induced buoyancy forces is clear from a common observation: an inverted candle does not behave as desired. A less obvious manifestation is the dependence of the vertical propagation limit (i.e. the dilution beyond which the mixture is not flammable) on the direction of propagation, upwards or downwards. Instability should play a role in such limits, since for upward propagation the light burnt gas underlies the heavy unburnt gas, so that a type of Taylor instability can be anticipated. (The characterization is not entirely accurate since the flame is not a material surface.)

A flame of characteristic dimension 1 cm and speed 45 cm/s has approximately equal gravity and dynamic heads. Faster or smaller flames are less influenced by gravity. Since flame thickness is typically much less than 1 mm (see section 2.4), gravity can be an important influence only in the hydrodynamic field of a nominally plane flame. The derivation of the basic equation (8.31) of SVFs is consistent with the conclusion, since it does not use vertical momentum balance, the only governing equation influenced by gravity (see section 1.2).

The unit of acceleration is $c_p M_r^3/\rho_c \lambda$, which is typically 2×10^5 cm/s^2. The dimensionless gravity force is then 5×10^{-3} and we may reasonably consider it to be $O(\theta^{-1})$. The momentum equation (8.33*b*) now gains a term $-\rho(g, 0)$ on its right side, where g is positive for downward propagation, so that terms $-g\chi$ and $-\sigma^{-1}g\chi$ are added to the pressures in front of, and behind, the flame. The perturbations $p_{1\mathrm{f}}$ and $p_{1\mathrm{b}}$ in equation (7) are now the sums of terms coming from equation (8), as before, and terms

$$-gF_1 = -gA\,\mathrm{e}^{\tilde{\alpha}\tau + \mathrm{i}\tilde{k}\eta}, \quad -\sigma^{-1}gF_1 = -\sigma^{-1}gA\,\mathrm{e}^{\tilde{\alpha}\tau + \mathrm{i}\tilde{k}\eta};$$

the condition (9) is, therefore, replaced by

$$\sigma[\tilde{\alpha} + (2\sigma - 1)|\tilde{k}|]P_\mathrm{f} + (\tilde{\alpha} - \sigma|\tilde{k}|)P_\mathrm{b} - (\sigma - 1)(2\sigma\tilde{\alpha} - g)A = 0. \tag{18}$$

Clearly there is no limiting effect on the diffusional–thermal mode, but there is on the hydrodynamic modes, the roots (12) becoming

$$\tilde{\alpha} = \sigma|\tilde{k}|[-1 \pm \sqrt{1 + (\sigma - \sigma^{-1})(1 - g\sigma^{-1}|\tilde{k}|^{-1})}]/(\sigma + 1). \tag{19}$$

Since g is negative for upward propagation, we immediately see that gravity has a destabilizing effect when the hot gas underlies the cold. When the reverse is true, disturbances for which $|\tilde{k}| < \sigma^{-1}g$ are stable: for a flame propagating downwards, modes of sufficiently long wavelength are stabilized.

That these conclusions do hold is seen from the cubic (14) with terms $(\sigma - 1)g|\tilde{k}|(\tilde{\alpha} + |\tilde{k}| - 2b^{-1})/(\sigma + 1)$ added to account for gravity. In either of the limits $\tilde{k}$ or $b \to 0$ the cubic becomes

$$(\tilde{\alpha} - 2b^{-1})[(\sigma + 1)\tilde{\alpha}^2 + 2\sigma|\tilde{k}|\tilde{\alpha} - \sigma(\sigma - 1)\tilde{k}^2 + (\sigma - 1)g|\tilde{k}|] = 0, \tag{20}$$

so that the diffusional–thermal mode is unaffected, but the modifications (19) of the hydrodynamic modes are obtained. For the latter, the stabilizing effect is most clearly seen for long waves, when

$$\tilde{\alpha} = \pm \mathrm{i}\sqrt{\frac{(\sigma - 1)g}{\sigma + 1}}|\tilde{k}|^{\frac{1}{2}} - \frac{\sigma}{\sigma + 1}|\tilde{k}| + O(\tilde{k}^{\frac{3}{2}}). \tag{21}$$

We conclude that, for $\mathscr{L} > 1$, gravity cannot completely stabilize even long waves, since the relevant root (5) is a result of purely one-dimensional disturbances, and these are unaffected by the hydrodynamics; but for $\mathscr{L} < 1$ it can be a stabilizer (Buckmaster 1977).

As $\tilde{k} \to \infty$, the cubic reduces to

$$(\tilde{\alpha}+|\tilde{k}|)[(\sigma+1)\tilde{\alpha}^2-\sigma(3\sigma-1)\tilde{k}^2]=0, \tag{22}$$

which is independent of g: short waves are unaffected by gravity, as expected. Note how instability persists in the limit: there is always a positive root, although the other two are negative.

We now examine the effect of curvature, but only for the diffusional–thermal model. (A full treatment is complicated, more so that for a plane flame.) Gravity stabilizes the hydrodynamical modes, at least for long wavelengths. Attention will be confined to a cylindrical source- or sink-flow

$$u=\pm R/r, \quad v=0, \tag{23}$$

where θr is radial distance and u and v are now polar components. The undisturbed flame is then circular, with radius θR and θ-multiplied curvature

$$\kappa=\mp R^{-1} \tag{24}$$

determined by the requirement that the mass flux through the unstretched flame is 1. (The treatment of cylindrical flames as hydrodynamic discontinuities is discussed at the end of section 8.1.)

Such a flame corresponds to the solution with

$$\boldsymbol{v}_{\perp f}=0, \quad W=1, \quad \text{and } \nabla_{\perp}\cdot\boldsymbol{v}_{\perp f}=-\kappa; \tag{25}$$

perturbations of this steady state satisfy

$$\dot{W}_1-\nabla_{\perp}\cdot\boldsymbol{v}_{1\perp f}-\kappa_1-(\kappa+2b^{-1})W_1=0. \tag{26}$$

Here $\boldsymbol{v}_{1\perp f}$ is due to the displacement of the flame in the source-or sink-flow; disturbances of the flow will be ignored. It is sufficient to consider plane disturbances, for which the equation of the flame may be written

$$r=R+F_1(\phi,\tau) \tag{27}$$

in polar coordinates r and ϕ. Then

$$W_1=-(R^{-1}F_1\pm F_{1\tau}), \quad \nabla_{\perp}\cdot\boldsymbol{v}_{1\perp f}=\mp 2R^{-2}F_1,$$
$$\kappa_1=\pm R^{-2}(F_1+F_{1\phi\phi}); \tag{28}$$

the behavior of the mode

$$F_1=A\,e^{\tilde{\alpha}\tau+in\phi} \tag{29}$$

is, therefore, determined by

$$\tilde{\alpha}^2-2(b^{-1}\mp R^{-1})\tilde{\alpha}-R^{-1}(\pm 2b^{-1}+n^2R^{-1})=0$$
$$\text{as } \tilde{\alpha}=b^{-1}\mp R^{-1}\pm b^{-1}\sqrt{1+(n^2+1)b^2R^{-2}}. \tag{30}$$

Clearly, the roots are always real.

To discuss the result we introduce the wave number

$$\tilde{k} = R^{-1} n \tag{31}$$

and note that the formula (4) is recovered in the limit $R \to \infty$ with $\tilde{k}$ fixed. The first root (30) corresponds to the larger root (4), associated with one-dimensional instability for $b > 0$; here also there is stability if, and only if, $b < 0$. The second root (30) is a generalization of the smaller root (4), for which there is two-dimensional instability for $b < 0$. Now there is stability if, and only if, $\pm R^{-1} > -\frac{1}{2} b \tilde{k}^2$. Thus, for a source flow and $\mathscr{L} < 1$, curvature is stabilizing for long wavelengths insofar as this mode is concerned. The same general conclusions may be drawn for spherical flow; in the case of one-dimensional disturbances they agree with the results of section 3.7.

4 Stability of plane NEFs

The conclusion that modes of short wavelength will always be unstable, as predicted by the theory of SVFs, contradicts experimental results. We therefore turn to the theory of NEFs for clarification, and here compromise is necessary. On the scale of the flame thickness, SVFs are almost plane and almost steady; because of these characteristics, the complete fluid mechanics, in particular variable density, can be incorporated without difficulty. By contrast, the continuity and momentum equations do not simplify when the flame is near-equidiffusional, so that even perturbations of the plane flame lead to nonconstant coefficients. For that reason most of the discussion of near-equidiffusional stability has relied on the constant-density approximation, as embodied in equations (8.55), and with the single exception of section 6 that is the framework adopted here.

For quiescent conditions upstream, the governing equations reduce to

$$\partial(T, h)/\partial t = \nabla^2(T, h + lT); \tag{32}$$

to these are added the jump conditions (8.49, 8.54). The steady plane wave corresponds to the solution (3.84), where $x_* = -t$ in the definition (3.21) of n (cf. equations (9.30)). Its linear stability is our first concern.

The class of disturbances is limited: $O(1)$ perturbations of T are admitted provided they are balanced by $O(1)$ perturbations of Y that keep the sum of order θ^{-1}. It follows that a finding of stability (in contrast to instability) is not definitive. Nevertheless, the conclusions are very convincing and it seems unlikely that an analysis dealing with a larger class of disturbances would alter them substantially.

Such perturbations, T_1 and h_1, also satisfy the equations (32) on both sides of the disturbed flame sheet

$$n = F_1(y, t). \tag{33}$$

The temperature perturbation vanishes behind the flame sheet so that the continuity of T and h, plus the jump conditions on their derivatives, yield

$$T_1 = -Y_f F_1, \quad \delta(h_1) = -lY_f F_1, \quad \partial T_1/\partial n = -Y_f F_1 + h_1/l_s,$$
$$\delta(\partial h_1/\partial n) = lh_1/l_s - 2lY_f F_1 \tag{34}$$

for $n = 0$, where l_s is the parameter (9.37b) and the right sides of the last two equations are to be evaluated for $n = 0+$, i.e. on the hot side of the flame sheet.

Variations in the z-direction may be ignored without loss in generality. The solution corresponding to

$$F_1 = A\,\mathrm{e}^{\alpha t + iky} \tag{35}$$

is

$$[T_1, h_1] = \begin{cases} [B, C + l(k_+^2 - k^2)Bn/(1 - 2k_+)]\,\mathrm{e}^{(\alpha t + iky + k_+ n)} \\ [0, D]\,\mathrm{e}^{(\alpha t + iky + k_- n)} \end{cases} \quad \text{for } n \lessgtr 0, \tag{36}$$

where

$$k_\pm = \tfrac{1}{2}[1 \pm \sqrt{1 + 4\alpha + 4k^2}] \quad \text{with } -\pi < \arg(1 + 4\alpha + 4k^2) \leqslant \pi. \tag{37}$$

To be admissible the solution must vanish far ahead of the flame sheet and be bounded far behind, i.e. $\mathrm{Re}(k_+)$ must be positive and $\mathrm{Re}(k_-)$ non-positive. For $\mathrm{Re}(\alpha) \geqslant 0$ (i.e. the unstable modes of interest and their neutral limits) these requirements are met automatically, so we need not be concerned further with them. Equations (34) now yield a homogeneous system for A, B, C, and D, which has a nontrivial solution only if

$$(1 - k_+)(1 - 2k_+)^2 + \bar{l}[(1 - k_+)^2 - k^2] = 0 \quad \text{with } \bar{l} = l/l_s, \tag{38}$$

a result first obtained by Sivashinsky (1977a).

This condition determines all those modes, i.e. values of α, corresponding to any given $\bar{l}$ and k. For $\mathrm{Re}(\alpha) \geqslant 0$ we can be sure that the mode is admissible, the boundaries of regions of instability $\mathrm{Re}(\alpha) > 0$ being curves $\mathrm{Re}(\alpha) = 0$. There are just three such curves, namely

$$\bar{l}\text{-axis: } k = 0; \quad \text{parabola P: } 4k^2 = -(\bar{l} + 1); \tag{39}$$
$$\text{curve } B\text{: } \bar{l} = 2(1 + 8k^2)[1 + \sqrt{3(1 + 8k^2)}]/(1 + 12k^2). \tag{40}$$

For the first two, $\mathrm{Im}(\alpha) = 0$ also.

The regions of instability can now be identified from explicit results for small k, namely,

$$\alpha = -(\bar{l} + 1)k^2 + \bar{l}^2(\bar{l} - 3)k^4 + O(k^6),$$
$$\tfrac{1}{32}[\bar{l}^2 - 4\bar{l} - 8 + \bar{l}\sqrt{\bar{l}(\bar{l} - 8)}] + O(k^2), \tag{41}$$

where both signs for the square root are implied for $0<\bar{l}<8$. Clearly there is a mode with $\mathrm{Re}(\alpha)>0$ only outside the range

$$-1\leqslant\bar{l}\leqslant 2(1+\sqrt{3}), \tag{42}$$

so that the stability regions are as marked in Figure 3. Note how misleading results, just as for an SVF, would be obtained by considering only one-dimensional disturbances: for $k=0$ the first of the values (41) vanishes and no instability is predicted for $\bar{l}<-1$. The one-dimensional stability condition $\bar{l}\leqslant 2(1+\sqrt{3})$ can be considered a refinement of $\mathscr{L}<1$ for slowly varying disturbances (section 2, cf. equation (5)]; the stability condition $\bar{l}\geqslant -1$ for two-dimensional disturbances is correspondingly a refinement of $\mathscr{L}>1$.

For each value of k there is a band of Lewis numbers, always including $-1\leqslant\bar{l}\leqslant\frac{16}{3}$ and, hence, $\mathscr{L}=1$, for which the flame is stable. The conclusion is in accord with that for slowly varying disturbances, where instability is found for all $\mathscr{L}\neq 1$ (section 2). Moreover, it provides a plausible explanation of why stable flames are observed and makes NEFs of great practical importance. In the limit $\theta\to\infty$ the stability band in $\mathscr{L}$ is vanishingly small, but in practice θ is only moderately large. Since Lewis numbers are usually close to 1, it is conceivable that all flames adequately described by activation-energy asymptotics may be considered NEFs.

3 Linear stability regions for plane NEFs; ---- maximum growth curve. The boundaries P and B are the curves (39*b*) and (40) respectively.

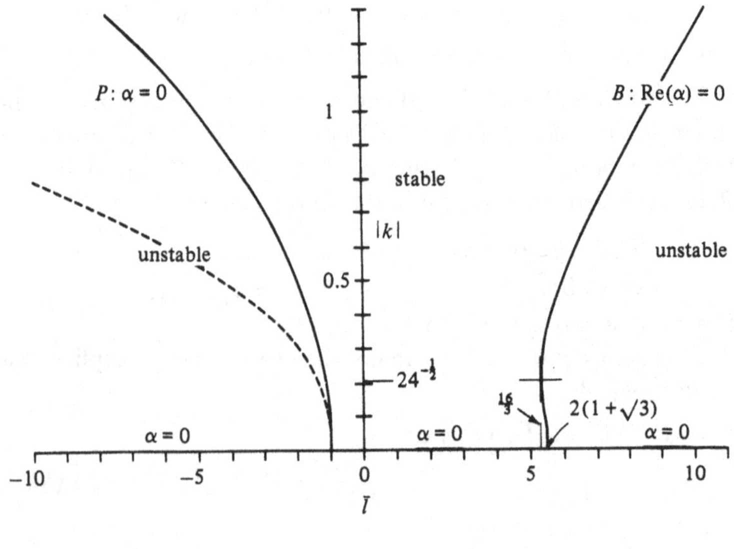

On the boundary B the neutral stability modes are oscillatory, i.e. $\mathrm{Im}(\alpha)\neq 0$. The boundary is relevant to mixtures whose deficient component is heavy and, therefore, has relatively small diffusivity (e.g. for lean, heavy hydrocarbon or rich hydrogen flames in air).

On the parabolic boundary and on the $\bar{l}$-axis the neutral modes are nonoscillatory, i.e. $\mathrm{Im}(\alpha)=0$. (In fact, all modes inside the parabola are also nonoscillatory.) For any $\bar{l}<-1$ and wavenumber in the range $0<|k|<\frac{1}{2}\sqrt{-(1+\bar{l})}$ there is a mode that will grow. Since $\mathrm{Re}(\alpha)$ vanishes at the ends of the range there must be an intermediate value, namely

$$k_{\mathrm{m}}=\{[4+15\bar{l}-9\bar{l}^2+(4-6\bar{l})\sqrt{1-3\bar{l}}]/(108\bar{l})\}^{\frac{1}{2}} \tag{43}$$

for which the growth is maximum (see Figure 3). Although the instability does not favor one wavelength to the exclusion of all others, an arbitrary disturbance can be expected to grow in such a way that the length k_{m}^{-1} plays a significant role. As we noted in section 1, it is characteristic of cellular flames, which are usually observed when the deficient component is light ($\bar{l}$ negative), for many modes to contribute to the structure; nevertheless, an average dimension can be assigned to the cells.

The notion that the stationary mode corresponding to the parabola has relevance to cellular flames is supported by the behavior of the flame-temperature perturbation

$$2T_{\mathrm{b}}^2(1-k_+)A\,\mathrm{e}^{\mathrm{i}ky}=T_{\mathrm{b}}^2(\sqrt{1+4k^2}-1)A\,\mathrm{e}^{\mathrm{i}(ky+\pi)}\quad\text{with } k=\tfrac{1}{2}\sqrt{-(1+\bar{l})}$$

corresponding to the displacement $F_1=A\exp\mathrm{i}ky$, with which it is exactly out of phase. The crests (as viewed from the burnt mixture) are colder than the troughs. We would, therefore, expect the crests to be the least luminous part of cellular flames, which is a well-known characteristic of them.

The stability boundary in the neighborhood of the point $\bar{l}=-1$, (i.e. $l=-l_{\mathrm{s}}$), $k=0$ can be explained by flame stretch. As noted in section 10.3, weak positive stretch decreases the speed of a flame for positive $1+\bar{l}$, and otherwise increases it. Thus, for $\bar{l}>-1$, the speed is decreased at the troughs of a corrugated flame (as viewed from the burnt mixture) where the stretch is positive, and increased at the crests. As a result, the corrugations are diminished in magnitude, a stabilizing effect. For $\bar{l}<-1$ the effect is destabilizing.

Section 3 describes the effects of buoyancy and curvature on the linear stability of an SVF. For an NEF, buoyancy effects can be accounted for if hydrodynamic effects are weak; details have been given by Matkowsky & Sivashinsky (1979*a*) for their constant-density approximation, and the buoyancy effect is incorporated into the discussion of section 6. The nonlinear analysis in section 7 covers the curvature effect here as a special case.

5 Cellular flames

The linear analysis of the previous section predicts that infinitesimal disturbances will grow without bound when $\bar{l}$ is less than -1, but there is strong evidence that the growth actually leads to cellular flames. In that case, nonlinearities must limit the growth; the nature of that interaction will now be investigated. Attention will be focused on the neighborhood of $\bar{l}=-1$ by setting

$$l=-l_s \mp \varepsilon, \tag{44}$$

where $\varepsilon>0$ is vanishingly small. Only the upper sign on ε is of interest here, corresponding to the linearly unstable waves in that neighborhood (i.e. the long ones); the lower sign will be used later. The neighborhood (44) can be treated by a perturbation analysis in ε. The stability boundary (39*b*) shows that the wavenumber k is $O(\varepsilon^{\frac{1}{2}})$ at most, while the result (41*a*) shows that disturbances grow no faster than $\alpha=O(\varepsilon^2)$. Accordingly, the slow variables

$$\eta=\varepsilon^{\frac{1}{2}}y, \quad \tau=\varepsilon^2 t \tag{45}$$

are appropriate for their description.

For fixed ε, an infinitesimal disturbance will grow in accordance with the linear analysis until its size makes some nonlinear term comparable to the linear terms. Its amplitude then turns out to be $O(\varepsilon)$ so that, following Sivashinsky (1977*b*), we take the flame sheet as

$$x=-t+\varepsilon F(\eta, \tau). \tag{46}$$

Nonlinearity has certainly intruded by then, as is seen from the resulting flame speed

$$W=1-\varepsilon^3(F_\tau+\tfrac{1}{2}F_\eta^2)+O(\varepsilon^6), \tag{47}$$

but it could conceivably intrude elsewhere for amplitude $o(\varepsilon)$; the consistency of the analysis stemming from the assumption (46) shows that it does not. If now the coordinate

$$n=x+t-\varepsilon F(\eta, \tau) \tag{48}$$

is introduced in place of x, so that the flame sheet is $n=0$, then derivatives in equations (32) become

$$\partial/\partial x=\partial/\partial n, \quad \partial/\partial y=-\varepsilon^{\frac{3}{2}}F_\eta\partial/\partial n+\varepsilon^{\frac{1}{2}}\partial/\partial\eta, \quad \partial/\partial t=(1-\varepsilon^3F_\tau)\partial/\partial n+\varepsilon^2\partial/\partial\tau$$

while the normal derivative in the jump conditions is

$$(1+\tfrac{1}{2}\varepsilon^3F_\eta^2)\partial/\partial n-\varepsilon^2F_\eta\partial/\partial\eta+O(\varepsilon^5).$$

Perturbation expansions in ε are now introduced for T, h, and F, the coefficients being functions of n, η, and τ. A sequence of problems then results, whose solutions must satisfy the requirements that: $T_1 = T_2 = T_3 = \ldots = 0$ for $n > 0$; conditions as $n \to -\infty$ are undisturbed; and exponential growth as $n \to +\infty$ is disallowed (though algebraic growth cannot be prevented). At the fourth problem, for T_3 and h_3, the nonlinear equation (54) for F_0 is obtained as a solubility condition.

The first problem leads to the steady plane wave (3.84), as might be expected, without using the jump condition on $\partial h_0/\partial n$; the condition is then satisfied automatically. Likewise, the solutions

$$T_1 = \begin{cases} 0 \\ 0 \end{cases}, \quad h_1 = \begin{cases} \pm Y_f n \, \mathrm{e}^n \\ 0 \end{cases} \quad \text{for } n \lessgtr 0 \tag{49}$$

and

$$T_2 = \begin{cases} Y_f F_{0\eta\eta} n \, \mathrm{e}^n \\ 0 \end{cases}, \quad h_2 = \begin{cases} l_s Y_f F_{0\eta\eta}(1+n^2) \, \mathrm{e}^n \\ l_s Y_f F_{0\eta\eta} \end{cases} \quad \text{for } n \lessgtr 0 \tag{50}$$

of the second and third problems are obtained without using the jump conditions on $\partial h_1/\partial n$ and $\partial h_2/\partial n$; the conditions are then satisfied automatically. The fourth problem has the solution

$$T_3 = \begin{cases} Y_f[An^2 - (2A+D)n] \, \mathrm{e}^n \\ 0 \end{cases},$$

$$h_3 = \begin{cases} Y_f[l_s A n^3 - (3l_s A - B + l_s D)n^2 + l_s(6A - C)n - l_s(2A+E)] \, \mathrm{e}^n \\ Y_f[-2l_s A n - l_s(2A+E)] \end{cases} \quad \text{for } n \lessgtr 0 \tag{51}$$

when the jump condition

$$\delta(\partial h_3/\partial n) = -h_3 - Y_f B + l_s Y_f(D-E) \quad \text{at } n=0 \tag{52}$$

is ignored; here

$$A = -\tfrac{1}{2}F_{0\eta\eta\eta\eta}, \quad B = \pm F_{0\eta\eta}, \quad C = F_{0\tau} + F_{0\eta}^2,$$
$$D = F_{0\tau} + F_{0\eta}^2 - F_{1\eta\eta}, \quad E = F_{0\tau} + \tfrac{1}{2}F_{0\eta}^2 - F_{1\eta\eta}. \tag{53}$$

Note that h_3 must be allowed to increase algebraically. The jump condition is not satisfied automatically but requires $-8A + B/l_s + C - D + E$ to vanish, i.e.

$$F_{0\tau} + 4F_{0\eta\eta\eta\eta} \pm F_{0\eta\eta}/l_s + \tfrac{1}{2}F_{0\eta}^2 = 0. \tag{54}$$

This result was found by Sivashinsky (1977*b*), who generalized it to

$$F_{0\tau} + 4\nabla^4 F_0 \pm \nabla^2 F_0/l_s + \tfrac{1}{2}|\nabla F_0|^2 = 0 \tag{55}$$

for variations in both the y- and z-directions.

The equation has an interesting implication for the flame speed. While the expression (47) is purely kinematical, its approximation

$$W = 1 + \varepsilon^3(4F_{0\eta\eta\eta\eta} \pm F_{0\eta\eta}/l_s) \tag{56}$$

is a profound consequence of the diffusional–thermal processes. Markstein (1964, p. 22), in an attempt to modify the Darrieus–Landau conclusion that all flames are hydrodynamically unstable, assumed that the flame-speed perturbation is proportional to curvature and so incorporated the term in $F_{0\eta\eta}$, but not that in $F_{0\eta\eta\eta\eta}$.

For $l < -l_s$ the linearized form of equation (54) has solutions proportional to $\exp(\alpha\tau/\varepsilon^2 + ik\eta/\varepsilon^{\frac{1}{2}})$ if

$$\alpha = -(1+\bar{l})k^2 - 4k^4, \tag{57}$$

which coincides with the first of the approximations (41) in the neighborhood of $l = -l_s$ and $k = 0$. This result is graphed in Figure 4. The maximum growth rate occurs for

$$|k_m| = \sqrt{-\tfrac{1}{8}(1+\bar{l})} \tag{58}$$

which is an approximation to the general formula (43). The connection of this maximum with the characteristic cell size has already been noted.

Discussion of the nonlinear equation (54) has been limited to numerical computations (Michelson & Sivashinsky 1977). Integration of the initial-value problem with periodic boundary conditions (using a period large compared to k_m^{-1}) leads to results such as those in Figure 5 when the plus sign is taken in the equation. (For the minus sign there is decay to zero, corresponding to linear stability.) The nonlinearity prevents unlimited growth of the disturbance and the resulting structure, which is quasi-

4 Growth rate α versus wavenumber k for $\bar{l}$ slightly less than -1, i.e. $\varepsilon = -(1+\bar{l})l_s$ small and positive.

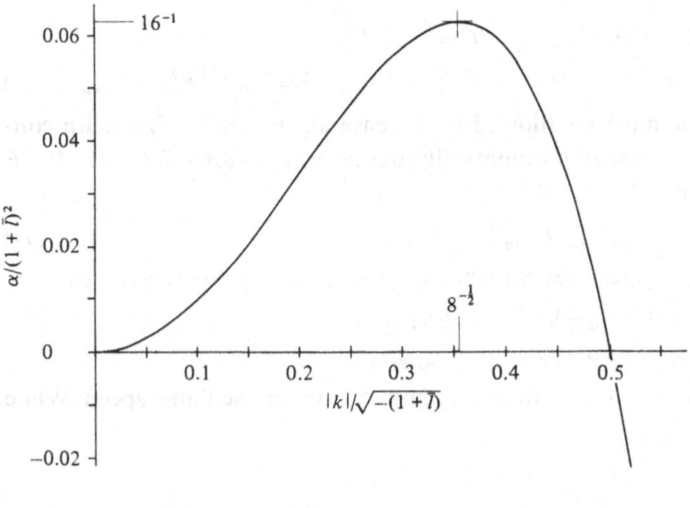

periodic with characteristic dimension k_m^{-1}, is highly suggestive of cellular flames.

6 Hydrodynamic effects

The analysis of section 5, leading as it does to a balance of small effects, provides a framework for discussing several generalizations when the additional effects are appropriately small. Here we shall consider weak interaction between the flame and the hydrodynamic field. A rigorous derivation of the governing equation replacing (54) requires analysis similar to that in section 5. A more transparent derivation can be made using plausible arguments.

The constant-density model is usually studied because it is simple and yet retains some of the key physical ingredients that govern flame behavior. Nevertheless, it is a rational asymptotic limit ($\sigma \to 1$, see section 8.6), albeit one of limited practical interest since density changes across flames are rarely small. When σ is close to 1, the hydrodynamic effects can be incorporated in a rational fashion as perturbations strong enough to influence the flame behavior described in section 5. Since we are only concerned with wavelengths $O(\varepsilon^{-\frac{1}{2}})$, the field outside the flame sheet has a dual structure similar to that of slowly varying flames (section 8.3): hydrodynamic disturbances described on the scale $n = O(\varepsilon^{-\frac{1}{2}})$ must be matched with those in the diffusion zone, where $n = O(1)$.

The flame speed (47) is now augmented by $(u_{1*} - \varepsilon^{\frac{1}{2}} F_\eta v_{1*})[1 + O(\varepsilon^3)]$, where u_{1*} and v_{1*} are the components of the velocity disturbance immediately ahead of the flame sheet. If this is to be a perturbation comparable to the existing one, then u_{1*} must be $O(\varepsilon^3)$; in which case v_{1*} (being of the same order) makes a smaller contribution and W is, to order ε^3, simply augmented by u_{1*}.

The problem now is to express u_{1*} in terms of F_0. To calculate the associated perturbation $u_{1\mathrm{f}}$ at the edge of the preheat zone it is necessary to make a hydrodynamic analysis, as in section 2. Thus, we obtain the result $P_\mathrm{f} = \frac{1}{2}(\sigma - 1)A$ by solving the jump conditions (9, 10) when k and α are appropriately small and σ is close to 1. This relates $u_{1\mathrm{f}}$ and F_0 when they are proportional to $\exp(\alpha t + \mathrm{i}ky)$ and, hence, it relates the Fourier transforms of $u_{1\mathrm{f}}$ and F_0 in general (since α is not involved). It follows that

$$u_{1\mathrm{f}}(\eta, \tau) = \frac{(\sigma - 1)\varepsilon}{4\pi} \int_{-\infty}^{\infty} \int_{-\infty}^{\infty} |k| \, \mathrm{e}^{\mathrm{i}k(\eta - \bar{\eta})} F_0(\bar{\eta}, \tau) \, \mathrm{d}\bar{\eta} \, \mathrm{d}k$$

$$= \frac{(\sigma - 1)\varepsilon^{\frac{1}{2}}}{2\pi} \fint_{-\infty}^{\infty} \frac{F_{0\eta}(\bar{\eta}, \tau)}{\eta - \bar{\eta}} \, \mathrm{d}\bar{\eta}, \qquad (59)$$

where $\sigma = 1 + O(\varepsilon^{\frac{1}{2}})$ if $u_{1\mathrm{f}}$ is to be $O(\varepsilon^3)$.

5 Nonlinear evolution of cellular structure for $\bar{l}$ slightly less than -1: integration of equation (54) with $+$ sign and initial conditions with period 97. Each part of the figure gives flame-sheet profiles at five equidistant instants in time:

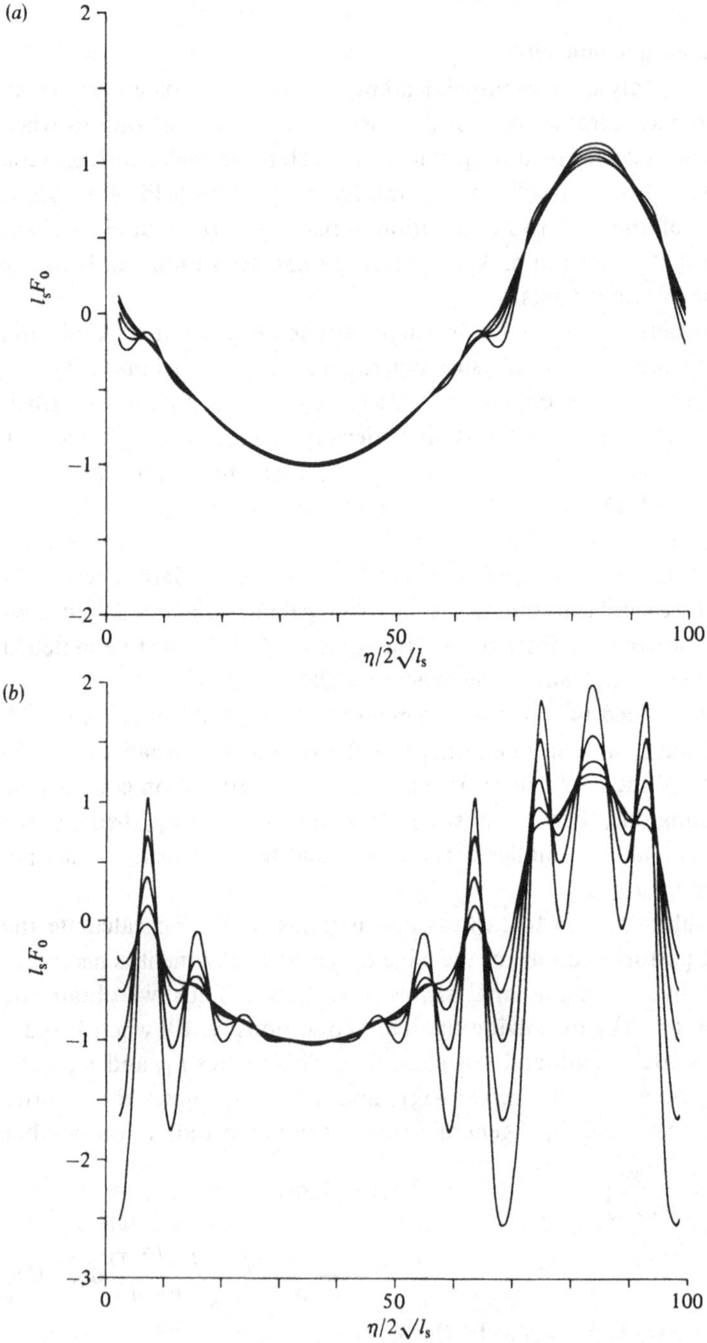

(*a*) early development; (*b*) formation of cells; (*c*) formation of quasi-periodic cellular structure; (*d*) continuing development of cellular structure. (Courtesy G. I. Sivashinsky). The fastest growing linear disturbance has wavelength $2\sqrt{2}\pi = 8.89$; note the changes in vertical scale.

(*c*)

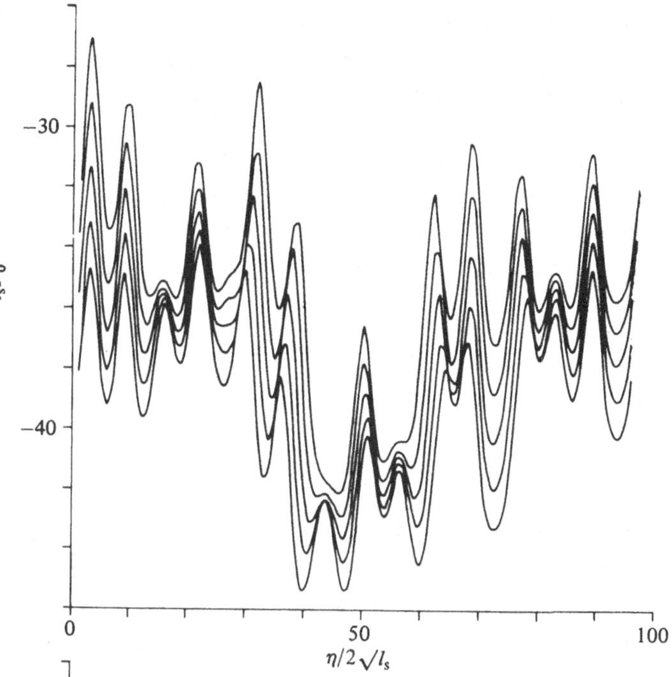

(*d*)

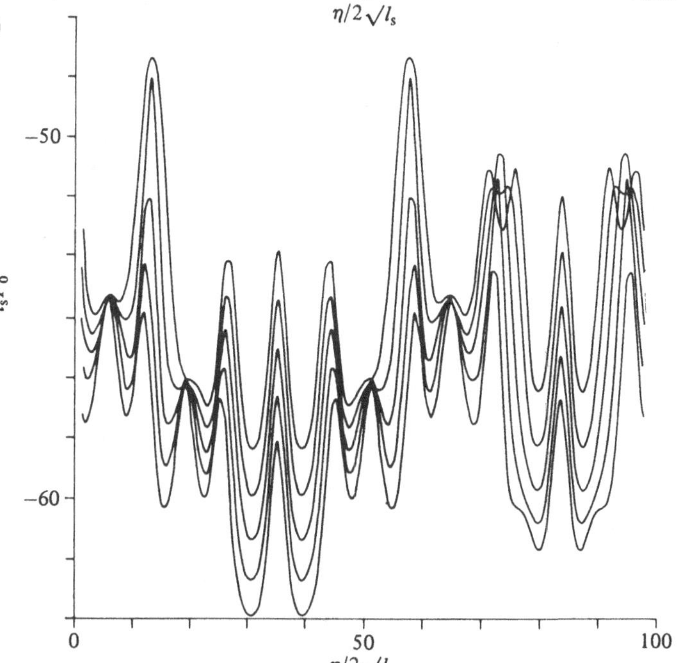

There are then two contributions to u_{1*}. One is simply an $O(\varepsilon^{3/2})$ constant generated by the density difference, reflecting the fact that the speed of the flame sheet is, in general, different from that of the hydrodynamic discontinuity. The constant is to be added to both (47) and (56), but may be ignored where the shape of the flame sheet is involved. The other contribution is (59), unaffected to leading order by the very weak density changes in the preheat zone.

So far we have been concerned with modifying the kinematical result (47) for the flame speed; now we consider the result (56) of the diffusional–thermal processes, which is found to be unchanged. It is plausible that such weak density changes do not affect these processes; that being the case, the generalization

$$F_{0\tau}+4F_{0\eta\eta\eta\eta}\pm F_{0\eta\eta}/l_s+\tfrac{1}{2}F_{0\eta}^2+\gamma \, \mathrm{P}\!\!\int_{-\infty}^{\infty} \frac{F_{0\eta}(\bar{\eta},\tau)}{\bar{\eta}-\eta}\,d\bar{\eta}=0 \quad \text{with } \gamma=(\sigma-1)/2\pi\varepsilon^{3/2} \qquad (60)$$

of equation (54) follows ($\sigma\to 1$ corresponds to $l_s\to\infty$ but a simple scaling can account for this).

The destabilizing effect of disturbances in the hydrodynamic field can be seen by considering small F_0. Under linearization, i.e. on dropping $\frac{1}{2}F_{0\eta}^2$, there are solutions $\exp(\tilde{\alpha}\tau+i\tilde{k}\eta)$ with

$$\tilde{\alpha}=-4\tilde{k}^4\pm\tilde{k}^2/l_s+\pi\gamma|\tilde{k}|; \qquad (61)$$

Figure 6 shows graphs of these curves. As may be expected from the last

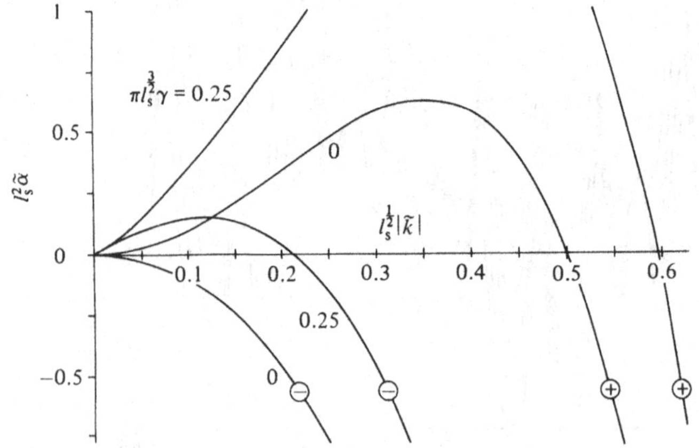

6 Destabilizing effect of hydrodynamic field: growth rate $\tilde{\alpha}$ versus wavenumber $\tilde{k}$ for $\bar{l}$ close to -1. The $\pm$ signs correspond with equation (61). Figure 4 is obtained from the $+$ curve with $\gamma=0$, since $\tilde{\alpha}=\alpha/\varepsilon^2$, $\tilde{k}=k/\varepsilon^{1/2}$ according to equations (45).

section, for $\gamma=0$ the lower sign gives stable modes only, but the upper sign involves unstable modes. For $\gamma>0$, i.e. when the hydrodynamic field is disturbed, both signs give unstable modes. The possibility of again stabilizing the modes by gravity can also be seen. For σ near 1, the unstable root (19) is $\frac{1}{2}|\tilde{k}|(\sigma-1)(1-g|\tilde{k}|^{-1})$, so that a term independent of $\tilde{k}$ must be added to the right side of equation (61). If g is positive and sufficiently large, the curves in Figure 6 are translated downwards into the quadrant $\tilde{\alpha}<0$.

Michelson & Sivashinsky (1977) have also treated the full equation (60), in particular, for the plus sign when the hydrodynamics is an extra destabilizing influence. Results similar to Figure 5 are obtained, except that now the large wrinkles contain several small wrinkles. We shall consider just two limits of the equation, both corresponding to significantly larger departures of l from $-l_s$ than of σ from 1, i.e. $\gamma \rightarrow 0$.

If changes remain on the scale of η and τ then the hydrodynamic term drops out and equation (54) is recovered in the limit. This tendency was found by Michelson and Sivashinsky in their numerical work.

If, however, changes occur on the slower scales $\gamma\eta$ and $\gamma^2\tau$, then the fourth derivative drops out and the equation reduces to

$$F_{0\tau} \pm F_{0\eta\eta}/l_s + \tfrac{1}{2}F_{0\eta}^2 + \gamma \,\mathrm{P}\!\!\int_{-\infty}^{\infty} \frac{F_{0\eta}(\bar{\eta},\tau)}{\bar{\eta}-\eta}\,\mathrm{d}\bar{\eta} = 0. \tag{62}$$

Interest now centers on the minus sign, i.e. the destabilizing effect of the hydrodynamics when there is stability otherwise. Michelson and Sivashinsky's computations show that from initial data with a long period (but otherwise arbitrary) there eventually emerges a steady progressive wave with the same period. For such a wave $F_{0\tau}$ is replaced by a constant $-V$ and, except near points of relatively large curvature, $F_{0\eta\eta}$ may be neglected. The shape is then described by a nonlinear integral equation for the slope $F_{0\eta}$, namely

$$\tfrac{1}{2}F_{0\eta}^2 + \gamma \,\mathrm{P}\!\!\int_{-\infty}^{\infty} \left(\frac{1}{\bar{\eta}-\eta} - \frac{1}{\bar{\eta}}\right) F_{0\eta}(\bar{\eta})\,\mathrm{d}\bar{\eta} = 0 \quad \text{with } V = \gamma \,\mathrm{P}\!\!\int_{-\infty}^{\infty} \frac{F_{0\eta}(\bar{\eta})}{\bar{\eta}}\,\mathrm{d}\bar{\eta}; \tag{63}$$

here η is measured from a point where the slope vanishes. The equation expresses constancy of the flame speed, as Darrieus and Landau assumed in their stability analyses.

A continuous periodic solution for F_0, obtained numerically (McConnaughey 1982), is shown in Figure 7; the value of V/γ^2 is 1.41. (A solution for any other period can be obtained from it by scaling η, without changing V.) The discontinuities in $F_{0\eta}$ at 0, $\pm 2\pi$ are smoothed out by the neglected $F_{0\eta\eta}$-term, whose inclusion perturbs the flame speed. These results suggest that, under more general circumstances, a possible manifestation of Darrieus–Landau instability is a well-behaved (non-chaotic), large-scale, stationary structure that does not obscure the underlying flame configuration. The local smoothing into small-scale wrinkles by diffusion–reaction effects is interesting in that fine wrinkles are sometimes observed on actual flames (Markstein 1964, p. 79). More recent numerical experiments of Sivashinsky & Michelson (1982) suggest, however, that the conclusion can only be valid for flames of moderate size. For larger flames, their results imply that the manifestations of Darrieus–Landau instability are chaotic. (Gaydon & Wolfhard (1970, p. 123) have noted that the formation of cusps from a sinusoidal deformation follows from elementary kinematics if the flame speed is constant and the induced flow field is neglected.)

7 Curved cellular flames

The cellular pattern in Figure 1 is formed by a flame located in a large (11 cm diameter) vertical tube; consistent with the results of section 5, it is observed to be highly unsteady. By contrast, cellular flames stabilized on burners can be quite steady, which suggests a classical bifurcation phenomenon similar to Taylor cells in a cylindrical Couette flow. At a critical value of some parameter, disturbances of a single finite wave-number become unstable and beyond that value there is an alternative (stable) steady state. Here we shall establish curvature as such a parameter leading to a steady cellular pattern.

Attention will be confined to the cylindrical source flow

$$u = r_*/r, \quad v = 0 \tag{64}$$

(cf. section 3, where sink flow was also considered): r is measured on the scale of the flame thickness. If disturbance of the hydrodynamic field is again neglected then equations (32) govern, provided the replacement

7 Stationary wrinkling of an otherwise stable plane flame due to hydrodynamic disturbances: approximate solution of equation (62) with − sign. Possible outcome of Darrieus–Landau instability.

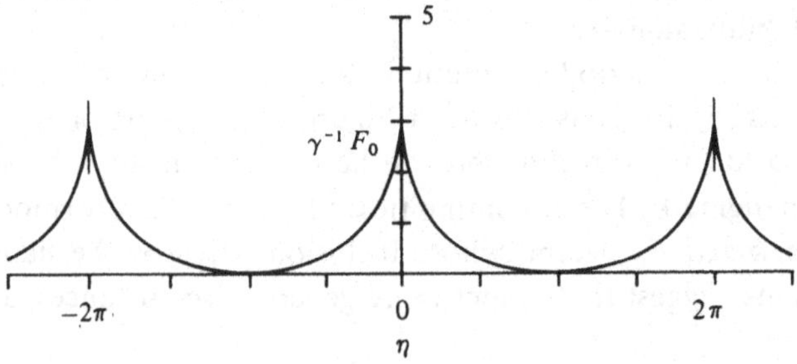

$\partial/\partial t \rightarrow \partial/\partial t + (r_*/r)\partial/\partial r$ is made. These possess a steady solution satisfying the jump conditions (8.49), (8.54) that reduces, for $l=0$, to the solution (7.10), when due account is paid to the difference in units.

Taking r_* to be the radius of the flame sheet implies that the dimensionless burning rate is 1, i.e. the dimensional burning rate is the same as that of an adiabatic plane flame. This is in apparent contradiction with previous sections, where we saw a change in burning rate with curvature, but an additional feature to consider here is flow divergence, which must have a canceling effect. Note that the stretch is 0, so that for large r_* the result is consistent with the observation in section 10.3 that changes in the speed of a weakly deformed plane flame are proportional to stretch.

The objective is to admit curvature disturbances just strong enough to modify equation (54). (By contrast, Matkowsky, Putnick & Sivashinsky (1980) allow l and r_* to be arbitrary without, however, uncovering any more features than this simpler analysis does.) Since the $F_{0\eta\eta}$-term corresponds to curvature $O(\varepsilon^2)$, we are led to introduce

$$R = \varepsilon^2 r_* \tag{65}$$

as the $O(1)$ scaled radius of the undisturbed flame sheet. An $O(\varepsilon^{-\frac{1}{2}})$ change in y nearby then corresponds to a change $O(\varepsilon^{\frac{3}{2}})$ in the polar angle ϕ, so that we take the disturbed position of the flame sheet to be

$$r = \varepsilon^{-2}R + \varepsilon f(\bar{\phi}, \tau) \quad \text{with } \bar{\phi} = \varepsilon^{-\frac{3}{2}}\phi \tag{66}$$

in a polar representation. If we draw the tangent to the undisturbed flame at $\bar{\phi}=0$ (y then being measured along the tangent from the contact point and x radially outwards) the distance between the disturbed flame and this tangent is, in the notation of earlier sections,

$$F_0 = -\tfrac{1}{2}R\bar{\phi}^2 + f(\bar{\phi}, \tau) \quad \text{with } \bar{\phi} = \eta/R \tag{67}$$

to sufficient accuracy.

The requirement that the existing curvature should not influence the flame velocity suggests modifying the result (56) to read

$$W = 1 + \varepsilon^3[4F_{0\eta\eta\eta\eta} \pm (F_{0\eta\eta} + R^{-1})/l_s] = 1 + \varepsilon^3[4R^{-4}f_{\bar{\phi}\bar{\phi}\bar{\phi}\bar{\phi}} \pm R^{-2}f_{\bar{\phi}\bar{\phi}}/l_s], \tag{68}$$

thus accounting for weak flow divergence. To see that the modification agrees with section 10.3 we note that for a flat flame it reads

$$W = 1 \pm \varepsilon^3 R^{-1}/l_s = 1 - (1 + l/l_s)\partial v/\partial y \tag{69}$$

where v is now the velocity in the y-direction. The factor $(F_{0\eta\eta} + R^{-1})$ is simply proportional to the flame stretch, so that the modification is consistent with (10.15).

A modification of the kinematical result (47) arises because the flame sheet moves in a nonuniform velocity field (without disturbing the field).

The accompanying changes $(-\varepsilon^3 F_0/R, \varepsilon^{\frac{1}{4}}\eta/R)$ in fluid velocity at the sheet lead to the new approximation

$$W = 1 - \varepsilon^3[F_{0\tau} + \tfrac{1}{2}F_{0\eta}^2 + R^{-1}(F_0 - \eta F_{0\eta})] = 1 - \varepsilon^3(f_\tau + \tfrac{1}{2}R^{-2}f_{\bar{\phi}}^2 + R^{-1}f). \tag{70}$$

Comparing the results (68) and (70) establishes

$$f_\tau + 4R^{-4}f_{\bar{\phi}\bar{\phi}\bar{\phi}\bar{\phi}} \pm R^{-2}f_{\bar{\phi}\bar{\phi}}/l_s + R^{-1}f + \tfrac{1}{2}R^{-2}f_{\bar{\phi}}^2 = 0 \tag{71}$$

as the fundamental equation governing the radial perturbation f in the position of the flame sheet. The term $R^{-1}f$ arises from the perturbation in radial velocity at the flame, so that for nominally spherical flames (where the velocity is proportional to r^{-2}) the term is replaced by $2R^{-1}f$ (Sivashinsky 1979). We shall pursue the stability of the cylindrical flame; the spherical flame requires only minor changes, including generalization of the $\bar{\phi}$-derivatives (which is also needed here for variations in the z-direction).

For linear stability we set f proportional to $\exp(\tilde{\alpha}\tau + i\bar{n}\bar{\phi})$ and neglect the quadratic term to obtain

$$\tilde{\alpha} = \pm R^{-2}\bar{n}^2/l_s - 4R^{-4}\bar{n}^4 - R^{-1} \tag{72}$$

where, although it is strictly a discrete parameter (being $\varepsilon^{\frac{1}{4}}$ times an

8 Stabilizing effect of curvature: growth rate $\tilde{\alpha}$ versus wavenumber $\tilde{k}$ with $\bar{l}$ slightly less than -1. Figure 4 is obtained from the curve with $R = \infty$, since $\tilde{\alpha} = \alpha/\varepsilon^2$, $\tilde{k} = k/\varepsilon^{\frac{1}{4}}$ according to equations (45).

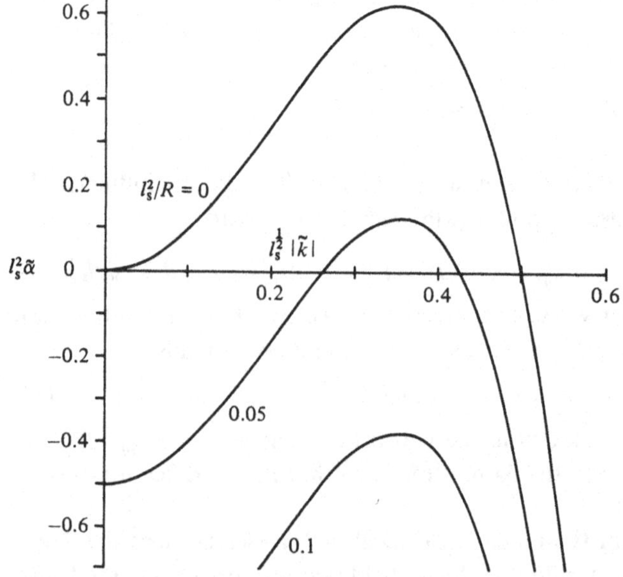

integer), $\bar{n}$ effectively takes all real values. (If variations in the z-direction are admitted, with wave number k, then $\bar{n}^2$ is replaced by $\bar{n}^2+R^2k^2$ and the conclusions are modified in an obvious way.) As $R\rightarrow\infty$, with

$$R^{-1}\bar{n}=\tilde{k} \qquad (73)$$

held fixed, the result (57) is recovered; and finite R clearly has a stabilizing effect. For $l>-l_s$ the curvature only augments the existing stability, but for $l<-l_s$ it can overcome the instability that occurs for small wave-numbers $\tilde{k}$ (see Figure 8). For all R the maximum growth rate is attained for the value

$$\tilde{k}_m^2=1/8l_s \qquad (74)$$

(cf. equation (58)) and it is zero when R has the critical value

$$R_c=16l_s^2. \qquad (75)$$

For $R\leqslant R_c$ there is stability, but for $R>R_c$ there is a band of unstable waves.

The stability regions in the plane of $l_s^{\frac{1}{2}}|\tilde{k}|$ and l_s^2/R are shown in Figure 9. as R increases through R_c, a small band of unstable wavenumbers appears, suggesting a classical bifurcation. The corresponding Landau equation (see Matkowsky 1970) is obtained in the usual way by setting

$$\delta^2=(R-R_c)/R_c, \quad f=\delta(f_0+\delta f_1+\delta^2 f_2+\ldots), \quad \bar{\tau}=\delta^2\tau, \qquad (76)$$

9 Stabilizing effect of curvature: the band of unstable wavenumbers $\tilde{k}$ for $\bar{l}$ slightly less than -1 shrinks as the curvature R^{-1} increases and disappears beyond $R_c^{-1}=(16l_s^2)^{-1}$.

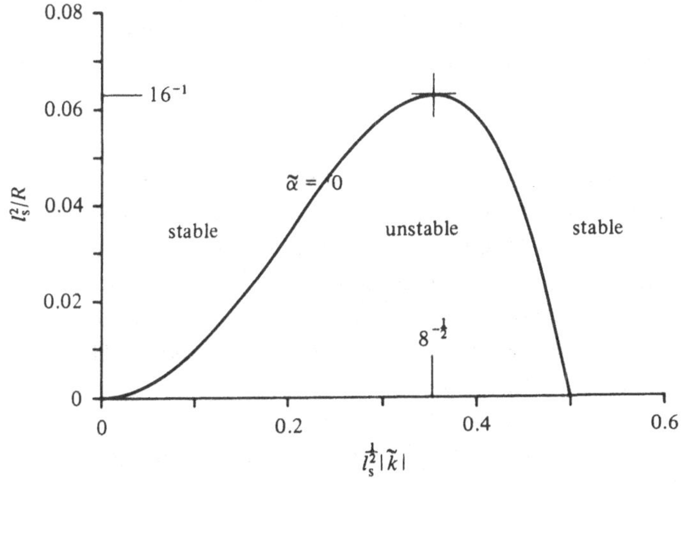

where

$$f = \delta A \cos(\bar{n}_c \bar{\phi}) \quad \text{for } \bar{\tau} = 0 \quad \text{with } \bar{n}_c = (32)^{\frac{1}{2}} l_s^{\frac{3}{2}}. \tag{77}$$

To leading order we find

$$f_0 = A_0(\bar{\tau}) \cos(\bar{n}_c \bar{\phi}), \quad A_0(0) = A; \tag{78}$$

then secular terms appear in f_2 unless

$$A_0' = A_0(9 - l_s^2 A_0^2)/144 l_s^2. \tag{79}$$

It follows that the amplitude asymptotes to the constant value

$$A_0 = \pm 3/l_s \quad \text{accordingly as } A \lessgtr 0. \tag{80}$$

We conclude that for small positive values of $R - R_c$ there is a stable cellular configuration, with amplitude proportional to $(R - R_c)^{\frac{1}{2}}$, to which the originally cylindrical flame sheet tends after disturbance.

It is characteristic of stationary cellular flames (including polyhedral flames) that the crests, i.e. the portions of the flame that are convex towards the burnt gas, are shaper than the troughs. To leading order the curvature is

$$f_{\bar{\phi}\bar{\phi}} = -32 \delta l_s^3 A_0 \cos(\bar{n}_c \bar{\phi}), \tag{81}$$

to which f_1 makes the (second-harmonic) correction $-(64\delta^2 l_s^4/9) A_0^2 \cos(2\bar{n}_c \bar{\phi})$. This increases the magnitude of the curvature at the crests and decreases it at the troughs, consistent with observations.

8 Delta-function models and the right stability boundary

The flames considered in this chapter can be characterized by a delta-function representation of the reaction term, whose strength is a rationally derived nonlinear function of the flame temperature. The strength, implied by the integral (3.11), is given by the formula

$$\delta(\partial T/\partial n) = -\sqrt{2\mathscr{L}\mathscr{D}}\, T_b^2\, e^{-\theta/2T_b}/M_r \theta. \tag{82}$$

This suggests a different starting point for a rational theory of flames in which such a delta function is used in place of the Arrhenius term, with θ treated as a finite parameter. In modeling the extremely complex processes occurring in a methane/air flame, for example, it is difficult to conceive of any *a priori* philosophical argument why such a substitution would render the approach substantially inferior to ours. The approach, known as the flame-sheet approximation, is not new (cf. Peskin & Wise 1966), but for the most part it lies outside the framework of our discussion. However, certain recent analyses of this type relate directly to the right stability boundary of Figure 3, so we shall briefly mention them.

Thermites are solids in which deflagrations can occur with the combustion products remaining solid. As will be shown (see sections 12.2 and 12.4), such deflagrations can be described by equations derived from those in Chapter 1 with $\mathscr{L} = \infty$ (no diffusion of species). Matkowsky & Sivashinsky (1978) then adopt (82) in place of the Arrhenius kinetics, setting $\mathscr{L} = 1$ (somewhat inconsistently). Steady plane solutions exist and their linear stability is examined for one-dimensional disturbances. Not surprisingly, in view of the results in sections 2 and 4, instability is found if θ is large enough; but for smaller values the flame is stable. For the critical value of the activation energy (corresponding to neutral stability) the flame oscillates; this should be compared with the fact that Im $(\alpha) \neq 0$ on the right boundary in Figure 3.

Numerical calculations using Arrhenius kinetics, reported by Shkadinsky, Khaikin, & Merzhanov (1971), are in substantial agreement with these predictions, suggesting that the model characterized by (82) is a good approximation for this specific problem. The implication for gaseous flames is that, as θ is decreased through finite values, the stability boundary moves to the right and eventually becomes inaccessible because $\mathscr{L}$ is necessarily finite. That oscillatory instabilities are observed in thermites (Merzhanov, Filonenko, & Borovinskaya 1973) suggests that the boundary has physical reality, despite the limitations under which it is derived.

Further evidence of the relevance of this boundary is provided by the work of Margolis (1980), who considers the stability of gaseous anchored flames subject to boundary conditions at the flameholder incompatible with an SVF or NEF (cf. section 5.6). An earlier paper of Joulin & Clavin (1979), on the stability of NEFs subject to distributed heat loss of the kind treated in section 3.4, shows a tendency for the right boundary to shift to the left, making it accessible to flames whose Lewis numbers do not differ too greatly from 1. Margolis reasons that heat loss to a flameholder could have a similar effect, thereby making the boundary relevant to actual flames. He attacks the question in two ways. First, he uses a delta-function model to find that, for $\mathscr{L}$ not too small, there is a critical value of the mass flux M below which oscillatory instability occurs. (As M decreases from its adiabatic value the flame moves in from infinity and the heat loss increases (see section 2.5).) Next, a complicated kinetic model of a hydrogen/oxygen flame is used in a numerical integration and again unstable oscillations are found. Margolis has also made unpublished reports of an experimental study in which oscillating flames are clearly identified. The oscillations are undulating rather than plane, suggesting that the instability has a finite wavelength, as in Figure 3.

These two investigations apparently provide the only concrete evidence that the right stability boundary is physically relevant. An investigation similar to that of Margolis needs to be done for the Arrhenius model using activation-energy asymptotics, to show that the oscillatory instabilities uncovered by him are unequivocally related to the boundary in Figure 3. A start on this question has been made by Matkowsky & Olagunju (1980).

12

IGNITION AND EXPLOSION

1 Synopsis

Ignition has been discussed in three contexts: burning of a linear condensate (Chapter 4), spherical diffusion flames (Chapter 6), and spherical premixed flames (Chapter 7). Based upon the S-shaped response curves determined by steady-state analyses it is argued that the burning rate will jump from a weak, almost extinguished, level to a vigorous one as the pressure (and, hence, the Damköhler number) is increased through some critical value (corresponding to the lower bend of the S). The occurrence of an intrinsically unsteady phenomenon, appropriately called ignition, is thereby inferred from results inherent in the steady state.

Such analyses have a fairly long history in combustion investigations, the simplest example being the thermal theory of spontaneous combustion identified with the name of its originator, Frank-Kamenetskii (1969). Section 2 presents a mathematical version of that theory, which shows an early appreciation of activation-energy asymptotics (though not in the formal sense of this monograph).

Even though steady-state analyses are undoubtedly useful, descriptions of the unsteady state are also needed; providing them is the main goal of this chapter. The three contexts already considered are too complicated for our purposes; we concentrate instead on a simpler problem containing the essential feature of ignition: the evolution of a deflagration wave from an unburnt state.

Consider a thermally insulated enclosure containing a mixture at sufficiently low temperature for the reaction to be very weak. Since the heat released cannot escape, the temperature must rise, albeit very slowly. After a long time the heat generated by the initially weak reaction will have raised the temperature of the mixture enough to excite a rapid reaction somewhere, and if the combustion field is nonhomogeneous a deflagration wave will sweep across it. Since the temperature rise to excitation is due to self-heating rather than external agencies, the process is called spontaneous combustion or auto-ignition. The term thermal explosion is used to describe the violence that follows excitation, also a feature of externally excited combustion. (Thermal refers to the temperature sensitivity of the reaction rate. Another type of explosion is due to a rapid chain reaction, initiated by conditions that permit the production of a few necessary radicals from one of the reactants.)

In section 3 the spatially homogeneous explosion, described by ordinary differential equations, provides insight into the mathematics of the rapid transient that is characteristic of the explosion process. A discussion of the spatially nonhomogeneous problem in section 4 is thereby facilitated.

A distinctive feature of auto-ignition is the very fast temperature increase, known as thermal runaway, that occurs at the end of a well-defined induction phase. The mathematical manifestation is unboundedness in the solution of an equation describing small perturbations of the initial state. (Evolution equations that generate unbounded solutions after a finite time have recently attracted the attention of mathematical analysts; see, for example, Payne (1975). Ayeni (1978) and Chandra & Davis (1980), amongst others, have considered the specific problems of this chapter, in particular for θ finite.) The subject of ignition by external agencies provides other examples, one of the simplest being a half space filled with a combustible mixture, at the boundary of which a positive heat flux is applied. After a certain time, ignition occurs at the boundary and a plane deflagration wave propagates into the interior, consuming the mixture in its passage. The early stages of the process are described in section 5.

The problems in sections 2–5 are concerned with premixed reactants; many such problems, to which activation-energy asymptotics could be profitably applied, are discussed by Merzhanov & Averson (1971). By contrast, section 6 considers a fuel and oxidant in separate half spaces that are brought into contact to interdiffuse and burn. Because of the many features that are in common with the spatially nonhomogeneous thermal explosion, it is possible to give a reasonably complete description of the combustion process by analogy.

2 Spontaneous combustion (auto-ignition)

Exothermic chemical reaction is a well-known hazard of holding certain bulk materials while in storage or transit. Joseph Conrad in his story *Youth* graphically describes the peril. Take one ship with a cargo of coal dampened down because of an earlier leak (now fixed); embark on a slow voyage to Bangkok; and before the China Sea is reached the legend 'Do or Die' proves ominous rather than brave.

Any moist volume of organic material (e.g. wood chips, wool, corn-cobs, compost) will, because of chemical reaction initiated by micro-organisms, get hot. (The reduction of spontaneous reactions in fruits and vegetables is a major concern of postharvest physiologists.) Certain inorganic materials, such as ammonium nitrate, lead azide, or a mixture of hydrogen sulphide and oxygen will also heat up because of chemical decomposition.

If the reaction is independent of temperature there is a maximum to which the temperature of a material will rise after a long time, dependent on its dimensions. (That maximum holds steady over an even longer period, but there is an eventual decay due to reactant depletion.) In the quasi-steady state the rate of generation of heat, which is proportional to a^3 for a volume of characteristic dimension a, is balanced by loss through the surface, which is proportional to a^2 with a factor of proportionality that is an increasing function of temperature. Clearly, the larger the volume, the hotter it tends to become. If the reaction does depend on temperature, proceeding much more vigorously at higher temperatures, there is a critical value of a beyond which the efflux at the surface is insufficient to dispose of the heat generated and unsteady combustion occurs. Our first task is to determine the steady states.

Any steady state is described by equation (1.58) with $\partial/\partial t = \mathbf{v} = 0$, i.e.

$$\nabla^2 T = -\Omega \quad \text{with } \Omega = D\,\mathrm{e}^{-\theta/T}, \tag{1}$$

the latter being appropriate when the length a is used in the nondimensionalization. For solids, the density has a constant value, which may be designated ρ_c to replace ρ by 1 in the definition (1.61) of D; pressure is not involved, which means that $\bar{p}_c$ should be replaced by $\rho_c QR/c_p m$ in that definition (thus eliminating the gas constant R from it); and α is usually taken to be 0. The only new assumption is that reactant depletion may be neglected (the constant mass fractions in Ω being absorbed into D) because the steady state is close to the initial state, a condition ensured *a posteriori* when θ is large. Application to gases requires us to accept the limitations of the constant-density approximation.

In the strictest sense, theories of the type presented in this chapter apply only to the so-called thermites (see Maksimov, Merzhanov, & Shkiro 1965), i.e. solids that burn internally to form solid products. Since their inception, however, they have been otherwise applied, in particular to the early propellants (section 4.1), whose products are mainly gaseous.

For an initially uniform temperature T_f such as we have so far envisaged, this purely thermal problem can be treated in the limit by setting $T_0 = T_f$. To first order we find

$$\nabla^2\phi = -\delta\, e^{\phi} \quad \text{with } \phi \equiv -(1/T)_1 = T_1/T_f^2, \quad \delta = D\theta\, e^{-\theta/T_f}/T_f^2. \qquad (2)$$

The sign of ϕ has been changed from the definition (2.18*c*) to keep it positive; the use of δ (rather than a variant of D) follows the convention established by Frank-Kamenetskii. Note that the steady state being sought is close to the initial state, so that the neglect of reactant depletion is justified. For natural organic materials, however, such as wood chips, the activation energy is usually not large and depletion is not small; in either circumstance the analysis is not appropriate.

The approximation is familiar from our earlier ignition studies (sections 6.4 and 7.5). In this context it was first used by Frank-Kamenetskii (1969), constituting an early example of activation-energy asymptotics. (References and a discussion of the solutions of equation (2) are given in his book.) On the other hand, linearization of the exponent θ/T in the Arrhenius factor has sometimes been adopted as a model not rationally consistent (in an asymptotic sense) with the underlying equation (1) for the specific problem being examined. The plane form of equation (2) has a long history (Stuart 1967), apparently originating with Liouville (1853), to whom it is sometimes attributed. The general solution is known, but there appears to be no way of fitting it to boundary conditions of interest. As a consequence, solutions of complete boundary-value problems have only been constructed in a few special cases.

Among these are problems depending solely on x. The corresponding problems for the exact equation (1) can then be solved also, providing useful checks for the asymptotics (cf. Shouman & Donaldson (1975)). A particularly simple example is an infinite slab $|x| \leq 1$ of combustible material whose surfaces are held at the temperature T_f, so that

$$\phi = 0 \quad \text{for } x = \pm 1. \qquad (3)$$

Such a surface condition corresponds to the limit of an efflux $k(T - T_f)$ per unit area as $k \to \infty$, and it is the ability of the material in the slab to conduct that heat to the surfaces that determines whether or not a steady state exists. By symmetry ϕ will attain its maximum, say ϕ_m, at $x = 0$; and in terms of it we have

$$\phi = 2\ln[e^{\frac{1}{2}\phi_m}\operatorname{sech}(cx)] \quad \text{with } c^2 = \tfrac{1}{2}\delta\, e^{\phi_m}. \qquad (4)$$

The boundary conditions (3) require, therefore,

$$\sqrt{\delta/2} = e^{-\frac{1}{2}\phi_m}\cosh^{-1}(e^{\frac{1}{2}\phi_m}), \qquad (5)$$

a relation fixing the maximum temperature in terms of δ. For a given material δ can be varied through a, the semithickness of the slab, or T_f, the temperature at its surfaces (see equation (2c) and the definition (1.61) of D).

Figure 1 shows the temperature distribution (4) and the response (5); the latter has the typical shape of an ignition curve. There is a critical value

$$\delta_c = 0.878 \tag{6}$$

above which no solution exists and below which there are two solutions. A value of δ greater than δ_c is said to be supercritical, while a smaller value is said to be subcritical. The implication is that thick, hot slabs (being supercritical) will burst into flames, but thin, cold slabs (being subcritical) will not; and this general picture has been confirmed experimentally for a number of substances (see Frank-Kamenetskii 1969, p. 412).

The symmetry of the boundary conditions makes the solution atypical, however, as we shall see from another example in which the surfaces $x = \pm 1$ are held at temperatures T_f and αT_f, respectively, where $0 < \alpha < 1$. Now the reaction is confined to a neighborhood of the hot boundary in the limit $\theta \to \infty$ instead of being spread across the slab, so that conduction of its heat to that surface is much easier and δ_c is much larger.

The initial temperature in the slab, due entirely to conduction, is

$$T = \tfrac{1}{2}[1 + \alpha + (1-\alpha)x]T_f; \tag{7}$$

we then seek a steady state within $O(\theta^{-1})$. For such a state the $O(1)$ temperature has its maximum at the hot boundary, so that reaction only occurs in a zone near there; the appropriate expansion variable is

$$\xi = \theta(x-1) \tag{8}$$

and $T_0 = T_f$. The structure of the reaction zone is found to be governed by

$$d^2\phi/d\xi^2 = -\tilde{\delta}\, e^{\phi} \quad \text{with } \phi = T_1/T_f^2, \quad \tilde{\delta} = \theta^{-2}\delta, \tag{9}$$

$$d\phi/d\xi = (1-\alpha)/2T_f + \ldots \quad \text{as } \xi \to -\infty, \quad \phi = 0 \quad \text{for } \xi = 0, \tag{10}$$

where the first boundary condition comes from matching with the temperature (7) outside the zone. The general solution of equation (9) is

$$\phi = 2\ln\{e^{\frac{1}{2}\phi_m}\operatorname{sech}[\tilde{c}(\xi - \xi_m)]\} \quad \text{with } \tilde{c}^2 = \tfrac{1}{2}\tilde{\delta}\, e^{\phi_m}, \tag{11}$$

where ξ_m is the location of the maximum ϕ_m; to satisfy the conditions (10) we must set

$$\tilde{\delta} = \tilde{\delta}_c e^{-\phi_m} \quad \text{and } \xi_m = \pm\tilde{c}^{-1}\operatorname{sech}^{-1}(e^{-\frac{1}{2}\phi_m}) \quad \text{with } \tilde{\delta}_c = (1-\alpha)^2/8T_f^2. \tag{12}$$

Since ϕ vanishes at the surface, ϕ_m ranges from 0 to ∞ (otherwise ξ_m is complex), so that $\tilde{\delta}$ cannot exceed the critical value $\tilde{\delta}_c$. The $\pm$ signs on ξ_m provide two solutions when $\tilde{\delta} < \tilde{\delta}_c$ (see Figure 2(*a*)); for the plus sign, ξ_m lies outside the slab and the maximum ϕ in the slab is 0 (its surface value) and not ϕ_m. The maximum inside the slab may, therefore, be written

$\mathscr{H}(-\xi_m)\phi_m$, where $\mathscr{H}$ is the unit-step function, and leads to the response shown in Figure 2(*b*), when the result (12*a*) is used.

Note that $\tilde{\delta}=\theta^{-2}\delta$ implies that the D corresponding to $\tilde{\delta}_c$ is much larger than that corresponding to δ_c. We conclude that raising the temperature of the cold boundary to that of the hot will greatly increase the risk of explosion. In fact, the formula (12*c*) suggests that the risk increases sharply as soon as α is within $O(\theta^{-1})$ of 1.

One further point should be noted. Nothing has been missed in either problem by looking for steady states within $O(\theta^{-1})$ of the initial state. Any steady state further away would correspond to a D of exponential order θ/T_m, where the maximum temperature T_m is greater than T_f. It would, therefore, correspond to the origin in Figures 1(*b*) and 2(*b*).

For a review of the subject, including other geometries and boundary

1 Steady state in slab with surfaces maintained at initial uniform temperature: (*a*) perturbation temperature profiles; (*b*) response determined by equation (5).

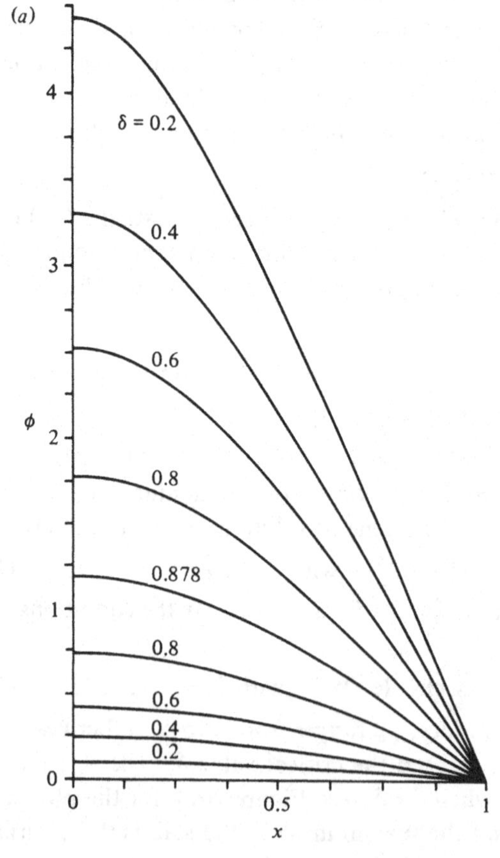

conditions, see Gray & Lee (1967). They do not use formal asymptotics; there is need for a comprehensive treatment doing so.

3 Adiabatic explosion

To describe the unsteady phase of the combustion, equation (1) is replaced by

$$\partial T/\partial t - \nabla^2 T = D\,e^{-\theta/T} \tag{13}$$

so that the perturbation equation (2) becomes

$$\partial \phi/\partial t - \nabla^2 \phi = \delta\,e^{\phi}. \tag{14}$$

The problem is completed by the initial condition

$$\phi = 0 \quad \text{for } t = 0 \tag{15}$$

and suitable boundary conditions.

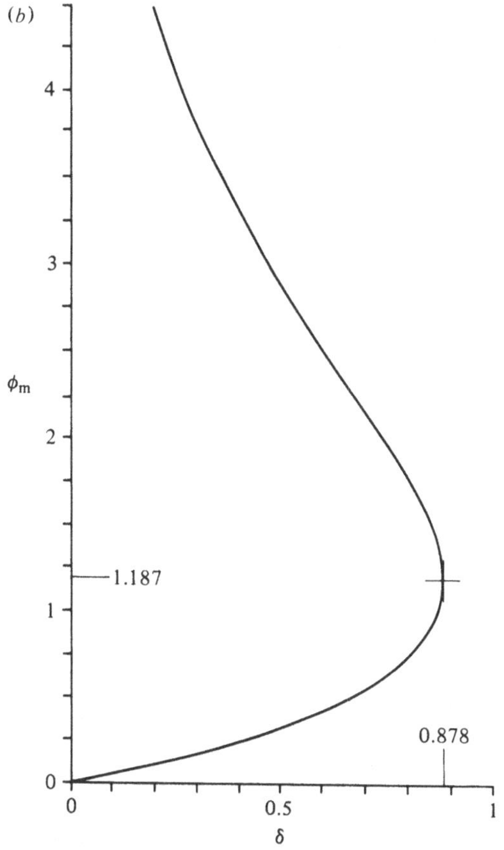

2 Steady state in slab with linear initial temperature and surface temperatures maintained: (*a*) perturbation temperature profiles, drawn for $1-\alpha=4T_f$; (*b*) response determined by equations (12).

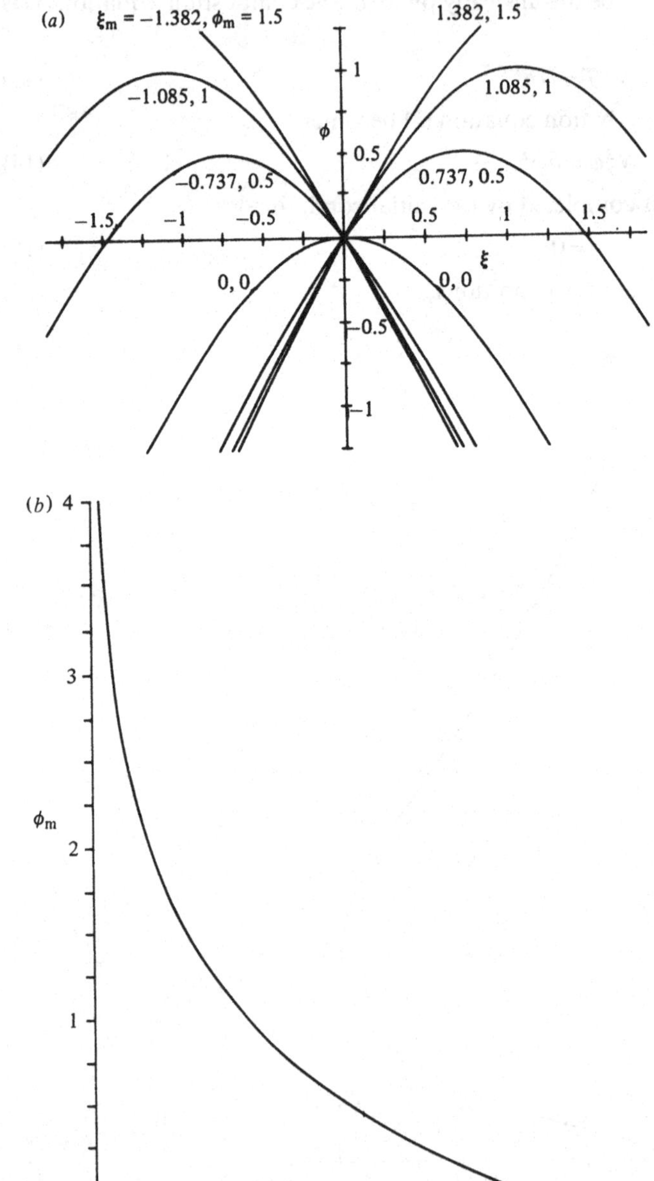

The lack of a steady-state solution for the slab with equally hot surfaces and supercritical δ implies that ϕ must increase without bound, i.e. that T goes further than $O(\theta^{-1})$ from T_f. This happens at a finite time, as we shall now show for the spatially homogeneous problem of a material thermally insulated at its boundary surface, when equation (13) reduces to

$$dT/dt = D\,e^{-\theta/T}. \tag{16}$$

Our object in this section is to gain insight into thermal explosions generally by examining those that are spatially homogeneous.

The corresponding perturbation equation is

$$d\phi/dt = \delta\,e^{\phi}, \tag{17}$$

which has the solution

$$\phi = -\ln(1-\tilde{t}) \quad \text{with } \tilde{t} = \delta t \tag{18}$$

satisfying the initial condition (15). At the finite time δ^{-1}, marking the end of the induction phase, the perturbation ϕ becomes unbounded, an event known as thermal runaway. Note that the phenomenon occurs for all values of δ in the spatially homogeneous problem because the heat generated by the reaction cannot escape, i.e. conditions are adiabatic. On the other hand, if a heat-loss term is introduced, then subcritical behavior can be restored (Kassoy & Poland 1975).

Clarification of the runaway process requires a return to equation (16). Beyond thermal runaway, however, a purely thermal model is of little practical significance since it predicts an unbounded increase in T, whereas the rise ceases when the reactant is consumed. Under the assumption that a single (decomposing) reactant is involved, its depletion can be incorporated by reinstating the mass fraction in equation (16) and adding a species equation, i.e. writing

$$dT/dt = -dY/dt = D(Y/Y_f)\,e^{-\theta/T}, \tag{19}$$

where Y_f is the initial fraction of reactant in the material. Clearly, the Shvab–Zeldovich relation (2.3) and the result (2.6) hold, where T_b is now the ultimate temperature. No change in our analysis of the induction phase (section 2) is necessary because Y then stays within $O(\theta^{-1})$ of Y_f and hence, may be replaced by it in equation (19a). We are led, therefore, to the problem

$$dT/dt = (T_f^2\delta/Y_f\theta)(T_b - T)\,e^{\theta/T_f-\theta/T}, \quad T = T_f \quad \text{for } t = 0. \tag{20}$$

Clearly, T is monotonically increasing but the factor $(T_b - T)$, corresponding to reactant depletion, ensures that it does not go beyond T_b; in fact, it rapidly approaches T_b (exponentially in time).

Kassoy (1975) has addressed the asymptotics of this problem directly; here we shall verify the essential features of his results from the exact solution. With the confidence so gained we shall attack spatially non-homogeneous problems in section 4, for which there are no exact solutions against which to check the asymptotics. The exact solution here, first given by Hermance (1975) and Parang & Jischke (1975), is

$$T_f^2\, e^{\theta/T_f}(\tilde{t}_e-\tilde{t})/Y_f\theta = e^{\theta/T_b}\mathrm{Ei}(\theta/T-\theta/T_b)-\mathrm{Ei}(\theta/T), \tag{21}$$

where

$$\tilde{t}_e = (Y_f\theta\, e^{-\theta/T_f}/T_f^2)[e^{\theta/T_b}\mathrm{Ei}(\theta/T_f-\theta/T_b)-\mathrm{Ei}(\theta/T_f)] \tag{22}$$

and the time variable (18b) is being used. Here $\mathrm{Ei}(x)$ is the exponential integral function (Gautschi & Cahill 1964); its asymptotic behavior as $x\to 0$ confirms that T approaches T_b exponentially as $\theta^{-1}\exp(\theta/T_f-\theta/T_b)(\tilde{t}_e-\tilde{t})\to\infty$.

Three phases can be distinguished when θ is large: induction, explosion, and relaxation. If T is bounded away from T_b, the arguments of both Ei-functions in the solution (21) become large so that

$$\tilde{t}_e-\tilde{t} = [Y_f T^2/T_f^2(T_b-T)]\, e^{\theta/T-\theta/T_f}[1+O(\theta^{-1})], \tag{23}$$

which reveals two of the phases. In the first phase T is still within $O(\theta^{-1})$ of T_f and the right side is $O(1)$, corresponding to induction. In fact, the result (18a) is recovered when

$$\tilde{t}_e = 1+T_f(2+T_f/Y_f)\theta^{-1}+O(\theta^{-2}) \tag{24}$$

is noted. In the second and most violent phase T is bounded from T_f (as well as from T_b) and the right side is exponentially small, showing that most of the change in T occurs within an exponentially small interval just before the time $\tilde{t}_e$. In that sense $\tilde{t}_e$ is the time at which the explosion takes place. The mathematical structure of the explosion is revealed by taking logarithms in equation (23) to obtain

$$\ln(\tilde{t}_e-\tilde{t}) = \theta(1/T-1/T_f)+O(1). \tag{25}$$

The asymptotic balance in this equation becomes explicit if the (fast) time τ is introduced by the nonlinear scaling

$$\tau = -\theta^{-1}\ln(\tilde{t}_e-\tilde{t}) \quad \text{so that } T_0 = T_f/(1-T_f\tau). \tag{26}$$

As τ ranges from 0 to Y_f/T_bT_f the temperature ranges from T_f to T_b, to leading order. While nonlinear scalings have been encountered before in an asymptotic analysis (Van Dyke 1975, p. 87), a logarithmic one (26a) appears to be novel.

The third phase, relaxation, corresponds to values of T within $O(\theta^{-1})$ of T_b; then the argument of the first Ei-function in the solution (21) is

$O(1)$ and we have

$$\tilde{t}_e - \tilde{t} = (Y_f\theta/T_f^2)\,e^{\theta/T_b - \theta/T_f}\,\mathrm{Ei}(\theta/T - \theta/T_b)[1 + O(\theta^{-1})]. \tag{27}$$

The final consumption of the reactant also takes place within an exponentially small interval which, since the Ei-function changes sign at some positive value of its argument, contains $\tilde{t}_e$.

We now turn to the question of deriving these asymptotic results directly from the governing equation (20*a*). Consider first the induction phase and set $T_0 = T_f$; we find the earlier result (18*a*) for T_1 and

$$T_2 = T_f^3\{\ln^2(1-\tilde{t}) - (2 + T_f/Y_f)[\ln(1-\tilde{t}) + \tilde{t}/(1-\tilde{t})]\}. \tag{28}$$

The logarithm in T_1 suggests that breakdown occurs when $\tilde{t}$ is exponentially close to 1, while the pole $(1-\tilde{t})^{-1}$ in T_2 suggests algebraical closeness; and correctly so, since the explosion time (24) is $O(\theta^{-1})$ away from 1. To generate a uniformly valid description it is only necessary to absorb the pole into T_1 by writing

$$T_1 = -T_f^2 \ln[1 + T_f(2 + T_f/Y_f)\theta^{-1} - \tilde{t}], \tag{29}$$

which amounts to introducing the $O(\theta^{-1})$ approximation to $\tilde{t}_e$. (The same end can be achieved formally by considering both T and $\tilde{t}$ to be functions of an auxiliary time variable.) In short, the time to explosion has been determined asymptotically by considering only the induction phase; it is reasonable to suppose that the same procedure will work in other problems.

The explosion phase can also be derived directly from equation (20*a*). It is only necessary to introduce the time variable τ, so that

$$\mathrm{d}T/\mathrm{d}\tau = (T_f^2/Y_f)(T_b - T)\,e^{\theta/T_f - \theta/T - \theta\tau}, \tag{30}$$

and the formula (26*b*) is recovered. Identification of the time scale τ is one of the key steps in unraveling the nonhomogeneous problem, as we shall see in section 4.

For the final relaxation phase we write $T_0 = T_b$ in equation (20*a*), which shows that the appropriate time variable is

$$\bar{\tau} = (\tilde{t} - \tilde{t}_e)/\varepsilon \quad \text{with } \varepsilon = (Y_f\theta/T_f^2)\,e^{-\theta Y_f/T_bT_f}. \tag{31}$$

Then

$$\mathrm{d}\phi/\mathrm{d}\bar{\tau} = -\phi\,e^{\phi} \quad \text{with } \phi = T_1/T_b^2, \tag{32}$$

the general solution of which is

$$\bar{\tau} - \bar{\tau}^0 = \mathrm{Ei}(1) - \mathrm{Ei}(-\phi); \tag{33}$$

here $\bar{\tau}^0$ is just the value of $\bar{\tau}$ when $\phi = -1$. The requirement that $\phi \to 0$ as $\bar{\tau} \to +\infty$ (i.e. all reactant is consumed) is satisfied automatically, as is that of matching with the approximation (26*b*). For the latter, note that introducing $\bar{\tau}$ into the approximation (23) and then keeping it fixed under expansion gives the result $\bar{\tau}=\mathrm{e}^{-\phi}/\phi$. Comparison with the solution (33) shows that the integration constant $\bar{\tau}^0$ is left undetermined and that it cannot be determined by finite asymptotic development; no matter how many terms are retained in the expansion (23) the Ei-functions of the exact solution will never be recaptured; it is those functions that determine ϕ. It is, therefore, fortunate that results analogous to

$$\bar{\tau}^0 = -\mathrm{Ei}(1), \tag{34}$$

here obtained from the exact solution, will not be needed in the sequel.

4 Explosion with heat loss

Kassoy (1977) has incorporated nonadiabaticity into the spatially homogeneous problem by inserting a bulk or lumped heat-loss term proportional to $T-T_f$ into equations (19). This model, introduced by Semenov (1928), has received considerable attention. Apart from the lack of a Shvab–Zeldovich relation, the new features are that the reaction cannot produce the adiabatic temperature T_b and there is an eventual decay to T_f. To obtain the essentially new phenomena of hot-spot formation and subsequent flash-through it is necessary to consider spatial nonhomogeneity, and that can be done without sacrificing the Shvab–Zeldovich relation.

Consider once more the infinite slab with equally hot surfaces. The equations (19) are now replaced by

$$\partial T/\partial t - \partial^2 T/\partial x^2 = -(\partial Y/\partial t - \mathscr{L}^{-1}\partial^2 Y/\partial x^2) = D(Y/Y_f)\,\mathrm{e}^{-\theta/T} \tag{35}$$

(cf. equation (13)). For a solid, where the reactant is unable to diffuse, the Lewis number $\mathscr{L}$ is infinite. Nevertheless, we shall take $\mathscr{L}=1$ to have the simplification of a Shvab–Zeldovich integral (2.3). Nothing spurious is obtained, and the results are justified for a gaseous reactant. Of more concern is the constancy of Y at the surfaces, assumed in reaching the integral. Rather than maintaining the reactant concentration it would be more practical to prevent its flux. We shall see, however, that the boundary condition on Y does not influence the solution until after the main events have taken place. The problem we shall treat is, therefore:

$$\partial T/\partial t - \partial^2 T/\partial x^2 = (T_f^2\delta/Y_f\theta)(T_b - T)\,\mathrm{e}^{\theta/T_f - \theta/T} \quad \text{with } \delta > \delta_c, \tag{36}$$

$$T = T_f \quad \text{for } t=0 \quad \text{and for } x = \pm 1. \tag{37}$$

By symmetry, the condition at $x=-1$ can be replaced by

$$\partial T/\partial x = 0 \quad \text{for } x=0, \tag{38}$$

with only the half-slab $0 \leqslant x \leqslant 1$ considered. A related problem, with emphasis on the ignition implications of the steady-state response, starts with the temperature distribution of the weak steady state for $\delta<\delta_c$ and considers the effect of a slow increase in δ beyond the critical value (Kapila 1981*b*).

In addition to the induction, explosion, and relaxation phases found in the spatially homogeneous problem (20) there are now transition and propagation phases to consider. These arise because the explosion is confined to a single location and then propagates elsewhere, involving a transition from the formation of a hot-spot to the flash-through of a deflagration wave. These two new phases postpone the relaxation phase, which is the only feature affected by the boundary condition on Y.

As for the homogeneous problem, the induction phase is described by a perturbation of T_f, but now it also depends on x, satisfying

$$\partial\phi/\partial t-\partial^2\phi/\partial x^2=\delta\, e^{\phi} \quad \text{with } \phi=T_1/T_f^2, \tag{39}$$

$$\partial\phi/\partial x=0 \quad \text{for } x=0, \quad \phi=0 \quad \text{for } x=1 \quad \text{and for } t=0. \tag{40}$$

If δ is greater than δ_c, the solution cannot tend to a steady limit. Moreover, the solution of the corresponding homogeneous problem becomes unbounded in a finite time, so that it is not surprising to find that the same is true here. Thermal runaway can no longer occur everywhere when the surfaces are held at T_f, but is localized at the symmetry plane $x=0$. The runaway time is no longer fixed on the $\tilde{t}$-scale, but is a function of δ; this was first determined (numerically) by Poland (1979) (see also Kassoy & Poland (1980)), and then independently by Kapila (1980). As $\delta\downarrow\delta_c$ the time to runaway becomes indefinitely long.

While ϕ cannot be calculated analytically, its form as $t\to t_e=\delta^{-1}\tilde{t}_e(\delta)$ can be determined, a step that is crucial for the subsequent development. Integration of a parabolic equation backward from a singularity is no novelty; more than thirty years ago Goldstein (1948) encountered such a question in his discussion of Prandtl's boundary-layer equations near a separation point. He showed that the similarity variable, here

$$\eta=x/(t_e-t)^{\frac{1}{2}}, \tag{41}$$

plays a central role near the singularity. In terms of η and $\tilde{t}$ our equation becomes

$$\partial\phi/\partial\tilde{t}-(\tilde{t}_e-\tilde{t})^{-1}[\partial^2\phi/\partial\eta^2-(\tfrac{1}{2}\eta)\partial\phi/\partial\eta]=e^{\phi}. \tag{42}$$

Because of the result (18) for the homogeneous problem, a uniform asymptotic approximation

$$\phi=-\ln(\tilde{t}_e-\tilde{t})+\psi(\eta)+\ldots \quad \text{as } \tilde{t}\uparrow\tilde{t}_e \tag{43}$$

is sought for η bounded. We find

$$\psi''-(\tfrac{1}{2}\eta)\psi'+e^{\psi}=1 \quad \text{with } \psi'(0)=0 \tag{44}$$

according to the symmetry condition (38); the other boundary conditions cannot be applied. But to determine ψ another boundary condition is needed and that comes from matching with the x, t-solution. Since ψ cannot be exponentially large, it has the asymptotic form

$$\psi = -2\ln\eta + A + \ldots \quad \text{as } \eta\to\infty, \tag{45}$$

which provides the needed condition once the constant A has been determined by matching. To effect the matching we write the explicit terms (45) as functions of x and $\tilde{t}$, and expand in $\tilde{t}_e - \tilde{t}$ to obtain

$$\psi = -2\ln x - \ln\delta + A + \ldots \quad \text{as } \tilde{t}\uparrow\tilde{t}_e \tag{46}$$

for the behavior of the induction solution with x bounded away from 0. The numerics mentioned do exhibit such a behavior, so that $A(\delta)$ can be determined (Poland 1979, Kassoy & Poland 1980, Kapila 1980). The procedure fails if δ is subcritical because the numerical integration does not yield the required form (46) but instead tends to a steady state.

The expansion (43) exhibits a focusing effect in that constant values of ψ move towards $x=0$ as $\tilde{t}\uparrow\tilde{t}_e$, i.e. the peak of the temperature distribution for $\tilde{t}$ constant becomes sharper as thermal runaway is approached. This feature persists into the explosion phase, as we shall see shortly, so that a well-defined hot-spot is formed. A combination of asymptotics and numerics, therefore, provides a complete picture for $\tilde{t}<\tilde{t}_e$.

Induction is immediately followed by explosion (discussed by Kapila (1980) and by Kassoy & Poland (1981)) for which the time variable (26*a*) is appropriate. The other variable for the explosion phase is η, which is not surprising in view of its role in thermal runaway. In terms of τ and η the governing equation becomes

$$\partial T/\partial\tau + \theta[(\tfrac{1}{2}\eta)\partial T/\partial\eta - \partial^2 T/\partial\eta^2] = (T_f^2/Y_f)(T_b - T)\exp(\theta/T_f - \theta/T - \theta\tau). \tag{47}$$

Since $x = \delta^{-\frac{1}{2}}\eta\, e^{-\frac{1}{2}\theta\tau}$, the $O(1)$ values of τ and η for which the equation is valid correspond to an exponentially thin region, the hot-spot, which rapidly gets thinner as time increases. The focusing effect as thermal runaway is approached continues during the explosion stage.

If the reaction is to play a role in equation (47) we must use the leading term (26*b*). The variable τ only appears as a parameter in the equation for T_1, so that the solution may be written

$$T_1 = T_f^2\{\psi(\eta) - \ln[(1-T_f\tau)(Y_f - T_b T_f\tau)/Y_f]\}/(1-T_f\tau)^2, \tag{48}$$

where ψ not only satisfies the differential equation (44*a*) and the initial condition (44*b*), but is also subject to the asymptotic condition (45), as can be seen from matching. It is, therefore, identical to the function determined there. It follows that the boundary layer at $x=0$ in the explosion phase is completely determined by the focusing effect in the induction phase and inherits the spatial structure of that effect.

The explosion is so rapid that the temperature in the slab away from (but close to) $x=0$ stays frozen at the value

$$T=T_f+\theta^{-1}T_f^2(-2\ln x-\ln\delta+A)+\dots \tag{49}$$

attained during the induction phase. Clearly, this does not match the boundary-layer expansion, even to leading order. An intermediate expansion is needed to describe the structure left behind by the rapidly shrinking hot-spot. The breakdown of the approximation (49) when $\ln x=O(\theta)$ suggests that

$$\chi=-2\theta^{-1}\ln x=\tau-\theta^{-1}\ln(\eta^2/\delta) \tag{50}$$

is an appropriate variable to use with τ. This intermediate region is thin for $\chi>0$, but is much thicker than the hot-spot for $\chi<\tau$. The governing equation is

$$\frac{\delta}{\theta}e^{\theta(\tau-\chi)}\frac{\partial T}{\partial\tau}-\frac{4}{\theta^2}\left(\frac{\theta}{2}\frac{\partial T}{\partial\chi}+\frac{\partial^2 T}{\partial\chi^2}\right)=\frac{T_f^2\delta}{Y_f\theta}(T_b-T)\exp\left(\frac{\theta}{T_f}-\frac{\theta}{T}-\theta\chi\right), \tag{51}$$

the solution of which is

$$T_0=T_f/(1-T_f\chi) \tag{52}$$

to leading order; the result is stationary, as are all algebraic perturbations (Kapila 1980). It ranges between the leading terms (26*b*) and (49) as χ decreases from τ to 0. (Note that, in this chapter, χ is a fast variable as are its variants, introduced later.)

Within the hot-spot the perturbation (48) is valid until τ approaches Y_f/T_bT_f, i.e. T approaches T_b. If a gaseous reactant is being considered and T_b differs appreciably from T_f, as it often does in practice, the fluid mechanics can no longer be ignored by adopting the constant-density approximation (Kassoy & Poland 1981). Otherwise one continues, as does Kapila (1980), to the subsequent phase where the time variable (31*a*) is appropriate; that is not suggested by the perturbation (48), but by the next term. The corresponding space variable is

$$\bar{\chi}=\delta^{\frac{1}{2}}x/\varepsilon^{\frac{1}{2}}, \tag{53}$$

since that ensures a diffusion term in the equation

$$\partial\phi/\partial\bar{\tau}-\partial^2\phi/\partial\bar{\chi}^2=-\phi\,e^{\phi} \tag{54}$$

governing the perturbation $T_1(\bar{\chi},\bar{\tau})$ of T_b. The initial condition

$$\phi=-\ln(-\bar{\tau})-\ln\ln(-\bar{\tau})+\psi(\eta)+\dots$$
$$\text{as } \bar{\tau}\to-\infty \quad \text{with } \eta=\bar{\chi}/(-\bar{\tau})^{\frac{1}{2}} \text{ fixed} \tag{55}$$

comes from matching with the explosion phase, while the boundary condition

$$\phi = -\ln(\bar{\chi}^2) - \ln\ln(\bar{\chi}^2) + A + \dots \quad \text{as } \bar{\chi} \to \infty \quad \text{with } \bar{\tau} \text{ fixed} \tag{56}$$

comes from matching with the intermediate expansion, neither of which will be derived here; in addition

$$\partial\phi/\partial\bar{\chi} = 0 \quad \text{for } \bar{\chi} = 0. \tag{57}$$

These conditions leave the origin of $\bar{\tau}$ undetermined, as in the homogeneous problem; but otherwise a unique solution is ensured.

Figure 3 shows the picture that emerges from the numerical solution (Kapila 1980). At first the curves have a plateau on the left that is continually eroded as residual focusing occurs. Eventually the erosion is stopped and the plateau begins to re-establish itself over ever-increasing distances, i.e. an incipient deflagration wave is formed. The focusing effect is then opposed by reactant depletion, and the hot-spot, now fast approaching the temperature T_b, begins to spread as the reaction zone detaches itself from the center and starts to propagate across the partially burnt intermediate zone, leaving behind almost depleted reactant. Most of what was the final relaxation phase in the homogeneous problem now consists of this transition from hot-spot to flash-through.

The remainder and more is taken up by the flash-through or propagation phase, which takes much longer than the transition phase, but is still

3 Transition phase, from hot-spot to flash-through: temperature profiles show the focusing effect giving way to formation of a deflagration wave as the fast time $\bar{\tau}$ increases. (Courtesy A. K. Kapila.)

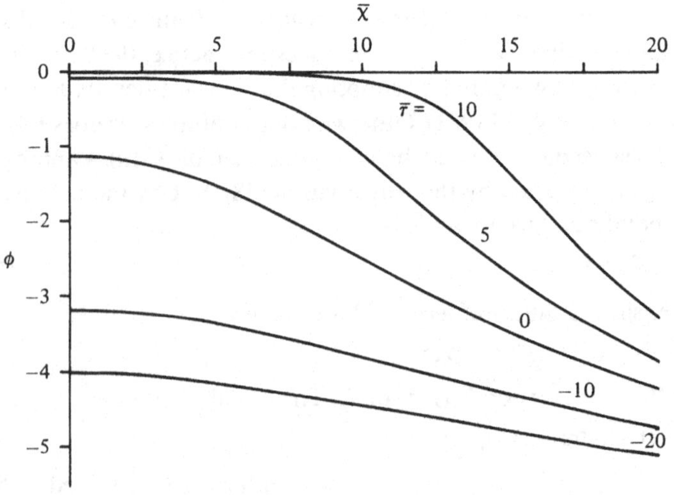

very brief. Considerations leading to the result (2.22) show that, for an $O(\theta^{-1}\,e^{\theta/T_f})$ value of D(i.e. δ of order 1) a deflagration wave has a thickness $O(\theta\varepsilon^{\frac{1}{2}})$ and takes a time $O(\theta\varepsilon^{\frac{1}{2}})$ to reach the surface of the slab. This suggests using the variables

$$\vec{\chi}=[x_*(\vec{\tau})-x]/\theta\varepsilon^{\frac{1}{2}}, \quad \vec{\tau}=(\tilde{t}-\tilde{t}_e)/\theta\varepsilon^{\frac{1}{2}} \tag{58}$$

to investigate the flash-through, where $\vec{\chi}$ is measured from the location x_* of the flame front. The fact that $O(\theta^{-1})$ changes in $\vec{\chi}$ measure the same distance as $O(1)$ changes in $\bar{\chi}$, i.e. that the flame-sheet structure will be described on the scale of the transition phase, supports this suggestion.

To trace the progress of the deflagration wave through the intermediate region into the main part of the slab, note that the region is both stationary and much thicker than the preheat zone of the wave. Far ahead of the flame sheet (on the scale of the preheat zone), when it reaches the position χ, the temperature (52) holds, while everywhere behind it the temperature is T_b. In the variables (58) the structure is quasi-steady, and so from the result (2.22) for the steady deflagration wave the flame velocity is seen to be given by

$$\dot{x}_*(\vec{\tau})=\sqrt{2/\delta}\,T_b^2(1-T_f\chi)/(Y_f-T_bT_f\chi)$$

corresponding to

$$W=\sqrt{2\delta}\frac{T_b^2T_f}{Y_f^{\frac{1}{2}}\theta^{\frac{1}{2}}}\,e^{Y_f\theta/2T_bT_f}\left(\frac{1-T_f\chi}{Y_f-T_bT_f\chi}\right) \tag{59}$$

on the x, t-scale. As χ decreases from Y_f/T_bT_f to 0 through the intermediate region, the speed decreases rapidly from infinity to $\sqrt{2\delta}(T_b^2T_f/Y_f^{\frac{1}{2}}\theta^{\frac{1}{2}})\times e^{Y_f\theta/2T_bT_f}$, at which it flashes through the main part of the slab. (The flame moves more rapidly the closer the unburnt mixture is to the flame temperature T_b because the heat of combustion is able to preheat a larger amount of material.)

The propagation phase ends when the preheat zone reaches the surface. In the ensuing relaxation phase the wave rapidly decelerates to 0, thereby establishing a steady state with the reaction zone just inside the slab. This is precisely the problem treated in section 10.2 when $\omega=\infty$ (see Figure 10.3 for $\bar{l}=0$, noting that χ is a time-like variable). In the present context Kapila (1980) has shown that the relaxation phase takes a very short time $O(\theta\varepsilon^{\frac{1}{2}})$ to be completed (cf. the $O(\varepsilon)$ relaxation time for the homogeneous explosion).

We do not go into details of this relaxation phase because it is a result of the nonphysical boundary condition on Y. If instead of maintaining the reactant at the surface it is prevented from diffusing across the surface, then nothing changes until the relaxation phase occurs, since $T+Y$ is constant to leading order everywhere. At that time the wave interacts with the cold wall along the lines described in section 9.5 and, assuming that quenching occurs, leaves behind reactant-depleted material at temperature T_b except for a small amount of unburnt reactant near the surface. Since the boundary continues to be maintained at T_f, there follows a relaxation phase of duration $O(1)$ during which essentially pure heat conduction again lowers the temperature everywhere to T_f and the residual reactant spreads across the slab, being slowly burnt in the process.

Our analysis describes the formation of a hot-spot from which emerges a deflagration wave that rapidly consumes the combustible material. These are characteristics of actual explosions; the model has clearly retained the essential physics. For a solid, modifications are necessary to account for an infinite Lewis number but these only change details. For a gas, the neglect of the fluid mechanics involved has undoubtedly excluded significant effects (Kassoy & Poland 1981), possibly the evolution of the deflagration wave into a detonation wave. The real value of the discussion, however, lies in its elucidation of the nature of the extremely rapid transients typical of the explosion process, one of the most challenging aspects of combustion theory.

5 Ignition by heat flux

Auto-ignition is triggered in a combustible material by the build-up of heat that cannot escape in sufficient amounts through the boundary. Ignition can also be caused by some external source of energy. For example, part or all of the boundary may suddenly be raised to a high temperature so that reaction starts nearby. Alternatively, a radiative heat flux may be applied, a mechanism that is sometimes a factor in the spread of fires.

Thermal runaway followed by explosion is once again a fundamental feature but, as is clear from section 4, if the goal is limited to calculating the time to runaway (as it is here), then only the induction phase need be considered. Our intention is not only to draw attention to ignition problems (that only make sense for large activation energy), but also to disclose another way in which runaway can occur.

The specific problem we shall examine is that of ignition of a half space by a constant energy flux (Liñán & Williams 1971). Modifications for a slab with its other surface insulated, for example, are minor and will be obvious. Since the sequel to thermal runaway is not considered, reactant depletion may be neglected; only the temperature equation (13) need be considered, with $\nabla^2 \equiv \partial^2/\partial x^2$ in this geometry. The boundary and initial conditions are

$$\partial T/\partial x = T_s' > 0 \quad \text{for } x=0, \quad T=T_f \quad \text{for } t=0 \quad \text{and as } x \to -\infty \tag{60}$$

if the half space lies in $-\infty < x \leqslant 0$. The constant T_s' is assumed to be $O(1)$.

In principle, an asymptotic development is possible for any value of D. If D is large enough self-heating will be dominant and the evolution will consist primarily of a homogeneous explosion far from the boundary. For smaller values of D, ignition will first occur at the surface because of the enhanced temperatures generated there by the heat flux. Since we are concerned with the latter situation it is appropriate to define T_r, a temperature characterizing the reactivity of the material, by

$$D = A(\theta)\, e^{\theta/T_r} \quad \text{with } T_r > T_f. \tag{61}$$

The algebraic factor $A(\theta)$ will be chosen to ensure simplicity in the analysis; different choices of A correspond to slightly different definitions of T_r.

The equation governing T is then

$$\partial T/\partial t - \partial^2 T/\partial x^2 = A(\theta)\, e^{\theta/T_r - \theta/T} \tag{62}$$

where the right side is negligible at times sufficiently small for T to be less than T_r everywhere. The solution

$$T = T_f + T_s'[2\sqrt{t/\pi}\, e^{-x^2/4t} + x\, \mathrm{erfc}(|x|/2\sqrt{t})] \tag{63}$$

satisfying the conditions (60) increases monotonically in t, but decreases monotonically in $|x|$, so that the value T_r is first attained at $x=0$, the time being

$$t_r = \pi(T_r - T_f)^2/4T_s'^2. \tag{64}$$

At that instant the reaction term must be reinstated. (To obtain t_r in terms of D, correct to leading order, insert $\theta/\ln D$ for T_r.)

Demonstration of runaway and an accurate estimate of the runaway time must involve details of the reaction (as in section 4). Nevertheless, the notion that the temperature rises purely by heat conduction until it reaches some specific value determined by the magnitude of D, whereupon reaction produces runaway within a time $O(\theta^{-1})$, is central to the development of the solution.

For times close to t_r, the reactionless solution (63) is only valid away from $x=0$, i.e. it becomes an outer expansion. Since the reaction requires T to be within $O(\theta^{-1})$ of T_r and runaway is expected to occur within a time $O(\theta^{-1})$ of t_r, we introduce an inner expansion

$$T = T_r + \theta^{-1} T_1(\xi, \tau) + \theta^{-\frac{3}{2}} T_2(\xi, \tau) + \ldots \quad \text{with } \xi = \theta x, \quad \tau = \theta(t - t_r). \tag{65}$$

The choice of scale for ξ is dictated by the requirement that $\partial T_1/\partial\xi$ be $O(1)$ to accommodate the imposed heat flux, namely

$$\partial T_1/\partial\xi = T_s' \quad \text{at } \xi=0; \tag{66}$$

the term in $\theta^{-\frac{1}{2}}$ is induced by an intermediate expansion that must be introduced later. (We use ξ rather than χ or a variant of it because it has been used consistently for a θ-multiplied spatial coordinate.)

The equation for T_1 is

$$\partial^2 T_1/\partial\xi^2 = -\lim_{\theta\to\infty} [\theta^{-1}A(\theta)]\, e^{T_1/T_1^2}$$

while $\partial^2 T_1/\partial\xi^2 \to 0$ as $\tau\to-\infty$ comes from matching with the solution (63). If A is not $o(\theta)$ this development soon leads to a contradiction so that, in view of the condition (66),

$$T_1 = T_{1s} + T_s'\xi \quad \text{with } T_{1s}(\tau) = T_s'\tau/\sqrt{\pi t_r} + o(1) \quad \text{as } \tau\to-\infty; \tag{67}$$

T_{1s} is otherwise undetermined at this stage. The equation for T_2 is

$$\partial^2 T_2/\partial\xi^2 = -e^{T_1/T_r^2} \tag{68}$$

(a steady balance of diffusion and reaction) if the choice

$$A = \theta^{\frac{1}{2}} + \ldots \tag{69}$$

is made; the requirement that $\partial^2 T_2/\partial\xi^2 \to 0$ as $\tau\to-\infty$ is met automatically. The solution with vanishing derivative for $\xi=0$ has the property

$$\partial T_2/\partial\xi = (T_r^2/T_s')\, e^{T_{1s}/T_r^2} + \ldots \quad \text{as } \xi\to-\infty, \tag{70}$$

which is all we shall need to know about it. So far there is no reason to suspect runaway: the fact that T_{1s} becomes infinite after a finite time follows from investigation of the intermediate region, now to be considered.

The inner approximation must break down when $\xi = O(\theta^{\frac{1}{2}})$ since on that scale the time derivative $\partial/\partial\tau$ is as important as $\partial^2/\partial\xi^2$. The inner expansion suggests that the temperature differs from T_r by $O(\theta^{-\frac{1}{2}})$ for such values of ξ, so that we write the intermediate expansion

$$T = T_r + \theta^{-\frac{1}{2}} T_1(\chi,\tau) + \theta^{-1} T_2(\chi,\tau) + \ldots \quad \text{with } \chi = \theta^{\frac{1}{2}} x. \tag{71}$$

The equation for T_1 is that of pure conduction, i.e.

$$(\partial/\partial\tau - \partial^2/\partial\chi^2) T_1 = 0, \tag{72}$$

the temperature being too far from T_r for there to be any reaction (at any order). Matchings with the inner and outer expansions require

$$T_1 = 0, \quad \partial T_1/\partial\chi = T_s' \quad \text{for } \chi=0,$$

$$T_1 = T_s'\chi + o(1) \quad \text{as } \chi\to-\infty \quad \text{and as } \tau\to-\infty. \tag{73}$$

There is one boundary condition too many but that does not prevent finding a solution $T_1 = T_s'\chi$. Continuing to T_2 yields

$$(\partial/\partial\tau-\partial^2/\partial\chi^2)T_2=0, \tag{74}$$

$$T_2=T_{1s},\quad \partial T_2/\partial\chi=(T_r^2/T_s')\,e^{T_{1s}/T_r^2}\quad \text{for } \chi=0,$$

$$T_2=T_s'(\chi^2+2\tau)/2\sqrt{\pi t_r}+o(1)\quad \text{as } \chi\to-\infty\quad \text{and as } \tau\to-\infty; \tag{75}$$

unlike the one for T_1, this problem is not overdetermined since T_{1s} has yet to be found. Eliminating it from the boundary conditions (75*a*, *b*) gives the single nonlinear condition

$$\partial T_2/\partial\chi=(T_r^2/T_s')\,e^{T_2/T_r^2}\quad \text{for } \chi=0. \tag{76}$$

All parameters can be purged by writing

$$\tilde{T}_2=T_2/T_r^2-T_s'(\chi^2+2\tau)/2\sqrt{\pi t_r}\,T_r^2,$$

$$\tilde{\chi}=(T_s'^2/\pi t_r T_r^4)^{\frac{1}{4}}\chi,\quad \tilde{\tau}=(T_s'^2/\pi t_r T_r^4)^{\frac{1}{2}}\tau-\ln(T_s'^{\frac{1}{2}}/\pi^{\frac{1}{4}}t_r^{\frac{1}{4}}T_r) \tag{77}$$

to obtain

$$(\partial/\partial\tilde{\tau}-\partial^2/\partial\tilde{\chi}^2)\tilde{T}_2=0, \tag{78}$$

$$\partial\tilde{T}_2/\partial\tilde{X}=e^{\tilde{\tau}+T_2}\quad \text{for } \tilde{\chi}=0,\quad T_2=o(1)\quad \text{as } \tilde{\chi}\to-\infty\quad \text{and as } \tilde{\tau}\to-\infty. \tag{79}$$

Numerical solution of this problem shows that $\tilde{T}_2$ becomes unbounded at a definite time $\tilde{\tau}=-0.431$ (Liñán & Williams 1971). In terms of the original time variable, runaway occurs, therefore, at the time

$$t_e=t_r+\theta^{-1}(\pi t_r T_r^4/T_s'^2)^{\frac{1}{2}}[-0.431+\tfrac{1}{4}\ln(T_s'^6/\pi t_r T_r^4)]. \tag{80}$$

The mathematical problem of runaway assumes different forms. For the auto-ignition problem of the last section, the nonlinear driving term is in the governing equation (39) and acts on $O(1)$ time and length scales. Here it is in the boundary condition (79) and acts over a time $O(\theta^{-1})$ at distances $O(\theta^{-\frac{1}{2}})$. Following runaway a hot-spot forms at the surface, from which emerges a deflagration wave that sweeps into the interior. Kapila (1981*a*) has given a detailed description of the process.

6 Auto-ignition and explosion of separated reactants

If two parallel streams are allowed to come into contact, one a combustible mixture and the other a hot inert, eventually the mixture will be ignited by the inert. This steady problem is considered by Marble & Adamson (1954). Here we shall treat a somewhat different problem with some of the same features but closer to the thermal explosion investigated in section 4. Mathematically it is identical to a special case of the unsteady mixing and explosion of two initially separated reactants treated by Liñán & Crespo (1972). Retaining the notation of the unsteady problem will enable results from section 4 to be used without translation. One feature, missing from the Marble–Adamson problem, is the diffusion flame that manifests itself as a Burke–Schuman flame sheet sufficiently far downstream.

Consider a semi-infinite flat plate (Figure 4) separating fast parallel streams containing oxidant X and fuel Y that come into contact at the trailing edge and then interdiffuse to form a combustible mixture downstream. To keep matters simple the speeds, reactant concentrations, and temperatures of the streams will be taken equal. Liñán and Crespo's problem is equivalent to one allowing for unequal concentrations and temperatures.) For small values of t no significant reaction occurs; for large values there is chemical equilibrium, with a Burke–Schumann flame sheet separating a region in which X is zero from one in which Y is zero. How the combustion field evolves from the initial regime to the final one as t increases is the main question to consider. Because such a transition arises in many combustion processes (another example is the unsteady ignition of a fuel drop) a mathematical description is of considerable interest. Experiments have been carried out under conditions similar to those in Figure 4 (Liebman, Corry, & Perlee (1970)).

Essential details of the combustion field can be determined using activation-energy asymptotics, as was shown by Liñán & Crespo (1972). Their discussion is intuitive since asymptotic methods were not well developed at that date; nevertheless, they accurately describe the main features of their more general problem. Our discussion is more systematic and the description more detailed; only for the sake of simplicity is the problem specialized.

The high velocity of the streams enables longitudinal diffusion to be neglected (cf. section 8.6); in addition, we adopt the constant-density

4 Combustion of initially separated reactants. * denotes explosion.

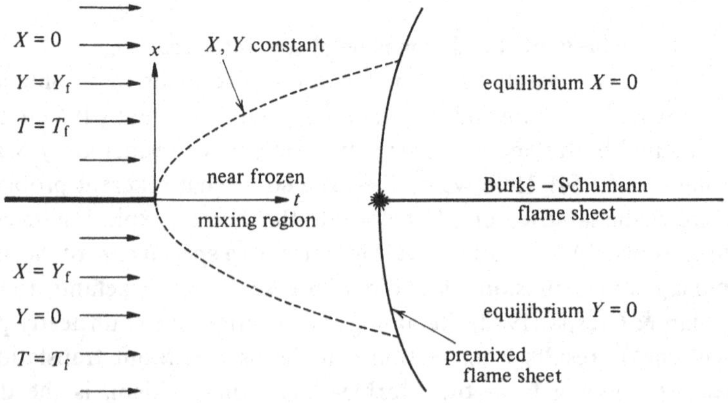

approximation. Taking unit Lewis numbers then leads to the governing equations

$$(\partial/\partial t-\partial^2/\partial x^2)T=-2(\partial/\partial t-\partial^2/\partial x^2)X$$
$$=-2(\partial/\partial t-\partial^2/\partial x^2)Y=D(XY/Y_f^2)\,e^{-\theta/T} \quad (81)$$

if the reaction is first-order in both fuel and oxidant. Such equations, being parabolic, require only initial conditions; we shall take

$$T=T_f,\quad X=\begin{cases}X_f\\0\end{cases},\quad Y=\begin{cases}0\\Y_f\end{cases},\quad \text{for } x\lessgtr 0 \quad \text{at } t=0 \quad \text{with } X_f=Y_f \quad (82)$$

to guarantee symmetry about the t-axis, a simplification that does not exclude any essential features. The variable t is time-like, so that the corresponding terminology will be adopted; when t is interpreted as time, equations (81) govern Liñán and Crespo's one-dimensional unsteady problem of two half spaces of fuel and oxidant brought suddenly into contact at the initial instant. (Their generalization consists of different temperatures for $x\lessgtr 0$ and $X_f\neq Y_f$; they also take variable density into account by means of the Dorodnitsyn–Howarth transformation (Stewartson 1964, p. 29).)

The linear combinations $T+2X$ and $T+2Y$ satisfy pure heat equations everywhere with step functions for initial values; the Shvab–Zeldovich relations follow:

$$T+2X=T_b-Y_f\operatorname{erf}\zeta,\quad T+2Y=T_b+Y_f\operatorname{erf}\zeta$$
$$\text{with}\quad T_b=H_f\equiv T_f+Y_f,\quad \zeta=x/2\sqrt{t},\quad (83)$$

where H_f is the available enthalpy after the reactants have mixed completely ($X=Y=\frac{1}{2}Y_f$) and the similarity variable ζ has arisen naturally. We now need only consider the temperature equation which, in terms of t and ζ, becomes

$$4t\partial T/\partial t-2\zeta\partial T/\partial\zeta-\partial^2T/\partial\zeta^2=4tD(XY/Y_f^2)\,e^{-\theta/T} \quad (84)$$

where X and Y have just been determined as functions of T and ζ. Behavior for small and large values of t can be inferred from this form of the equation, for any value of θ, by noting that the effective Damköhler number is tD: when t is small the combustion is nearly frozen, whereas when t is large it is near equilibrium.

In the limit $t\to 0$ we find

$$T=T_f,\quad X=\tfrac{1}{2}Y_f(1-\operatorname{erf}\zeta),\quad Y=\tfrac{1}{2}Y_f(1+\operatorname{erf}\zeta). \quad (85)$$

At any finite value of x when t is small there is a combustible mixture formed by interdiffusion of the reactants, in which no significant reaction has taken place. We may, therefore, expect a thermal explosion similar in many respects to the spatially nonhomogeneous one discussed in section 4.

In the limit $t \to \infty$ we must have the equilibrium (6.21); then the relations (83) show that

$$T=\begin{cases} T_b+Y_f \operatorname{erf}\zeta \\ T_b-Y_f \operatorname{erf}\zeta \end{cases}, \quad X=\begin{cases} -Y_f \operatorname{erf}\zeta \\ 0 \end{cases}, \quad Y=\begin{cases} 0 \\ Y_f \operatorname{erf}\zeta \end{cases} \quad \text{for } x \lessgtr 0. \tag{86}$$

There is a Burke–Schumann flame sheet, corresponding to an infinite Damköhler number (cf. section 6.2), located on the t-axis at a temperature T_b. On the scale of ζ the flame sheet is a discontinuity; to investigate its structure we introduce the variables

$$\xi = t^{\frac{1}{4}}\zeta = x/2t^{\frac{1}{4}}, \quad T_1=\sqrt{\pi}t^{\frac{1}{4}}(T_b-T)/2Y_f \tag{87}$$

and seek a solution for which T_1 is a function of ξ alone. This choice of variables ensures a meaningful balance of diffusion and reaction, i.e. a distinguished limit of equation (84). The limit is

$$d^2T_1/d\xi^2 = C(T_1^2-\xi^2) \quad \text{with } C \doteq 2D\,e^{-\theta/T_b}/\sqrt{\pi}Y_f; \tag{88}$$

matching with the outer solutions (86) imposes the boundary conditions

$$dT_1/d\xi = \pm 1 + \dots \quad \text{as } \xi \to \pm\infty. \tag{89}$$

The problem (88), (89) is that for the classical Burke–Schumann structure (see the problem (6.27), (6.28) with T_1 and ξ replaced by linear functions of themselves).

Evolution from the frozen to the equilibrium regime is remarkably simple in the limit $\theta \to \infty$. Thermal runaway occurs in the diffusion-generated mixture at a point on the line of symmetry and a hot-spot develops just as for the spatially nonhomogeneous explosion of section 4. As the temperature in the hot-spot approaches its ultimate value T_b (corresponding to complete consumption of the reactants) a symmetrical pair of deflagration waves is formed; they penetrate the large temperature gradients left behind by the focusing hot-spot and flash through the layers of frozen mixture consuming all of the deficient reactant. This happens within a range of t (including the runaway point) that is exponentially smaller than the scale of mixing.

A detailed description starts with the induction phase, during which the temperature rises from the initial value T_f by only an $O(\theta^{-1})$ amount. The corresponding small-disturbance equation

$$\partial\phi/\partial t - \partial^2\phi/\partial x^2 = \delta(1-\operatorname{erf}^2\zeta)\,e^{\phi} \tag{90}$$

differs from (39) for an initially uniform slab only in having an extra factor on the right side; if we now restrict the discussion to the region $x>0$, the boundary conditions

$$\partial\phi/\partial x = 0 \quad \text{for } x = 0, \quad \phi = 0 \quad \text{for } t = 0 \tag{91}$$

are the same. As there, the solution becomes unbounded at a point $t = t_e$ on $x = 0$ that can be determined numerically (Liñán & Crespo 1972).

The behavior of ϕ as $t \to t_e$ is given by equation (43) since ζ is $O(t_e - t)^{\frac{1}{2}}$ in the variables $\tilde{t}$ and η. The function ψ is changed because the constant A in its asymptotic expansion (45) is different, being determined by the asymptotic form (46) of the solution to a different problem.

The explosion phase is governed by the equation (47) with the right side replaced by

$$(T_f^2/4Y_f^2)(T_b - T)^2 \exp(\theta/T_f - \theta/T - \theta\tau) \tag{92}$$

since ζ is effectively 0 in the hot-spot. The reactant concentrations are changed from $\frac{1}{2}Y_f$ to 0 as T increases from T_f to T_b. The leading term (26*b*) is still valid, but now the perturbation is

$$T_1 = T_f^2\{\psi(\eta) - 2\ln[(Y_f - T_b T_f \tau)/2Y_f]\}/(1 - T_f\tau)^2. \tag{93}$$

Between the hot-spot described by this expansion and the essentially frozen combustion field on either side there is a stationary intermediate structure on the scale (50), with leading term (52).

The transition phase that follows is governed by the modification

$$\partial\phi/\partial\bar{\tau} - \partial^2\phi/\partial\bar{\chi}^2 = \phi^2 e^\phi \tag{94}$$

of the equation (54), provided the definition (31*b*) is replaced by

$$\varepsilon = (4Y_f^2\theta^2/T_b^2 T_f^2)\, e^{-\theta Y_f/T_b T_f}. \tag{95}$$

The modification arises because the hot-spot contains the reactants in stoichiometric proportion; this results in a truly second-order reaction, rather than an essentially first-order one. The initial condition (55) is replaced by

$$\phi = -\ln(-\bar{\tau}) - 2\ln\ln(-\bar{\tau}) + \psi(\eta) + \ldots$$
$$\text{as } \bar{\tau} \to -\infty \quad \text{with } \eta = \bar{\chi}/(-\bar{\tau})^{\frac{1}{2}} \text{ fixed}, \tag{96}$$

while the boundary condition (56) is replaced by

$$\phi = -\ln(\bar{\chi}^2) - 2\ln\ln(\bar{\chi}^2) + A + \ldots \quad \text{as } \bar{\chi} \to \infty \quad \text{with } \bar{\tau} \text{ fixed}. \tag{97}$$

Numerical investigation of the problem (94), (96), (97) by Kapila (unpublished) yields a transition from the focusing hot-spot to an incipient deflagration wave similar to that displayed by the problem (54), (55), (56).

The wave must first traverse the intermediate structure left by the shrinking hot-spot. The situation is the same as that in section 4, except that ahead there is a mixture of fuel and oxidant in stoichiometric proportion, i.e. $X = Y$ (cf. equations (83) with $\zeta = 0$), instead of a single reactant. It follows that the variables (58) are again appropriate, if the new definition

(95) of ε is used, but that the velocity (59) must be replaced by

$$W = \sqrt{\delta}\, \frac{T_b^3 T_f}{Y_f \theta^2}\, e^{Y_f\theta/2T_bT_f} \left(\frac{1 - T_f\chi}{Y_f - T_bT_f\chi} \right) \tag{98}$$

in accordance with the result (2.55). As χ decreases from Y_f/T_bT_f to 0 through the intermediate region the wave decelerates rapidly from an infinite speed to

$$W = \sqrt{\delta}(T_b^3 T_f / Y_f^2 \theta^2)\, e^{Y_f\theta/2T_bT_f}, \tag{99}$$

the speed with which it enters the near-frozen external field.

For the slab considered in section 4, the wave then continues at the same speed to the surface, but here it encounters nonuniformities in available enthalpy that produce $O(1)$ changes in its temperature and hence exponentially large changes in its velocity, a situation excluded from the general discussion of unsteady flames in Chapter 8. These nonuniformities have been formed by interdiffusion of the reactants up to the time t_e, so that

$$T = T_f, \quad X = \tfrac{1}{2}Y_f(1 - \operatorname{erf}\zeta_e), \quad Y = \tfrac{1}{2}Y_f(1 + \operatorname{erf}\zeta_e) \quad \text{with } \zeta_e = x/2\sqrt{t_e}; \tag{100}$$

there is no time for them to change during transit of the wave.

Fortunately the situation is similar to that in the intermediate region: the combustion field is stationary and has a much larger scale than most of the preheat zone of the flame, so that plane steady results may be invoked. Far ahead of the wave on the latter scale the temperature and mass fractions have the distributions (100) at the instant the wave reaches the position x, while behind they have the constant values

$$T = T_b - Y_f \operatorname{erf}\zeta_e, \quad X = 0, \quad Y = Y_f \operatorname{erf}\zeta_e \quad \text{for } x > 0. \tag{101}$$

To determine the velocity of the wave we use the result (2.55) which (as was pointed out) holds both near and far from stoichiometry. Thus,

$$W = \sqrt{\delta}(T_f / Y_f^2 \theta^2)(1 - \operatorname{erf}\zeta_e)^{-1}(T_b - Y_f \operatorname{erf}\zeta_e) \times [(T_b - Y_f \operatorname{erf}\zeta_e)^2 + \theta Y_f \operatorname{erf}\zeta_e]^{\frac{1}{2}}\, e^{\alpha\theta} \tag{102}$$

where

$$\alpha = Y_f(1 - \operatorname{erf}\zeta_e)/2T_f(T_b - Y_f \operatorname{erf}\zeta_e) \tag{103}$$

is the wave speed, which clearly reduces to the value (99) for $\zeta_e = 0$. The result cannot be valid as $\zeta_e \to \infty$ since the propagation is then associated with significant changes in t; well before then, however, the formula (2.55) fails because J_s (i.e. the oxidant flux) becomes small. The correct result has not yet been obtained, perhaps because it would add very little to the picture. Note that equilibrium with a Burke–Schumann flame sheet holds at any finite x immediately after the explosion: the combustion field (101) is precisely of the ultimate form (86) for $x > 0$.

The picture that has emerged is remarkably simple; it consists of regions of frozen chemistry and equilibrium with reaction zones in between, the hallmark of activation-energy asymptotics.

TEXT REFERENCES

Adams, G. K. & Wiseman, L. A. (1954). The combustion of double-based propellants. In *Selected Combustion Problems; Fundamentals and Aeronautical Applications*, ed. W. R. Hawthorne & J. Fabri, pp. 277–88. London: Butterworths Scientific Publications.

Aly, S. L. & Hermance, C. E. (1981). A two-dimensional theory of laminar flame quenching. *Combustion and Flame*, **40**, 173–85.

Arrhenius, S. (1889). Über die Reaktionsgeschwindigkeit bei der Inversion von Rohrzucher durch Saüren. *Zeitschrift für Physikalische Chemie*, **4**, 226–48.

Ayeni, R. O. (1978). *Thermal Runaway*. Cornell University: Ph.D. thesis.

Batchelor, G. K. (1967). *An Introduction to Fluid Dynamics*. Cambridge: Cambridge University Press.

Berman, V. S. & Riazantsev, Iu. S. (1972). Analysis of the problem of thermal flame propagation by the method of matched asymptotic expansions. *Journal of Applied Mathematics and Mechanics (PMM)*, **36**, 622–8.

Blythe, P. A. (1964). Asymptotic solutions in nonequilibrium nozzle flow. *Journal of Fluid Mechanics*, **20**, 243–72.

Bowen, R. M. (1976). Theory of mixtures. In *Continuum Physics*, vol. III, part I, ed. A. C. Eringen, pp. 1–127. New York: Academic Press.

Buckmaster, J. (1975*a*). A new large Damköhler number theory of fuel droplet burning. *Combustion and Flame*, **24**, 79–88.

Buckmaster, J. (1975*b*). Combustion of a liquid fuel drop. *Letters in Applied and Engineering Sciences*, **3**, 365–72.

Buckmaster, J. (1976). The quenching of deflagration waves. *Combustion and Flame*, **26**, 151–62.

Buckmaster, J. (1977). Slowly varying laminar flames. *Combustion and Flame*, **28**, 225–39.

Buckmaster, J. (1979*a*). The quenching of two-dimensional premixed flames. *Acta Astronautica*, **6**, 741–69.

Buckmaster, J. (1979*b*). The quenching of a deflagration wave held in front of a bluff body. In *Seventeenth International Symposium on Combustion*, pp. 835–42. Pittsburgh: The Combustion Institute.

Buckmaster, J. (1979*c*). A mathematical description of open and closed flame tips. *Combustion Science and Technology*, **20**, 33–40.

Buckmaster, J. (1981). Two examples of a stretched flame. *Quarterly Journal of Mechanics and Applied Mathematics* (in press).

Buckmaster, J. & Crowley, A. (1982). The fluid mechanics of flame tips.

Buckmaster, J. D., Kapila, A. K., & Ludford, G. S. S. (1976). Linear condensate deflagration for large activation energy. *Acta Astronautica*, **3**, 593–614.

Buckmaster, J. D., Kapila, A. K., & Ludford, G. S. S. (1978). Monopropellant decomposition in a reactive atmosphere. *Combustion Science and Technology*, **17**, 227–36.

Buckmaster, J. & Mikolaitis, D. (1982). The premixed flame in a counterflow. *Combustion and Flame*. (in press).

Buckmaster, J. & Nachman, A. (1981). Propagation of an unsteady flame in a duct of varying cross-section. *Quarterly Journal of Mechanics and Applied Mathematics* (in press).

Buckmaster, J. & Takeno, T. (1981). Blow-off and flashback of an excess enthalpy flame. *Combustion Science and Technology* **25**, 153–9.

Burke, S. P. & Schumann, T. E. W. (1928). Diffusion flames. *Industrial and Engineering Chemistry*, **20**, 998–1004.

Bush, W. B. & Fendell, F. E. (1970). Asymptotic analysis of laminar flame propagation for general Lewis numbers. *Combustion Science and Technology*, **1**, 421–8.

Carrier, G. F., Fendell, F. E., & Bush, W. B. (1978). Stoichiometry and flame-holder effects on a one-dimensional flame. *Combustion Science and Technology*, **18**, 33–46.

Chandra, J. & Davis, P. W. (1980). Rigorous bounds and relations among spatial and temporal approximations in the theory of combustion. *Combustion Science and Technology*, **23**, 153–62.

Chu, B. T. (1956). Stability of systems containing a heat source – the Rayleigh criterion. *National Advisory Committee for Aeronautics Research Memorandum 56D27.*

Clarke, J. F. (1975). The pre-mixed flame with large activation energy and variable mixture strength: elementary asymptotic analysis. *Combustion Science and Technology*, **10**, 189–94.

Clarke, J. F. & McIntosh, A. C. (1980). The influence of a flameholder on a plane flame, including its static stability. *Proceedings of the Royal Society of London*, **A372**, 367–92.

Culick, F. E. C. (1968). A review of calculations for unsteady burning of a solid propellant. *American Institute of Aeronautics and Astronautics Journal*, **6**, 2241–55.

Darrieus, G. (1938). Propagation d'un front de flame. Essai de theorie de vitesses anomales de deflagration par developpement spontane de la turbulence. (Unpublished typescript of paper given at the *Sixth International Congress of Applied Mechanics, Paris, 1946*; mentioned in Darrieus, G. (1939). La mécanique des fluides–quelques progrès récents. *La Technique Moderne Supplément*, **XXXI** (15), IX–XVII.

Denison, M. R. & Baum, E. (1961). A simplified model of unstable burning in solid propellants, *American Rocket Society Journal*, **31**, 1112–22.

Deshaies, B. & Joulin, G. (1980). Asymptotic study of an excess-enthalpy flame. *Combustion Science and Technology*, **22**, 281–5.

Eckhaus, W. (1961). Theory of flame-front stability. *Journal of Fluid Mechanics*, **10**, 80–100.

Emmons, H. W. (1958). Flow discontinuities associated with combustion. In *High Speed Aerodynamics and Jet Propulsion*, vol. III, ed. H. W. Emmons, pp. 584–607. Princeton: Princeton University Press.

Emmons, H. W. (1971). Fluid mechanics and combustion. In *Thirteenth International Symposium on Combustion*, pp. 1–18, Pittsburgh: The Combustion Institute.

Fendell, F. E. (1969). Quasi-steady spherico-symmetric monopropellant decomposition in inert and reactive environments. *Combustion Science and Technology*, **1**, 131–45.

Fendell, F. E. (1972). Asymptotic analysis of premixed burning with large activation energy. *Journal of Fluid Mechanics*, **56**, 81–95.

Ferguson, C. R. & Keck, J. C. (1979). Stand-off distances on a flat flame burner. *Combustion and Flame*, **34**, 85–98.

Fick, A. (1855). Über diffusion. *Annalen der Physik und Chemie*, **94**, 59–86.

Frank-Kamenetskii, D. A. (1969). *Diffusion and Heat Transfer in Chemical Kinetics*, New York: Plenum Press.

Fristrom, R. M. & Westenberg, A. A. (1965). *Flame Structure*. New York: McGraw-Hill.
Gautschi, W. & Cahill, W. F. (1964). Exponential integral and related functions. In *Handbook of Mathematical Functions with Formulas, Graphs and Mathematical Tables*, ed. M. Abramowitz and I. A. Stegun, pp. 227–51. National Bureau of Standards, Applied Mathematics Series 55. Washington, D.C.: Superintendent of Documents, U.S. Government Printing Office.
Gaydon, A. G. & Wolfhard, H. G. (1970). *Flames. Their Structure, Radiation and Temperature*. London: Chapman and Hall.
Goldstein, S. (1948). On boundary-layer flow near a position of separation. *Quarterly Journal of Mechanics and Applied Mathematics*, **1**, 43–69.
Gray, P. & Lee, P. R. (1967). Thermal explosion theory. *Oxidation and Combustion Reviews*, **2**, 0–183.
Guirao, C. & Williams, F. A. (1971). A model for Ammonium Perchlorate deflagration between 20 and 100 atm. *American Institute of Aeronautics and Astronautics Journal*, **9**, 1345–56.
Hermance, C. E. (1975). Implications concerning general ignition processes from the analysis of homogeneous, thermal explosions. *Combustion Science and Technology*, **10**, 261–5.
Higgins, B. (1802). Letter in *Nicholson's Journal*, **1**, 130–1.
Hirschfelder, J. O. & Curtiss, C. F. (1949). The theory of flame propagation. *The Journal of Chemical Physics*, **17**, 1076–81.
Hirschfelder, J. O., Curtiss, C. F., & Campbell, D. E. (1953). The theory of flames and detonations. In *Fourth International Symposium on Combustion*, pp. 190–211. Baltimore: Williams and Wilkins.
Holmes, P. (1982). On a second order boundary value problem arising in combustion theory. *Quarterly of Applied Mathematics* (in press).
Istratov, A. G. & Librovich, V. B. (1969). On the stability of gasdynamic discontinuities associated with chemical reactions. The case of a spherical flame. *Astronautica Acta*, **14**, 453–67.
Janssen, R. D. (1982). *Drop Combustion*. Cornell University: Ph.D. thesis.
Johnson, W. E. (1963). On a first-order boundary value problem from laminar flame theory. *Archive for Rational Mechanics and Analysis*, **13**, 46–54.
Johnson, W. E. & Nachbar, W. (1962). Deflagration limits in the steady linear burning of a monopropellant with applications to ammonium perchlorate. In *Eighth International Symposium on Combustion*, pp. 678–89. Baltimore: Williams and Wilkins.
Johnson, W. E. & Nachbar, W. (1963). Laminar flame theory and the steady linear burning of a monopropellant. *Archive for Rational Mechanics and Analysis*, **12**, 58–91.
Joulin, G. (1981). Asymptotic analysis of non-adiabatic flames: heat losses towards small inert particles. In *Eighteenth International Symposium on Combustion*, pp. 1395–404. Pittsburgh: The Combustion Institute.
Joulin, G. & Clavin, P. (1976). Analyse asymptotique des conditions d'extinction des flammes laminaires. *Acta Astronautica*, **3**, 223–40.
Joulin, G. & Clavin, P. (1979). Linear stability analysis of non-adiabatic flames: diffusional-thermal model. *Combustion and Flame*, **35**, 139–53.
Kanury, A. M. (1975). *Introduction to Combustion Phenomena*. New York: Gordon and Breach.
Kapila, A. K. (1978). Homogeneous branched-chain explosion: initiation to completion. *Journal of Engineering Mathematics*, **12**, 221–35.
Kapila, A. K. (1980). Reactive-diffusive system with Arrhenius kinetics: dynamics of ignition. *SIAM Journal on Applied Mathematics*, **39**, 21–36.
Kapila, A. K. (1981*a*). Evolution of deflagration in a cold combustible subjected to a uniform energy flux. *International Journal of Engineering Science*, **19**, 495–509.
Kapila, A. K. (1981*b*). Arrhenius systems: dynamics of jump due to slow passage through criticality. *SIAM Journal on Applied Mathematics*, **41**, 29–42.

Kapila, A. K. & Ludford, G. S. S. (1977). Deflagration of gasifying condensates with distributed heat loss. *Acta Astronautica*, **4**, 279–89.

Kapila, A. K., Ludford, G. S. S., & Buckmaster, J. D. (1975). Ignition and extinction of a monopropellant droplet in an inert atmosphere. *Combustion and Flame*, **25**, 361–8.

Karlovitz, B., Denniston, D. W., Knapschaefer, D. H., & Wells, F. E. (1953). Studies on turbulent flames. In *Fourth International Symposium on Combustion*, pp. 613–20. Baltimore: Williams and Wilkins.

Kaskan, W. E. (1957). The dependence of flame temperature on mass burning velocity. In *Sixth International Symposium on Combustion*, pp. 134–43. New York: Reinhold.

Kassoy, D. R. (1975). A theory of adiabatic, homogeneous explosion from initiation to completion. *Combustion Science and Technology*, **10**, 27–35.

Kassoy, D. R. (1977). The supercritical spatially homogeneous thermal explosion: initiation to completion. *Quarterly Journal of Mechanics and Applied Mathematics*, **30**, 71–89.

Kassoy, D. R. & Poland, J. (1975). The subcritical spatially homogeneous explosion; initiation to completion. *Combustion Science and Technology*, **11**, 147–52.

Kassoy, D. R. & Poland, J. (1980). The thermal explosion confined by a constant temperature boundary: I – the induction period solution. *SIAM Journal on Applied Mathematics*, **39**, 412–30.

Kassoy, D. R. & Poland, J. (1981). The thermal explosion confined by a constant temperature boundary: II – the extremely rapid transient. *SIAM Journal on Applied Mathematics*, **41**, 231–46.

Kassoy, D. R. & Williams, F. A. (1968). Effects of chemical kinetics on near equilibrium combustion in nonpremixed systems. *Physics of Fluids*, **11**, 1343–51.

Knight, B. & Williams, F. A. (1980). Observations on the burning of droplets in the absence of buoyancy. *Combustion and Flame*, **38**, 111–9.

Kontrat'ev, V. N. (1972). *Rate Constants of Gas Phase Reactions*, translated by L. J. Holtschlag, ed. R. M. Fristrom. National Bureau of Standards COM-72-10014. Springfield, Va.: National Technical Information Service.

Landau, L. (1944). On the theory of slow combustion. *Acta Physicochimica* (*URSS*), **XIX**, 77–85. (See Landau, L. D. (1965). *Collected Papers* (ed. D. ter Haar), No. 54, pp. 396–403, New York: Gordon and Breach, Pergamon. Also Landau, L. D. & Lifshitz, E. M. (1959). *Fluid Mechanics*, p. 478. New York: Pergamon Press.)

Law, C. K. (1975). Asymptotic theory for ignition and extinction in droplet burning. *Combustion and Flame*, **24**, 89–98.

Levy, J. & Friedman, R. (1962). Further studies of pure ammonium perchlorate deflagration. In *Eighth International Symposium on Combustion*, pp. 663–72. Baltimore: Williams and Wilkins.

Lewis, B. & von Elbe, G. (1961). *Combustion, Flames and Explosions of Gases*. New York: Academic Press.

Libby, P. & Williams, F. A. (1982). Structure of laminar flamelets in premixed turbulent flames. *Combustion and Flame*, **44**, 287–303.

Liebman, I., Corry, J., & Perlee, H. E. (1970). Flame propagation in layered methane–air systems. *Combustion Science and Technology*, **1**, 257–67.

Liñán, A. (1971). A theoretical analysis of premixed flame propagation with an isothermal chain reaction. *Instituto Nacional de Tecnica Aeroespacial "Esteban Terradas"* (*Madrid*), USAFOSR Contract No. EOOAR 68–0031, Technical Report No. 1.

Liñán, A. (1974). The asymptotic structure of counterflow diffusion flames for large activation energies. *Acta Astronautica*, **1**, 1007–39.

Liñán, A. (1975). Monopropellant droplet decomposition for large activation energies. *Acta Astronautica*, **2**, 1009–29.

Liñán, A. & Crespo, A. (1972). An asymptotic analysis of unsteady diffusion flames for large activation energies. *Instituto Nacional de Tecnica Aeroespacial "Estaban*

Terradas" (*Madrid*). ARO-INTA Subcontracts 71-14-TS/OMD & 72-13-TS/OMD, Technical Report No. 1. Also (1976), *Combustion Science and Technology* **14**, 94–117.
Liñán, A. & Williams, F. A. (1971). Theory of ignition of a reactive solid by a constant energy flux. *Combustion Science and Technology*, **3**, 91–8.
Liouville, J. (1853). Sur l'équation aux différences partielles $d^2 \log \lambda/dudv \pm \lambda/2a^2 = 0$. *Journal de Mathématiques Pures et Appliquées*, **18**, 71–2.
Lu, G. C. (1981). *Diffusion Flames in a Chamber and Plane Steady Detonations for Large Activation Energy*. Cornell University: Ph.D. thesis.
Ludford, G. S. S. (1976). Combustion for large activation energy. *Letters in Applied and Engineering Sciences*, **4**, 49–62.
Ludford, G. S. S. (1980). Premixed cylindrical flames. In *Transactions of the 26th Conference of Army Mathematicians*. ARO Report No. 80–1, pp. 155–60. Research Triangle Park, N.C.: U.S. Army Research Office.
Ludford, G. S. S. & Sen, A. K. (1981). Burning rate maximum of a plane premixed flame. *Progress in Astronautics and Aeronautics*, **76**, 427–36. (*Combustion in Reactive Systems*, ed. J. Ray Bowen, N. Manson, Antoni K. Oppenheim, and R. I. Soloukhin.)
Ludford, G. S. S., Yannitell, D. W., & Buckmaster, J. D. (1976a). The decomposition of a cold monopropellant in an inert atmosphere. *Combustion Science and Technology*, **14**, 133–46.
Ludford, G. S. S., Yannitell, D. W., & Buckmaster, J. D. (1976b). The decomposition of a hot monopropellant in an inert atmosphere. *Combustion Science and Technology*, **14**, 125–31.
Maksimov, E. I., Merzhanov, A. G., & Shkiro, V. M. (1965). Gasless compositions as a simple model for the combustion of nonvolatile condensed systems. *Combustion, Explosion, and Shock Waves*, **1**(4), 15–18.
Marble, F. E. & Adamson, R. C., Jr. (1954). Ignition and combustion in a laminar mixing zone. *Jet Propulsion*, **24**, 85–94.
Margolis, S. B. (1980). Bifurcation phenomena in burner-stabilized premixed flames. *Combustion Science and Technology*, **22**, 143–69.
Markstein, G. H. (1964). *Nonsteady Flame Propagation*. AGARDOgraph no. 75. New York: MacMillan.
Matalon, M. & Ludford, G. S. S. (1980). On the near-ignition stability of diffusion flames. *International Journal of Engineering Science*, **18**, 1017–26.
Matalon, M., Ludford, G. S. S., & Buckmaster, J. (1979). Diffusion flames in a chamber. *Acta Astronautica*, **6**, 943–59.
Matkowsky, B. J. (1970). Nonlinear dynamic stability: a formal theory. *SIAM Journal on Applied Mathematics*, **18**, 872–83.
Matkowsky, B. J. & Olagunju, D. O. (1981). Pulsations in a burner-stabilized premixed plane flame. *SIAM Journal on Applied Mathematics*, **40**, 551–62.
Matkowsky, B. J., Putnick, L. J., & Sivashinsky, G. I. (1980). A nonlinear theory of cellular flames. *SIAM Journal on Applied Mathematics*, **38**, 489–504.
Matkowsky, B. J. & Sivashinsky, G. I. (1978). Propagation of a pulsating reaction front in solid fuel combustion. *SIAM Journal on Applied Mathematics*, **35**, 465–78.
Matkowsky, B. J. & Sivashinsky, G. I. (1979a). Acceleration effects on the stability of flame propagation. *SIAM Journal on Applied Mathematics*, **37**, 669–85.
Matkowsky, B. J. & Sivashinsky, G. I. (1979b). An asymptotic derivation of two models in flame theory associated with the constant density approximation. *SIAM Journal on Applied Mathematics*, **37**, 686–99.
McConnaughey, H. V. (1982). *Three Topics in Combustion Theory*. Cornell University: Ph.D. thesis.
Merzhanov, A. G. & Averson, A. E. (1971). The present state of the thermal ignition theory: an invited review. *Combustion and Flame* **16**, 89–124.
Merzhanov, A. G., Filonenko, A. K., & Borovinskaya, I. P. (1973). New phenomena in

combustion of condensed systems. *Akademii Nauk SSR, Physical chemistry Doklady*, **208**. 122–5.

Michelson, D. M. & Sivashinsky, G. I. (1977). Nonlinear analysis of hydrodynamic instability in laminar flames – II. Numerical experiments. *Acta Astronautica*, **4**, 1207–21.

Mitani, T. (1980). Propagation velocities of two-reactant flames. *Combustion Science and Technology*, **21**, 175–7.

Morse, P. M. & Feshbach, H. (1953). *Methods of Theoretical Physics*. New York: McGraw-Hill.

Müller, I. (1977). Thermodynamik von Mischungen als Modellfall der rationalen Thermodynamik. *Zeitschrift für angewandte Mathematik und Mechanik*, **57**, T36–42.

Nir, E. C. (1973). An experimental study of the low pressure limit for steady deflagration of Ammonium Perchlorate. *Combustion and Flame*, **20**, 419–35.

Normandia, M. J. & Ludford, G. S. S. (1980). Surface equilibrium in drop combustion. *SIAM Journal on Applied Mathematics*, **38**, 326–39.

Onsager, L. (1945). Theories and problems of liquid diffusions. *Annals of the New York Academy of Sciences*, **46**, 241–65.

Parang, M. & Jischke, M. C. (1975). Adiabatic ignition of homogeneous systems. *American Institute of Aeronautics and Astronautics Journal*, **13**, 405–8.

Payne, L. E. (1975). *Improperly Posed Problems in Partial Differential Equations*. Regional Conference Series in Applied Mathematics. Philadelphia: Society for Industrial and Applied Mathematics.

Peskin, R. L. & Wise, H. (1966). A theory for ignition and deflagration of fuel drops. *American Institute of Aeronautics and Astronautics Journal*, **4**, 1646–50.

Peters, N. (1978). On the stability of Liñán's "premixed flame regime". *Combustion and Flame*, **33**, 315–18.

Peters, N. & Warnatz, J. (eds) (1982). *Numerical Methods in Laminar Flame Propagation*. Vol. 5 in *Notes on Numerical Fluid Mechanics*. Wiesbaden: Vieweg & Sohn.

Poland, J. (1979). *Transient Phenomena in Thermal Explosions in Confined Gases*. University of Colorado: Ph.D. thesis.

Potter, J. M. & Riley, N. (1980). Free convection over a burning sphere. *Combustion and Flame*, **39**, 83–96.

Semenov, N. N. (1928). Zur Theorie der Verbrennungsprozesses. *Zeitschrift für Physik*, **48**, 571–82.

Sen, A. K. & Ludford, G. S. S. (1979). The near-stoichiometric behavior of combustible mixtures. Part I: Diffusion of the reactants. *Combustion Science and Technology*, **21**, 15–23.

Sen, A. K. & Ludford, G. S. S. (1981*a*). Effects of mass diffusion on the burning rate of non-dilute mixtures. In *Eighteenth International Symposium on Combustion*, pp. 417–24. Pittsburgh: The Combustion Institute.

Sen, A. K. & Ludford, G. S. S. (1981*b*). The near-stoichiometric behavior of combustible mixtures. Part II: Dissociation of the products. *Combustion Science and Technology*, **26**, 183–91.

Shkadinsky, K. G., Khaikin, B. I., & Merzhanov, A. G. (1971). Propagation of a pulsating exothermic reaction front in the condensed phase. *Combustion, Explosion, and Shock Waves*, **7**, 15–22.

Shouman, A. R. & Donaldson, A. B. (1975). The stationary problem of thermal ignition in a reactive slab with unsymmetric boundary temperatures. *Combustion and Flame*, **24**, 203–10.

Sivashinsky, G. I. (1973). On a steady corrugated flame front. *Astronautica Acta*, **18**, 253–60.

Sivashinsky, G. I. (1974*a*). On a converging spherical flame front. *International Journal of Heat and Mass Transfer* (Transactions of the ASME), **17**, 1499–506.

Sivashinsky, G. I. (1974*b*). The diffusion stratification effect in bunsen flames. *Journal of Heat and Mass Transfer*, **11**, 530–5.

Sivashinsky, G. I. (1975). Structure of Bunsen flames. *The Journal of Chemical Physics*, **62**, 638–43.
Sivashinsky, G. I. (1976). On the distorted flame front as a hydrodynamic discontinuity. *Acta Astronautica*, **3**, 889–918.
Sivashinsky, G. I. (1977*a*). Diffusional-thermal theory of cellular flames. *Combustion Science and Technology*, **15**, 137–45.
Sivashinsky, G. I. (1977*b*). Nonlinear analysis of hydrodynamic instability in laminar flames – I. Derivation of basic equations. *Acta Astronautica*, **4**, 1177–206.
Sivashinsky, G. I. (1979). On self-turbulization of a laminar flame. *Acta Astronautica*, **6**, 569–91.
Sivashinsky, G. I. & Gutfinger, C. (1975). Applications of asymptotic methods to laminar flame theory. In *Current Topics in Heat and Mass Transfer*, ed. C. Gutfinger, pp. 555–607. Washington, D.C.: Hemispheric Publishing Corporation.
Sivashinsky, G. I. & Michelson, D. M. (1982). Thermal-expansion induced cellular flames. *Combustion and Flame* (in press).
Spalding, D. B. (1960). The theory of burning of solid and liquid propellants. *Combustion and Flame*, **4**, 59–67.
Spalding, D. B. & Jain, V. K. (1959). Theory of the burning of monopropellant droplets. *Aeronautical Research Council Current Paper no. 447*. Technical Report 20 176.
Spalding, D. B. & Yumlu, V. S. (1959). Experimental demonstration of the existence of two flame speeds. *Combustion and Flame*, **3**, 553–6.
Stewartson, K. (1964). *The Theory of Laminar Boundary Layers in Compressible Fluids*. Oxford: Clarendon Press.
Stuart, J. T. (1967). On finite amplitude oscillations in laminar mixing layers. *Journal of Fluid Mechanics*, **29**, 417–40.
Takeno, T. & Sato, K. (1981). A theoretical and experimental study on an excess enthalpy flame. *Progress in Astronautics and Aeronautics*, **76**, 596–610. (*Combustion in Reactive Systems*, eds J. Ray Bowen, N. Manson, Antoni K. Oppenheim, and R. I. Soloukhin.)
Taliaferro, S., Buckmaster, J. & Nachman, A. (1981). The near-extinction stability of fuel-drop combustion. *Quarterly Journal of Mechanics and Applied Mathematics* (in press).
Titchmarsh, E. C. (1946). *Eigenfunction Expansions Associated with Second-Order Differential Equations*. Oxford: Clarendon Press.
Truesdell, C. A. (1965). On the foundations of mechanics and energetics. In *Continuum Mechanics II: The Rational Mechanics of Materials*, ed. C. Truesdell, pp. 292–305. New York: Gordon and Breach.
Tsuji, M. & Yamaoka, I. (1981). An experimental study of extinction of near-limit flames in a stagnation flow. In *First International Specialists Meeting of the Combustion Institute*, pp. 111–16. Section Française du Combustion Institute.
Turing, A. M. (1953). The chemical basis of morphogenesis. *Philosophical* Transactions *of the Royal Society of London*, **B237**, 37–72.
Van Dyke, M. (1975). *Perturbation Methods in Fluid Mechanics*. Stanford: The Parabolic Press.
von Kármán, Th. & Millán, G. (1953). The thermal theory of constant pressure deflagration. In *Anniversary Volume on Applied Mechanics Dedicated to C. B. Biezeno*, pp. 58–69. Haarlem: N. V. de Techniche Uitgerverij H. Stam.
Westbrook, C. E., Adamczyk, A. A., & Lavoie, G. A. (1981). A numerical study of laminar flame wall quenching. *Combustion and Flame*, **40**, 81–99.
Williams, F. A. (1962). Response of a burning solid to small amplitude pressure oscillations. *Journal of Applied Physics*, **33**, 3153–66.
Williams, F. A. (1965). *Combustion Theory*. Reading, Massachusetts: Addison-Wesley.
Williams, F. A. (1969). Theories of steady-state propellant combustion. In *Fundamental Aspects of Solid Propellant Rockets*, by M. Barrère, N. C. Huang, & Forman

Williams, pp. 333–93. AGARDograph no. 116. Slough, England: Technivision Services.

Williams, F. A. (1971). Theory of combustion in laminar flows. *Annual Review of Fluid Mechanics*, **3**, 171–88.

Williams, F. A. (1973). Quasi-steady gas-phase flame theory in unsteady burning of a homogeneous solid propellant. *American Institute of Aeronautics and Astronautics Journal*, **11**, 1328–30.

Williams, F. A. (1975). A review of some theoretical considerations of turbulent flame structure. In *Analytical and Numerical Methods for Investigation of Flow Fields, with Chemical Reactions, Especially Related to Combustion*. AGARD Conference Proceedings No. 164, pp. II 1-1–II 1-25. Neuilly sur Seine: Advisory Group for Aerospace Research and Development, NATO.

Zeldovich, Y. B. & Barenblatt, G. I. (1959). Theory of flame propagation. *Combustion and Flame*, **3**, 61–74.

FURTHER REFERENCES

Allison, R. A. & Clarke, J. F. (1981). Theory of a hydrogen-oxygen diffusion flame. Part II: Large activation energy asymptotics. *Combustion Science and Technology*, **25**, 97–108.

Berman, V. S. & Riazantsev, Iu. S. (1972). Use of the method of matched asymptotic expansions to calculate the steady thermal propagation of the front of an exothermic reaction in a condensed medium. *Journal of Applied Mechanics and Technical Physics* (*PMTF*), No. 5, 688–92.

Berman, V. S. & Riazantsev, Iu. S. (1973*a*). Asymptotic analysis of stationary distribution of the front of a two-stage consecutive exothermic reaction in a condensed medium. *Journal of Applied Mechanics and Technical Physics* (*PMTF*), No. 1, 60–70.

Berman, V. S. & Riazantsev, Iu. S. (1973*b*). Asymptotic analysis of stationary propagation of the front of a two-stage exothermic reaction in a gas. *Journal of Applied Mathematics and Mechanics* (*PMM*), **37**, 995–1004.

Berman, V. S. & Riazantsev, Iu. S. (1975). Asymptotic analysis of stationary propagation of the front of a parallel exothermic reaction. *Journal of Applied Mathematics and Mechanics* (*PMM*), **39**, 286–96.

Berman, V. S. & Riazantsev, Iu. S. (1976). Igniting a homogeneous reacting medium by heat source with finite heat content. *Journal of Applied Mathematics and Mechanics* (*PMM*), **40**, 1008–12.

Berman, V. S. & Riazantsev, Iu. S. (1977). Asymptotic analysis of gas ignition by a glowing surface. *Journal of Applied Mechanics and Technical Physics* (*PMTF*), No. 1, 57–62.

Berman, V. S. & Riazantsev, Iu. S. (1978). Ignition of a gas in a boundary layer at a heated plate. *Journal of Fluid Mechanics*, **12**, 758–64.

Berman, V. S., Riazantsev, Iu. S. & Shevtsova, V. M. (1979). On igniting a reacting gas by a heat source of finite heat capacity. *Journal of Applied Mathematics and Mechanics* (*PMM*), **43**, 79–87.

Berman, V. S., Riazantsev, Iu. S. & Shevtsova, V. M. (1980). Asymptotic analysis of the process of igniting a combustible gas mixture by thermal inhomogeneity. *Journal of Applied Mathematics and Mechanics* (*PMM*), **44**, 60–4.

Blythe, P. A. (1978). Wave propagation and ignition in a combustible mixture. In *Seventeenth International Symposium on Combustion*, pp. 909–16, Pittsburgh: The Combustion Institute.

Bobrova, N. R., Burkina, R. S. & Vilyunov, V. N. (1981). Stationary combustion in one-dimensional gas flows. *Combustion, Explosion, and Shock Waves*, **16**, 292–7.

Broido, A. & Williams, F. A. (1973). Use of asymptotic analysis of the large activation-energy limit to compare graphical methods of treating thermogravimetry data. *Thermochimika Acta*, **6**, 245–53.

Buckmaster, J. (1975), Combustion of a fuel drop immersed in a hot atmosphere. *International Journal of Engineering Science*, **13**, 915–21.
Buckmaster, J. & Ludford, G. S. S. (1977). Activation energy asymptotics and unsteady flames. In *Transactions of the 22nd Conference of Army Mathematicians.* ARO Report 77-1, pp. 183–201, Research Triangle Park, N.C.: U.S. Army Research Office.
Burkina, R. S. & Vilyunov, V. N. (1976). Asymptotic analysis of the problem of reactive substance ignition by a heated surface. *Journal of Applied Mechanics and Technical Physics (PMTF)*, No. 6, 829–35.
Bush, W. B. (1979). Asymptotic analysis of laminar flame propagation: review and extension. *International Journal of Engineering Science*, **17**, 597–613.
Bush, W. B. (1980). Stoichiometry effects on a one-dimensional flame. *Combustion Science and Technology*, **23**, 263–6.
Bush, W. B. & Fendell, F. E. (1971). Asymptotic analysis of the structure of a steady-planar detonation. *Combustion Science and Technology*, **2**, 271–85.
Bush, W. B., Fendell, F. E. & Fink, S. F. (1980). Effect of boundary thermal constraint on planar premixed-flame/wall interaction. *Combustion Science and Technology*, **24**, 53–70.
Bush, W. B. & Williams, F. A. (1975). Radiant ignition of a surface-cooled reactive solid. *Acta Astronautica*, **2**, 445–62.
Bush, W. B. & Williams, F. A. (1976). Influence of strong conductive gas-phase cooling on radiant ignition of a reactive solid. *Combustion and Flame*, **27**, 321–9.
Carrier, G. F., Fendell, F. E. & Bush, W. B. (1979). Interaction of a planar premixed flame with a parallel adiabatic end wall. *Combustion Science and Technology*, **20**, 195–207.
Cheng, H. K. & Lee, R. S. (1967). Non-equilibrium-transition patterns of a quasi-one-dimensional dissociating-gas flow. *American Institute of Aeronautics and Astronautics Journal*, **5**, 1686–9.
Cheng, H. K. & Lee, R. S. (1968). Freezing of dissociation and recombination in supersonic nozzle flows: Part I. General discussion and special analysis. *American Institute of Aeronautics and Astronautics Journal*, **6**, 823–30.
Clarke, J. F. (1975). Parameter perturbations in flame theory. *Progress in Aerospace Sciences*, **16**, 3–29.
Clarke, J. F. (1979). On the evolution of compressible pulses in an exploding atmosphere; initial behavior. *Journal of Fluid Mechanics*, **94**, 195–208.
Clarke, J. F. (1981). Propagation of gasdynamic disturbances in an explosive atmosphere. *Progress in Astronautics and Aeronautics*, **76**, 383–402. (*Combustion in Reactive Systems*, ed. J. Ray Bowen, N. Manson, Antoni K. Oppenheim and R. I. Soloukhin.)
Clavin, P. & Pelce, P. (1981). Dynamical properties of premixed flame fronts. In *First International Specialists Meeting of the Combustion Institute*, pp. 130–5, Section Francaise du Combustion Institute.
Clavin, P. & Williams, F. A. (1979). Theory of premixed flame propagation in large-scale turbulence. *Journal of Fluid Mechanics*, **90**, 589–604.
Clavin, P. & Williams, F. A. (1981). Effects of Lewis number on propagation of wrinkled flames in turbulent flow. *Progress in Astronautics and Aeronautics*, **76**, 403–11. (*Combustion in Reactive Systems*, ed. J. Ray Bowen, N. Manson, Antoni K. Oppenheim and R. I. Soloukhin.)
Garcia-Ybarra, P. L. & Clavin, P. (1981). Cross-transport effects in non-adiabatic premixed flames. *Progress in Astronautics and Aeronautics*, **76**, 463–81. (*Combustion in Reactive Systems*, ed. J. Ray Bowen, N. Manson, Antoni K. Oppenheim and R. I. Solonkhin.)
Hermance, C. E. (1973). Ignition analysis of adiabatic, homogeneous systems including reactant consumption. *American Institute of Aeronautics and Astronautics Journal*, **11**, 1728–31.
Ibiricu, M. M. & Williams, F. A. (1975). Influence of externally applied thermal radiation

on the burning rates of homogeneous solid propellants. *Combustion and Flame*, **24**, 185–98.

Janicka, J. & Peters, N. (1980). Asymptotic evaluation of the mean NO-production rate in turbulent diffusion flames. *Combustion Science and Technology*, **22**, 93–6.

Joulin, G. (1981). Sur l'interaction d'une flamme plissée et d'un bruleur plan. In *First International Specialists Meeting of the Combustion Institute*, pp. 142–145. Section Française du Combustion Institute.

Joulin, G. & Clavin, P. (1975*a*). Etude théorique des flammes engendrées par des réactions en chaine. In 2ème *Symposium Européen sur la Combustion*, pp. 185–9 Section Française du Combustion Institute.

Joulin, G. & Clavin, P. (1975*b*). Asymptotic analysis of a premixed laminar flame governed by a two-step reaction. *Combustion and Flame*, **25**, 399–92.

Joulin, G. & Mitani, T. (1981). Linear stability analysis of two-reactant flames. *Combustion and Flame*, **40**, 235–46.

Kapila, A. K. & Ludford, G. S. S. (1977). Two-step sequential reactions for large activation energy. *Combustion and Flame*, **29**, 167–76.

Kapila, A. K. & Matkowsky, B. J. (1979). Reactive-diffusive system with Arrhenius kinetics: multiple solutions, ignition and extinction. *SIAM Journal on Applied Mathematics*, **36**, 373–89.

Kapila, A. K. & Matkowsky, B. J. (1980). Reactive-diffusive system with Arrhenius kinetics: the Robin problem. *SIAM Journal on Applied Mathematics*, **39**,391–401.

Kapila, A. K., Matkowsky, B. J. & Vega, J. (1980). Reactive-diffusive system with A Arrhenius kinetics: peculiarities of the spherical geometry. *SIAM Journal on Applied Mathematics*, **38**, 382–401.

Kapila, A. K. & Poore, A. B. (1981). The steady response of a nonadiabatic tubular reactor: new multiplicities. *Chemical Engineering Science*, **37**, 57–68.

Kassoy, D. R. (1975), Perturbation methods for mathematical models of explosion and ignition phenomena. *Quarterly Journal of Mechanics and Applied Mathematics*, **28**, 63–74.

Kassoy, D. R. (1976). Extremely rapid transient phenomena in combustion, ignition and explosion. *SIAM-AMS Proceedings*, **10**, 61–72. (*Asymptotic Methods and Singular Perturbations*, ed. R. E. O'Malley, Jr.)

Kassoy, D. R. & Liñán, A. (1978). The influence of reactant consumption on the critical conditions for homogeneous thermal explosions. *Quarterly Journal of Mechanics and Applied Mathematics*, **31**, 99–112.

Kindelán, M. & Liñán, A. (1978). Gasification effects in the heterogeneous ignition of a condensed fuel by a hot gas. *Acta Astronautica*, **5**, 1199–211.

Kindelán, M. & Williams, F. A. (1975*a*). Theory for endothermic gasification of a solid by a constant energy flux. *Combustion Science and Technology*, **10**, 1–19.

Kindelán, M. & Williams, F. A. (1975*b*). Radiant ignition of a combustible solid with gas-phase exothermicity. *Acta Astronautica*, **2**, 955–79.

Kindelán, M. & Williams, F. A. (1977). Gas-phase ignition of a solid with in-depth absorption of radiation. *Combustion Science and Technology*, **16**, 47–58.

Krishnamurthy, L. (1976). On gas-phase ignition of diffusion flame in the stagnation-point boundary layer. *Acta Astronautica*, **3**, 935–42.

Krishnamurthy, L., Williams, F. A. & Seshadri, K. (1976). Asymptotic theory of diffusion-flame extinction in the stagnation-point boundary layer. *Combustion and Flame*, **26**, 363–77.

Law, C. K. (1978*a*). Deflagration and extinction of fuel droplet in a weakly-reactive atmosphere. *The Journal of Chemical Physics*, **68**, 4218–21.

Law, C. K. (1978*b*). On the stagnation-point ignition of a premixed combustible. *International Journal of Heat and Mass Transfer*, **21**, 1363–8.

Law, C. K. (1978*c*). Ignition of a combustible mixture by a hot particle. *American Institute of Aeronautics and Astronautics Journal*, **16**, 628–30.
Law, C. K. (1979). Transient ignition of a combustible by stationary isothermal bodies. *Combustion Science and Technology*, **19**, 237–42.
Law, C. K. & Chung, S. H. (1980). An ignition criterion for droplets in sprays. *Combustion Science and Technology*, **22**, 17–26.
Law, C. K. & Law, H. K. (1979). Thermal ignition analysis in boundary layer flows. *Journal of Fluid Mechanics*, **92**, 97–108.
Law, C. K. & Law, H. K. (1981). Flat-plate ignition with reactant consumption. *Combustion Science and Technology*, **25**, 1–8.
Lengelle, G. (1970). Thermal degradation kinetics and surface pyrolysis of vinyl polymers. *American Institute of Aeronautics and Astronautics Journal*, **8**, 1989–97.
Libby, A. D. (1980). Ignition, combustion and extinction of carbon particles. *Combustion and Flame*, **38**, 285–300.
Liñán, A. & Kindelán, M. (1981). Ignition of a reactive solid by an inert hot spot. *Progress in Astronautics and Aeronautics*, **76**, 412–26. (*Combustion in Reactive Systems*, ed. J. Ray Bowen, N. Manson, Antoni K. Oppenheim and R. I. Soloukhin,)
Liñán, A. & Williams, F. A. (1972). Radiant ignition of a reactive solid with in-depth absorption. *Combustion and Flame*, **18**, 85–97.
Liñán, A. & Williams, F. A. (1979). Ignition of a reactive solid exposed to a step in surface temperature. *SIAM Journal on Applied Mathematics*, **36**, 587–603.
Ludford, G. S. S. (1977*a*). Combustion: basic equations and peculiar asymptotics. *Journal de Méchanique*, **16**, 531–51.
Ludford, G. S. S. (1977*b*). The premixed plane flame. *Journal de Mécanique*, **16**, 553–73.
Ludford, G. S. S. & Buckmaster, J. (1980). Activation-energy asymptotics in turbulent combustion. In *Fluid Mechanics in Energy Conversion*, pp. 263–75. Philadelphia: SIAM.
Ludford, G. S. S. & Stewart, D. S. (1980). Mathematical questions from combustion theory. In *Transactions of the 27th Conference of Army Mathematicians*. ARO Report No. 81-1, pp. 53–66, Research Triangle Park, N.C.: U.S. Army Research Office.
Ludford, G. S. S. & Stewart, D. S. (1981). A theory of deflagrations and its application. In *First International Specialists Meeting of the Combustion Institute*, pp. 158–61. Section Française du Combustion Institute.
Ludford, G. S. S. & Stewart, D. S. (1982). Deflagration to detonation transition. In *Transactions of the 27th Conference of Army Mathematicians*. ARO Report No. 82-1, pp. 563–72. Research Triangle Park, N.C.: U.S. Army Research Office.
Margolis, S. B. (1981). Effects of selective diffusion on the stability of burner-stabilized premixed flames. In *Eighteenth International Symposium on Combustion*, pp. 679–93. Pittsburgh: The Combustion Institute.
Margolis, S. B. & Matkowsky, B. J. (1982). Flame propagation with multiple fuels. In *First International Specialists Meeting of the Combustion Institute*, pp. 146–51. Section Française du Combustion Institute.
Matalon, M. (1980). Complete burning and extinction of a carbon particle in an oxidizing atmosphere. *Combustion Science and Technology*, **24**, 115–27.
Matalon, M. (1981). Weak burning and gas-phase ignition about a carbon particle in an oxidizing atmosphere. *Combustion Science and Technology*, **25**, 43–8.
Matalon, M. & Ludford, G. S. S. (1979*a*). The asymptotic derivation of isola and S responses for chambered diffusion flames. *SIAM Journal on Applied Mathematics*, **37**, 107–23.
Matalon, M. & Ludford, G. S. S. (1979*b*). Chambered diffusion flames for different supply temperatures. *Acta Astronautica*, **6**, 1377–86.
Matkowsky, B. J. & Olagunju, D. O. (1980). Propagation of a pulsating flame front in a

gaseous combustible mixture. *SIAM Journal on Applied Mathematics*, **39**, 290–300.

McIntosh, A. C. & Clarke, J. F. (1981). The time response of a premixed flame to changes in mixture strength. *Progress in Astronautics and Aeronautics*, **76**, 443–62. (*Combustion in Reactive Systems*, ed. J. Ray Bowen, N. Manson, Antoni K. Oppenheim and R. I. Soloukhin.)

Mikolaitis, D. & Buckmaster, J. (1981). Flame stabilization in a rear stagnation point flow. *Combustion Science and Technology*, **27**, 55–68.

Mitani, T. (1980). Asymptotic theory for extinction of curved flames with heat loss. *Combustion Science and Technology*, **23**, 93–101.

Mitani, T. (1981). A flame inhibition theory by inert dust and spray. *Combustion and Flame*, **43**, 243–53.

Niioka, T. (1981). Ignition time in the stretched-flow field. In *Eighteenth International Symposium on Combustion*, pp. 1807–13. Pittsburgh: The Combustion Institute.

Niioka, T. & Williams, F. A. (1977). Ignition of a reactive solid in a hot stagnation-point flow. *Combustion and Flame*, **29**, 43–54.

Normandia, M. J. & Buckmaster, J. (1977). Near-asphyxiated fuel-drop burning. *Combustion and Flame*, **29**, 277–82.

Peters, N. (1978*a*). Das Flammenzonenmodell und seine Anwendungen. *Verein Deutscher Inginieure-Berichte*, **315**, 321–342.

Peters, N. (1978*b*). An asymptotic analysis of nitric oxide formation in turbulent diffusion flames. *Combustion Science and Technology*, **19**, 39–49.

Peters, N. (1979). Premixed burning in diffusion flames—the flame zone model of Libby and Economos. *International Journal of Heat and Mass Transfer*, **22**, 691–703.

Peters, N., Hocks, W. & Mohiuddin, G. (1981). Turbulent mean reaction rates in the limit of large activation energies. *Journal of Fluid Mechanics*, **110**, 411–32.

Sivashinsky, G. I. (1980). On flame propagation under conditions of stoichiometry. *SIAM Journal on Applied Mathematics*, **39**, 67–82.

Sivashinsky, G. (1981*a*). On spinning propagation of combustion waves. *SIAM Journal on Applied Mathematics*, **40**, 432–8.

Sivashinsky, G. (1981*b*). On the structure and stability of laminar flames. In *First International Specialists Meeting of the Combustion Institute*, pp. 136–40. Section Française du Combustion Institute.

Sivashinsky, G. I. & Gutfinger, C. (1974). The extinction of spherical diffusion flame. In *Heat Transfer in Flames*, ed. N. H. Afgan and J. M. Beer, pp. 495–501. Washington, D.C.: Halstead Press.

Sivashinsky, G. I. & Matkowsky, B. J. (1981). On the stability of nonadiabatic flames. *SIAM Journal on Applied Mathematics*, **40**, 255–60.

Sohrab, S. & Williams, F. A. (1981). Extinction of planar diffusion flames adjacent to flat surfaces of burning polymers. *Journal of Polymer Science: Polymer Chemistry Edition*, **19**, 2955–76.

Stewart, D. S. & Ludford, G. S. S. (1982*a*). The equation governing the propagation of fast deflagration waves for small heat release. In *Transactions of the 27th Conference of Army Mathematicians*. ARO Report No. 82-1, pp. 185–99. Research Triangle Park, N.C.: U.S. Army Research Office.

Tromans, P. S. (1981). The interaction between strain fields and flames—a possible source of combustion variations in spark ignition engines. In *Fluid Mechanics of Combustion Systems*, ed. T. Morel, R. P. Lohmann and J. M. Rackley, pp. 201–6. New York: The American Society of Mechanical Engineers.

CITATION INDEX

Adamczyk 166
Adams 61
Adamson 241
Aly 45
Arrhenius 6
Averson 222
Ayeni 222

Barenblatt 47
Batchelor 175
Baum 81, 83
Berman 33
Blythe 114
Borovinskaya 219
Bowen 3
Boyer 195
Buckmaster 17, 34, 42, 46, 49, 55, 61, 70, 95, 103, 104, 117, 119, 125, 129, 130, 135, 141, 145, 148, 151, 152, 156, 163 165, 166, 168, 177, 179, 181, 183, 184, 185, 199, 200
Burke 13, 102
Bush 2, 25, 28, 34, 40, 93

Cahill 230
Campbell 23
Carrier 34, 93
Chandra 222
Chu 193
Clarke 34, 93
Clavin 42, 219
Conrad 222
Corry 242
Crespo 241, 242, 245
Crowley 156
Culick 91
Curtiss 23, 29
Darrieus 137, 197
Davis 222

Denison 81, 83
Denniston 146
Deshaies 46
Donaldson 224

Eckhaus 145, 199
Emmons 8, 45, 67, 199

Fendell 2, 25, 28, 34, 40, 93, 125, 128, 135
Ferguson 29
Feshbach 119
Fick 5
Filonenko 219
Frank-Kamenetskii 2, 221, 224, 225
Friedman 64
Fristrom 36

Gautschi 230
Gaydon 214
Goldstein 233
Gray 227
Guirao 70
Günther 155
Gutfinger 42

Hermance 45, 230
Higgins 193
Hirschfelder 23, 29
Holmes 102

Istratov 194

Jain 120
Janssen 76, 98, 115, 116, 135
Jischke 230
Johnson 23, 34, 61, 70, 73
Joulin 42, 46, 47, 219

Kanury 163

Kapila 18, 34, 61, 70, 71, 73, 76, 78, 125, 129, 135, 233, 234, 235, 236, 237, 241, 245
Karlovitz 146
Kaskan 29
Kassoy 8, 101, 102, 229, 230, 232, 233, 234, 235, 238
Keck 29
Khaikin 219
Knapschaefer 146
Knight 97
Kontrat'ev 17

Landau 137, 197
Lavoie 166
Law 108, 109, 111, 113
LeChatelier 199
Lee 227
Levy 64
Lewis 28, 42, 138, 146, 149, 155, 161, 166, 170, 171
Libby 183
Librovich 194
Liebman 242
Liñán 18, 96, 113, 125, 238, 241, 242, 245
Liouville 224
Lu 95
Ludford 17, 34, 61, 70, 71, 73, 76, 78, 95, 116, 117, 119, 120, 124, 125, 129, 130, 133, 135
Maksimov 223
Mallard 199
Marble 241
Margolis 94, 219
Markstein 11, 51, 136, 194, 208, 214
Matalon 17, 95, 119
Matkowsky 55, 141, 152, 205, 215, 217, 219, 220
McConnaughey 213
McIntosh 93
Merzhanov 219, 222, 223
Michelson 208, 213, 214
Mikolaitis 184
Milán 23
Mitani 34
Morse 119
Müller 5

Nachbar 34, 61, 70, 73
Nachman 55, 117, 119
Nir 61
Normandia 116, 117, 124, 133

Olagunju 220
Onsager 5

Parang 230
Payne 222
Perlee 242
Peskin 218
Peters 49, 119
Poland 229, 233, 234, 235, 238
Potter 97
Putnick 215

Riazantsev 33
Riley 97

Sabathier 194
Sato 45
Schumann 13, 102
Semenov 232
Sen 34
Shkadinsky 219
Shkiro 223
Shouman 224
Shvab 16
Sivashinsky 42, 43, 51, 55, 141, 142, 145, 152, 159, 190, 196, 199, 203, 205, 206, 207, 208, 211, 213, 214, 215, 216, 219
Spalding 50, 64, 78, 120
Stewartson 55, 172, 243
Stuart 224

Takeno 45, 46
Taliaferro 117, 119
Titchmarsh 118
Truesdell 3
Tsuji 183
Turing 11

Van Dyke 19, 157, 230
von Elbe 28, 42, 138, 146, 149, 161, 166, 170, 171
von Kármán 23

Warnatz 49
Wells 146
Westbrook 166
Westenberg 36
Williams 2, 5, 8, 12, 16, 23, 25, 34, 60, 62, 70, 78, 81, 89, 97, 101, 102, 172, 183, 238, 241
Wise 218
Wiseman 61
Wolfhard 214

Yamaoka 183
Yannitell 125, 130
Yumlu 50

Zeldovich 16, 47

SUBJECT INDEX

acoustic admittance 89, 91
acoustic waves
  amplification of 90
  burning of linear condensates, effect on 79, 84, 86, 87
activation energy 6, 15
  asymptotics 2, 17, 25, 103, 247
ammonium perchlorate 60
anchored flames 23
  stability of 92, 219
area change, effect on flame speed 51, 57 (*see also* stretch)
Arrhenius factor 5
  replacement by delta function 218
asymptotic expansions
  nonlinear scaling in 230
  uniform notation for 19
auto-ignition 222
  of initially separated reactants 241

blow-off of
  excess-enthalpy flames 47
  flames in front of bluff bodies 183
  tube flames 155
Buckmaster limit 103, 125
buoyancy (*see* gravity *and* gravity effect on)
Burke-Schumann limit 101, 125 (*see also* flame sheets)
burners 154
  , Bunsen 20, 95, 138
  for excess-enthalpy flames 45
  , flat 28, 92
burning rate (*see also* flame speed)
  of cylindrical flames 123
  of diffusion flames 99, 100, 101, 103, 106, 109, 111
  of linear condensates 62, 66, 69, 72, 74, 77, 82, 86
  maximum 36, 61
  of plane flames 27, 31, 35, 40
  , reference 41
  of spherical flames 125, 126, 128, 129, 130, 132, 133, 134
  cellular flames 194, 205, 206
chain reactions 18, 222
Charles's law 13
Clausius-Clapeyron relation (law) 74
cold-boundary difficulty 21, 28
combustion approximation 11
condensate burning 58, 79
conductance of burners 93
constant-density approximation 11
  for explosions 235
  for initially separated reactants 242
  for NEFs 55, 151, 171, 202
  for SVFs 159
continuum theory
  balances in 2
  constitutive equations for 3
convection-diffusion
  in different dimensions 121
  operator 15
  zone 31, 123, 129
corrugated flames 198 (*see also* wrinkled flames)
counterflow
  flames 180
  diffusion flames 96, 113
C-responses 44
  for auto-ignition of slabs 227, 228
  for diffusion flames 100, 115
  for linear condensates 66, 73, 78
  for spherical flames 127, 130
crests 205, 218
curvature
  cellular flames, effect on 214
  of flame sheet 143
  flame speed, effect on 51, 52
  NEF stability, effect on 205

curvature (*contd.*)
SVF stability, effect on 199
vorticity and temperature stratification due to 136
cylindrical flames 52, 120, 138
, diffusion 95
stability of 201, 214
, unsteady 173
Damköhler number 15, 18
asymptotics 98, 124
, effective 243
Darrieus-Landau instability 197, 208
Davy lamp 45
deflagration waves 20, 38
in auto-ignition of initially separated reactants 244
evolution into detonation waves 194, 238
, incipient 236
delta (δ-) function 26, 27
models 218
detonations 1, 194, 238
diffusion
, differential mass 7, 15
flame 95, 241
velocity 2, 5
drops
, fuel 96, 115
, monopropellant 133, 135
duct flames
NEFs 55
SVFs 50, 51

eigenvalue
for diffusion-flame stability 118, 119
, flame-speed 22, 23, 27, 35, 36, 40
energy flux 16
enthalpy 4
, available 16
balance of 22, 39, 42, 56, 62, 65, 71, 82, 84, 92, 104, 140
, conditional 16
small variations of 141
transfer in excess-enthalpy flame 45
entropy layer 79, 87
equidiffusion of mass and heat 15
equilibrium 7
limit (*see* Buckmaster *and* Burke-Schumann limits)
Euler's equations 136
evaporating liquids as
drops 96, 115, 133
linear condensates 24, 27, 60, 73, 76, 86 91
evolution equation of
cellular flames 207, 212, 216
SVFs 145

experiments on
anchored flames 29, 219
auto-ignition 225
cellular flames 194
excess-enthalpy flames 45
initially separated reactants 242
linear condensates 60, 64, 83, 91
near-stoichiometric burning rate 36
open and closed flame tips 155
porous spheres 97
stagnation-point extinction 183
streamlines at flames 155, 158, 171
explosions, thermal 222
exact solution for 230
, homogeneous and nonhomogeneous 222
of initially separated reactants 241
phases of 32, 222, 230, 233
explosive regime 32
extinction (*see also* quenching)
of diffusion flame 108
through heat loss 45
for linear condensates 60, 64, 67, 68, 71, 77, 78
of spherical flames 127
at stagnation points 181, 192

Fick's law 5
flame
area element of a 146
bluntness of a 174
characteristic surface of a 146
crests 205, 218
, diffusion 20, 95, 241
holders 51
, near-surface 31, 123, 125
position for flat burner 29, 32
, remote 31, 100, 123, 125
shape 138, 154, 158, 159, 163, 170
-sheet approximation 218
sheets 18, 26, 139
, surface 31, 103, 114, 123, 125, 131
thickness 26, 148
volume element of a 148
flame-sheet structure 26, 33, 36, 40, 122
, Buckmaster 103, 126
, Burke-Schumann 102, 126, 244
, diffusion 110, 112
, unsteady 118
flame speed 41, 46 (*see also* burning rate)
, adiabatic 22
when anchored 27
constancy of 137, 145
diffusional-thermal effect on 208
of excess-enthalpy flames 46
flame temperature, sensitivity to 38, 41, 140

in flash-through 237, 246
heat-loss effect on 44
of NEFs 57
, negative 183
when weakly stretched 183, 215
for straining flow 181, 184, 186, 190
of SVFs 48, 49, 50, 52, 145
flame temperature
, adiabatic 22, 29, 35, 40, 99, 125, 137
when anchored 24, 39
burning-rate, effect on 38, 41, 140
of cellular flames 205
of diffusion flames 102, 105, 109, 110, 112, 114
for heat loss 44, 49, 52, 71
for linear condensates 63, 65, 68, 71, 82, 84, 86
of multidimensional flames 140
of NEFs 56
of spherical flames 128, 129, 131
, stagnation point 181
of SVFs 42, 48, 49, 52, 144
at tips 163, 166
flame tips
, closed and open 155, 161, 165, 167
hydrodynamic analysis of 156, 170
of NEFs 165, 167, 170
, polyhedral 194
of SVFs 159, 161
flameless combustion 61, 69
flammability (propagation) limits 171, 193, 199
flashback 47, 154, 170
flash-through 236, 241, 246
frozen combustion 18, 100, 125

gasification 59 (*see also* evaporating liquid and pyrolysis)
gravity 3, 12
gravity effect on
cellular flames 213
fuel drops 97
SVFs 199
NEFs 205

heat exchange
, distributed 71, 76
in excess-enthalpy flames 45
, radiative 64, 67, 76
heat loss (*see also* heat exchange)
in ducts 43, 49, 51
in explosions 232
to inerts particles 47
to walls 168
heat of combustion 13
heat of evaporation (*see* latent heat)
heat of gasification 62
heat release 137, 152 (*see also* constant-density approximation)
hot-spots 18, 232, 244
focusing of 234, 244
hydrocarbon flames 37, 155, 161, 205
hydrodynamic discontinuity 136
limitations of concept 138
hydrogen flames 155, 161, 205, 219
ignition 221 (*see also* auto-ignition)
of diffusion flames 108
by heat fluxes 238
by hot inerts 241
of linear condensates 68
of spherical flames 127
induction (*see* explosions, thermal, phases of)
instability (*see* stability)

jump conditions at
flame sheets 54, 150
hydrodynamic discontinuities 137

Karlovitz stretch (*see* stretch)

latent heat 73
Lewis number 9, 15, 53, 139, 141
linear condensates 58, 79

Mach number 11, 14, 88, 136
mass-flux fraction 17
mass fraction 2
mixtures
continuum theory of 2
one-reactant model for 21
two-reactant model for 20, 34
molecular-mass equality 7
monotonic responses 58
for diffusion flames 106, 114, 117
for linear condensates 64, 71
for spherical flames 128, 130

near-surface flames (*see* flame, near-surface)
NEFs (near-equidiffusional flames) 38, 53, 139, 149
cold wall, effect of 168
as a restricted class 141
, sheared 176
stability of 202
, strained 179, 185
tips of 163, 170
nonuniformities 175, 185 (*see also* shear and strain)
one-reactant model (*see* mixtures, one-reactant model for)
partial burning 27, 112, 131
perturbations (*see also* stability)
defined 19
NEFs as 38, 53
SVFs as 38, 47

plane flames 20, 25, 39
polyhedral flames 194
preheat zone 26, 142
propagation limits (*see* flammability limits)
propellants 59
pyrolysis 60, 62, 64, 79, 80, 91
pyrolysis law 62

quasi-steady approximation 80, 97
quenching (*see also* extinction)
  by cold walls 168
  in explosions 238
  by shear 175, 177, 179, 189
  by strain 175, 187, 192
  , flame-tip 163, 166, 167
  hypothesis 166

radiation 43, 64, 67, 76
rate constant 6
reactant depletion 223, 229
reaction rate 6
reactions 5, 7
  of arbitrary order 33
  , chain 18, 222
  , exothermic 13
  , switch-on 23, 24, 28, 58
remote flames (*see* flame, remote)
responses 58, 74, 77 (*see also* C-responses, monotonic responses, *and* S-responses)
  , oval 78
Shvab-Zeldovich variables 15
shear 138, 147, 168, 175, 176, 187
similarity solutions 166, 169, 178, 179, 233, 243, 244
slow variations 82, 86, 98, 104, 127 (*see also* SVFs)
specific-heat equality 7
species flux (*see* mass-flux fraction)
spherical flames 52, 120, 124, 138
  , diffusion 95
spontaneous combustion 221 (*see also* auto-ignition)
spray combustion 59
S-responses for
  counterflow flames 184
  diffusion flames 100, 106, 112, 117
  linear condensates 68, 73
  spherical flames 128, 130
stability 193
  of anchored flames 92, 219
  of chambered diffusion flames 119
  , Darrieus-Landau 197
  of diffusion flames 103, 117
  diffusional-thermal effect on 196
  in ducts 52
  gravity effect on 199, 213
  for heat loss 50
  hydrodynamic effect on 209
  of linear condensates 66, 69, 75, 78, 79, 80, 82, 84, 86, 90
  of NEFs 202, 206, 209, 214, 218
  of spherical flames 129
  for stagnation-point flow 181
  of SVFs 50, 53, 196, 199
  of thermites 219
stagnation-point flow (*see* counterflow *and* strain)
Stephan problems 151
Stephan problems relating to
  cold walls 168
  shear 176
  strain 185
  tips 163
stoichiometric
  average mass 13
  coefficients 5, 7
stoichiometry 34
strain 152, 175, 179, 185, 189
stretch 146 (*see also* area change)
  for shear 177, 189
  stability implications of 205
  for strain 183, 187, 192
  universal effect when weak 183
sub- and supercritical 225
supply
  conditions 23, 39, 92, 98, 122, 124
  as heat sink 26, 39, 98, 122
surface flames (*see* flame, surface)
SVFs (slowly varying flames) 38, 47, 50, 51, 139, 142
  , anchored 92
  basic equation of 145, 148
  , sheared 187
  stability of 196
  , strained 189
  tips of 159

thermal expansion coefficient 152
thermal runaway for
  auto-ignition 222, 229, 233
  heat flux 239
  initially separated reactants 245
thermites 18, 218, 223
tips (*see* flame tips)
transport coefficients 172
turbulence 175, 183, 193
two-reactant model (*see* mixtures, two-reactant model for)

units 13, 41

vorticity 136, 157, 197

weak-burning limit 108, 116, 129 (*see also* frozen combustion)
wrinkled flames 214 (*see also* corrugated flames)